AF356142

Dietary Interventions for Human General and Oral Health and Disease Reduction

Dietary Interventions for Human General and Oral Health and Disease Reduction

Guest Editors

Theodoros Varzakas
Maria Antoniadou

Basel • Beijing • Wuhan • Barcelona • Belgrade • Novi Sad • Cluj • Manchester

Guest Editors

Theodoros Varzakas
University of Peloponnese
Kalamata
Greece

Maria Antoniadou
National and Kapodistrian
University of Athens
Athens
Greece

Editorial Office
MDPI AG
Grosspeteranlage 5
4052 Basel, Switzerland

This is a reprint of the Special Issue, published open access by the journal *Applied Sciences* (ISSN 2076-3417), freely accessible at: https://www.mdpi.com/journal/applsci/special_issues/Dietary_Interventions_Human_Health.

For citation purposes, cite each article independently as indicated on the article page online and using the guide below:

Lastname, A.A.; Lastname, B.B. Article Title. *Journal Name* **Year**, *Volume Number*, Page Range.

ISBN 978-3-7258-1485-5 (Hbk)
ISBN 978-3-7258-1486-2 (PDF)
https://doi.org/10.3390/books978-3-7258-1486-2

Contents

About the Editors

Theodoros Varzakas

Theodoros Varzakas has served as full professor at the Department of Food Science and Technology, the University of the Peloponnese, Greece, since 2019, specializing in issues of food technology, food processing/engineering, and food quality and safety. Varzakas is the Section Editor of the Food Security and Sustainability Section of Foods (2020 onwards). He was also formerly the Editor-in-Chief for Current Research in Nutrition and Food Science (2015–2019). Varzakas is a reviewer and member of the Editorial Boards of many international journals, such as the International Journal of Food Science and Technology, the Journal of Food Engineering, Waste Management, Critical Reviews in Food Science and Nutrition, the Italian Journal of Food Science, the Journal of Food Processing and Preservation, the Journal of Culinary Science and Technology, the Journal of Agricultural and Food Chemistry, the Journal of Food Quality, Foods, Microorganisms, and Current Research in Food Science. Varzakas has written more than 200 research papers and chapters in books and presented more than 160 papers and posters at national and international conferences. He has written and edited 14 books in Greek, as well as 16 in English on sweeteners, biosensors, food engineering, food processing, all published by CRC. He has participated in many European and national research programs as a coordinator or scientific member. According to Google Scholar, his h-index is 36 and no. of citations is approx. 5500. According to September 2022 data and October 2023-update for "Updated science-wide author databases of standardized citation indicators" Varzakas Theodoros is in the top 2% of citations worldwide.

Maria Antoniadou

Maria Antoniadou is currently an associate professor at the Department of Dentistry, School of Health Sciences, National and Kapodistrian University of Athens, with experience in the clinical and managerial practice of dentistry. She has previously worked as a research and clinical member at the prosthetics clinic of the University of Freiburg, Germany. She holds a Master's Degree in Aesthetic Dentistry from the University of Athens, while her doctoral dissertation was on the application of dental biomaterials in the modern practice of dentistry. She is also a Certified Systemic Analyst Professional (CSAP), having participated in the Executive Mastering Program in Systemic Management at the University of Piraeus, Greece. She specializes in dental psychology and patient coaching and management, professional counseling, performance enhancement and dental business initiative counseling, dental management, and marketing. She is also an ACSTH due to completing an HRE-certified ICF program and an AC Accredited Coach focused on dental business enhancement. She is the author of the book "Application of the Humanities and Basic Principles of Coaching in Health Sciences" (Athens 2021, Tsiotras Ed., ISBN:978-618-5495-45-9).

Preface

The aim of this Special Issue was to address Dietary Interventions for Human General and Oral Health and Disease Reduction. Hence, it enhanced scholarly understanding of nutrition as a critical component of human well-being, particularly in relation to various systemic and oral diseases. Moreover, another aim was to underline the impacts of nutritional and dietary interventions on disease reduction and overall oral and systemic health improvement. Finally, this Special Issue also aimed to address dental and oral health issues in relation to nutrition guidelines

The motivation for writing this scientific work was to address these issues regarding diet and its relationship with disease reduction and facilitate the beneficial effects of diet and nutrition on treatment and prevention. Hence, the audience consisted of nutritionists, food technologists, doctors of medicine, and dentists.

A total of 10 papers were accepted in this context from eminent professionals worldwide, and we would like to acknowledge their valuable contributions. Finally, an Editorial written by the Guest Editors highlighted the significance of diet and nutrition for the intersection between health and oral health.

Theodoros Varzakas and Maria Antoniadou
Guest Editors

Editorial

Dietary Interventions for Human General and Oral Health and Disease Reduction

Maria Antoniadou [1],* and Theodoros Varzakas [2],*

[1] Department of Dentistry, School of Health Sciences, National and Kapodistrian University of Athens, 11527 Athens, Greece

[2] Department of Food Science and Technology, University of Peloponnese, 24100 Kalamata, Greece

* Correspondence: mantonia@dent.uoa.gr (M.A.); t.varzakas@uop.gr (T.V.)

Citation: Antoniadou, M.; Varzakas, T. Dietary Interventions for Human General and Oral Health and Disease Reduction. *Appl. Sci.* **2024**, *14*, 5095. https://doi.org/10.3390/app14125095

Received: 29 May 2024

Accepted: 11 June 2024

Published: 12 June 2024

According to the World Health Organization (WHO), "a healthy diet is essential for good health and nutrition." [1]. This Special Issue is dedicated to enhancing the understanding of nutrition as a critical component of human well-being, particularly in relation to various systemic and oral diseases. The aim is to approach diet from the perspective of healthcare professionals, highlighting the significance of maintaining their own health in busy and stressful working environments. This Special Issue, which is comprised of ten papers, underlines the profound impact of nutritional and dietary interventions on disease reduction and overall oral and systemic health improvement.

In this context, the first contribution deals with diabetes, a widespread metabolic disease, and shows the linkage of nutritional recommendations with affected patients with type 2 diabetes mellitus (T2DM), illustrating how nutritional recommendations can substantially benefit these patients. Similarly, the third contribution explores obesity and the ketogenic diet, emphasizing the importance of reducing the intake of sweet and salty snacks for effective weight management. The second contribution presents innovative technological solutions, demonstrating the positive effects of incorporating probiotics, prebiotics, and various gelling agents into mulberry jellies.

Weight management and disease control during the COVID-19 pandemic are the focus of the fourth contribution, which examines the use of Clinical Decision Support Systems (CDSSs) for patients aiming to lose weight. Dietary supplementation is further explored in the fifth and sixth contributions, which assess the benefits of hydroxytyrosol from olive oil by-products for hypercholesterolemic individuals and the combined use of black garlic and pomegranate to regulate blood pressure, respectively. The eighth contribution discusses the role of mushrooms as functional foods in managing Ménière's disease. Additionally, the tenth paper investigates drug-food interactions within the Mediterranean diet, highlighting its relevance in disease management.

This Special Issue also addresses dental and oral health. The seventh contribution examines the impact of mental health, legal status, and discrimination on dental health among migrants and refugees in ten European countries, as well as the frequency of dental service utilization. Finally, the ninth contribution explores the relationship between vitamin D deficiency and the risk of dental caries in adults.

Oral health is a vital component of overall health, significantly impacting both quality of life and systemic health [2,3]. Poor oral health is linked to numerous systemic conditions, including cardiovascular disease, diabetes, respiratory infections, and adverse pregnancy outcomes [4,5]. Despite its importance, oral health is often neglected, particularly among underprivileged populations who face barriers to accessing dental care [6]. This neglect exacerbates health disparities and leads to a cycle of poor health outcomes [7]. Research on the link between oral health and systemic health in underprivileged populations is crucial for several reasons. First, it helps to identify specific risk factors and health behaviors that contribute to poor oral health in these communities [8,9]. Understanding these factors can inform targeted interventions and public health strategies designed to improve access to dental care and promote healthy practices [10]. Second, such research can illuminate the broader health implications of poor oral health, emphasizing the need for integrated healthcare approaches that address both dental and systemic health [11].

Overall, the importance of this Special Issue lies in its comprehensive exploration of how nutrition intersects with various health conditions and the above-mentioned concerns. Highlighting evidence-based dietary interventions and their effects on disease management and prevention as mentioned by all authors, this collection of papers reinforces the critical role of nutrition in enhancing overall health and well-being.

Conflicts of Interest: The authors declare no conflicts of interest.

List of Contributions

1. Gortzi, O.; Dimopoulou, M.; Androutsos, O.; Vraka, A.; Gousia, H.; Bargiota, A. Effectiveness of a Nutrition Education Program for Patients with Type 2 Diabetes Mellitus. *Appl. Sci.* **2024**, *14*, 2114. https://doi.org/10.3390/app14052114.
2. Szydłowska, A.; Zielińska, D.; Sionek, B.; Kołożyn-Krajewska, D. The Mulberry Juice Fermented by *Lactiplantibacillus plantarum* O21: The Functional Ingredient in the Formulations of Fruity Jellies Based on Different Gelling Agents. *Appl. Sci.* **2023**, *13*, 12780. https://doi.org/10.3390/app132312780.
3. Markovikj, G.; Knights, V.; Gajdoš Kljusurić, J. Body Weight Loss Efficiency in Overweight and Obese Adults in the Ketogenic Reduction Diet Program—Case Study. *Appl. Sci.* **2023**, *13*, 10704. https://doi.org/10.3390/app131910704.
4. Detopoulou, P.; Papandreou, P.; Papadopoulou, L.; Skouroliakou, M. Implementation of a Nutrition-Oriented Clinical Decision Support System (CDSS) for Weight Loss during the COVID-19 Epidemic in a Hospital Outpatient Clinic: A 3-Month Controlled Intervention Study. *Appl. Sci.* **2023**, *13*, 9448. https://doi.org/10.3390/app13169448.
5. Cicero, A.F.G.; Fogacci, F.; Di Micoli, A.; Veronesi, M.; Grandi, E.; Borghi, C. Hydroxytyrosol-Rich Olive Extract for Plasma Cholesterol Control. *Appl. Sci.* **2022**, *12*, 10086. https://doi.org/10.3390/app121910086.
6. Fogacci, F.; Di Micoli, A.; Grandi, E.; Fiorini, G.; Borghi, C.; Cicero, A.F.G. Black Garlic and Pomegranate Standardized Extracts for Blood Pressure Improvement: A Non-Randomized Diet-Controlled Study. *Appl. Sci.* **2022**, *12*, 9673. https://doi.org/10.3390/app12199673.
7. Karnaki, P.; Katsas, K.; Diamantis, D.V.; Riza, E.; Rosen, M.S.; Antoniadou, M.; Gil-Salmerón, A.; Grabovac, I.; Linou, A. Dental Health, Caries Perception and Sense of Discrimination among Migrants and Refugees in Europe: Results from the Mig-HealthCare Project. *Appl. Sci.* **2022**, *12*, 9294. https://doi.org/10.3390/app12189294.
8. Bell, V.; Fernandes, T.H. Mushrooms as Functional Foods for Ménière's Disease. *Appl. Sci.* **2023**, *13*, 12348. https://doi.org/10.3390/app132212348.
9. Peponis, M.; Antoniadou, M.; Pappa, E.; Rahiotis, C.; Varzakas, T. Vitamin D and Vitamin D Receptor Polymorphisms Relationship to Risk Level of Dental Caries. *Appl. Sci.* **2023**, *13*, 6014. https://doi.org/10.3390/app13106014.
10. Spanakis, M.; Patelarou, E.; Patelarou, A. Drug-Food Interactions with a Focus on Mediterranean Diet. *Appl. Sci.* **2022**, *12*, 10207. https://doi.org/10.3390/app122010207.

References

1. WHO Healthy Diet. Available online: https://www.who.int/initiatives/behealthy/healthy-diet#:~:text=A%20healthy%20diet%20is%20essential,are%20essential%20for%20healthy%20diet (accessed on 27 May 2024).
2. Oral Health in America: Advances and Challenges [Internet]. National Institute of Dental and Craniofacial Research: Bethesda, MD, USA, 2021 Dec. Section 1, Effect of Oral Health on the Community, Overall Well-Being, and the Economy. Available online: https://www.ncbi.nlm.nih.gov/books/NBK578297/ (accessed on 27 May 2024).
3. Baiju, R.M.; Peter, E.; Varghese, N.O.; Sivaram, R. Oral Health and Quality of Life: Current Concepts. *J. Clin. Diagn. Res.* **2017**, *11*, ZE21–ZE26. [CrossRef] [PubMed]
4. Cho, G.J.; Kim, S.Y.; Lee, H.C.; Kim, H.Y.; Lee, K.M.; Han, S.W.; Oh, M.J. Association between dental caries and adverse pregnancy outcomes. *Sci. Rep.* **2020**, *10*, 5309. [CrossRef] [PubMed]
5. Botelho, J.; Mascarenhas, P.; Viana, J.; Proença, L.; Orlandi, M.; Leira, Y.; Chambrone, L.; Mendes, J.J.; Machado, V. An umbrella review of the evidence linking oral health and systemic noncommunicable diseases. *Nat. Commun.* **2022**, *13*, 7614. [CrossRef] [PubMed]

6. Ghanbarzadegan, A.; Balasubramanian, M.; Luzzi, L.; Brennan, D.; Bastani, P. Inequality in dental services: A scoping review on the role of access toward achieving universal health coverage in oral health. *BMC Oral Health* **2021**, *21*, 404. [CrossRef] [PubMed]
7. National Academies of Sciences, Engineering, and Medicine; Health and Medicine Division; Board on Population Health and Public Health Practice; Committee on Community-Based Solutions to Promote Health Equity in the United States. The Root Causes of Health Inequity. In *Communities in Action: Pathways to Health Equity*; Weinstein, J.N., Geller, A., Negussie, Y., Baciu, A., Eds.; National Academies Press: Washington, DC, USA, 2017. Available online: https://www.ncbi.nlm.nih.gov/books/NBK425845/ (accessed on 27 May 2024).
8. Nazir, M.A.; Izhar, F.; Akhtar, K.; Almas, K. Dentists' awareness about the link between oral and systemic health. *J. Family Community Med.* **2019**, *26*, 206–212. [CrossRef] [PubMed]
9. Schwarz, C.; Hajdu, A.I.; Dumitrescu, R.; Sava-Rosianu, R.; Bolchis, V.; Anusca, D.; Hanghicel, A.; Fratila, A.D.; Oancea, R.; Jumanca, D.; et al. Link between Oral Health, Periodontal Disease, Smoking, and Systemic Diseases in Romanian Patients. *Healthcare* **2023**, *11*, 2354. [CrossRef] [PubMed]
10. de Abreu, M.H.N.G.; Cruz, A.J.S.; Borges-Oliveira, A.C.; Martins, R.C.; Mattos, F.F. Perspectives on Social and Environmental Determinants of Oral Health. *Int. J. Environ. Res. Public Health* **2021**, *18*, 13429. [CrossRef] [PubMed]
11. Nakre, P.D.; Harikiran, A.G. Effectiveness of oral health education programs: A systematic review. *J. Int. Soc. Prev. Community Dent.* **2013**, *3*, 103–115. [CrossRef] [PubMed]

applied sciences

Article

Effectiveness of a Nutrition Education Program for Patients with Type 2 Diabetes Mellitus

Olga Gortzi [1,*], Maria Dimopoulou [1], Odysseas Androutsos [2], Anna Vraka [1], Helen Gousia [1] and Alexandra Bargiota [3,*]

[1] Department of Agriculture Crop Production and Rural Environment, School of Agriculture Sciences, University of Thessaly, 38446 Volos, Greece; mdimopoulou@uth.gr (M.D.); annavraka@hotmail.com (A.V.); elgousia1995@gmail.com (H.G.)

[2] Laboratory of Clinical Nutrition and Dietetics, Department of Nutrition and Dietetics, University of Thessaly, 42132 Trikala, Greece; oandroutsos@uth.gr

[3] Department of Endocrinology and Metabolic Diseases, Faculty of Medicine, School of Health Sciences, University Hospital of Larissa, University of Thessaly, 41334 Larissa, Greece

* Correspondence: olgagortzi@uth.gr (O.G.); abargio@med.uth.gr (A.B.); Tel.: +30-2421093289 (O.G.); +30-2413502879 (A.B.)

Abstract: Diabetes is a metabolic disease that is a major health problem globally. Dietary interventions contribute to the management of the disease and the improvement in patients' quality of life. The aim of the present study was to assess the effects of a nutrition and lifestyle education intervention on a sample of patients with diabetes. The duration of the intervention was 3 months, and it focused on the promotion of the Mediterranean diet through information pamphlets, diet plans and healthy lifestyle guidelines, which were provided in addition to patients' standard medical treatment. Patients were enrolled in the outpatient clinic of the University Hospital of Larissa (Greece). Anthropometric and biochemical parameters were recorded at baseline and follow-up using standardized equipment and methods. The intervention improved patients' body mass index, body composition, fasting glucose, postprandial glucose, triglycerides, HDL/LDL cholesterol and cholesterol. For smoking status, alcohol consumption and physical activity categorization, physical activity improved but not the other two indices. The results of this study show that patient education should be provided according to the nutritional recommendations for T2DM plus a more individually structured intervention. It is therefore necessary to direct the attention of doctors to the need for continuous and detailed discussions with patients in relation to both the standards of a healthy diet and the benefits it brings. Patients, for their part, need to commit to following an appropriate, healthy diet.

Keywords: diabetes; nutrition; weight loss; education

Citation: Gortzi, O.; Dimopoulou, M.; Androutsos, O.; Vraka, A.; Gousia, H.; Bargiota, A. Effectiveness of a Nutrition Education Program for Patients with Type 2 Diabetes Mellitus. *Appl. Sci.* **2024**, *14*, 2114. https://doi.org/ 10.3390/app14052114

Academic Editor: Wojciech Kolanowski

Received: 14 January 2024
Revised: 22 February 2024
Accepted: 23 February 2024
Published: 4 March 2024

1. Introduction

The prevalence of type 2 diabetes mellitus (T2DM) worldwide is increasing at epidemic proportions [1]. Chronic hyperglycemia is considered a major risk factor for cardiovascular and kidney disease, retinopathy and neuropathy [2]. The Center for Disease Control (CDC) estimates that almost 33% of adults in the U.S. have prediabetes; therefore, preventing or delaying T2DM is a public health imperative to help extend and improve the lives of millions of people [3].

The increase in the prevalence of T2DM is paralleled with the increase in overweight/obesity, and it has become particularly evident in the last decade [4]. Undoubtedly, as a person's body mass index (BMI) increases, the risk of developing T2DM increases in a "dose-dependent" manner. The prevalence of T2DM is 3–7 times higher in obese than in normal-weight adults [4], and people with a BMI > 35 kg/m^2 are 20 times more likely to develop T2DM than those with a BMI between 18.5 and 24.9 kg/m^2 [5]. Obesity also complicates the management of T2DM by increasing insulin resistance and blood glucose concentrations [6]. Obesity is also an independent risk factor for dyslipidemia, hypertension and cardiovascular complications and cardiovascular mortality in patients with T2DM [2].

Attaining and maintaining a healthy body weight is a major therapeutic target in the management of T2DM. Governments are looking to identify the most effective services to support overweight/obese patients with T2DM to lose weight and improve their health status and quality of life [7]. Data from the Diabetes Prevention Program (DPP) showed that weight loss (7% of weight lost in the first year), increased physical activity (150 min of walking per week) and improvements in other lifestyle behaviors (e.g., dietary behavior) decreased the 4-year incidence of T2DM by 58% in men and women with glucose tolerance disorders [8].

Some strategies associated with successful long-term weight loss include lifestyle modification, specifically the adoption of low-caloric diets, frequent body weight monitoring, and participation in regular physical activity. Successful weight loss is accompanied by a reduction in the portion size of meals, foods and snacks, daily breakfast consumption, and 3 or less hours of screen-watching (television, computer, tablets/smartphone) per week on average [9,10]. The link between overweight and obesity and many non-communicable diseases is well known [11]. Weight loss has a major impact on improving glycemic control and reducing the risk of cardiovascular disease [12].

Dietary recommendations need to be based on personal choices, access to food and the patient's culture and ability to make behavioral changes [13]. Interventions in the patient's lifestyle to increase physical activity and reduce caloric intake aim to reduce body weight by 5%, as this can improve HbA1c and cholesterol and reduce cardiovascular risk [14]. The American Diabetes Association emphasizes the importance of educating the patient to make conscious food choices that take into account personal preferences, culture and religion and individualized metabolic goals. The diet patterns with the most beneficial effect on the metabolic profile have been mentioned as the Mediterranean diet and the vegetarian diet [15,16].

Long-term improvements to lifestyle, especially nutrition and physical activity, are challenging for most patients. The role of healthcare professionals working with diabetic patients is to encourage, monitor and support them in this effort [17]. Several techniques can be used to promote behavioral change. First, behaviors need to be screened, and personalized goals need to be set by the healthcare professional in agreement with the patient. Setting realistic and achievable goals allows patients to achieve success, which can be a starting point for further lifestyle change [18,19]. Strategies such as self-monitoring, avoidance of situations that trigger food intake and problem-solving (to the extent possible) may support self-regulation over time. Frequent communication and educational meetings (e.g., every fortnight) are associated with better long-term maintenance of weight loss [20].

The purpose of this study was to investigate the effectiveness of a nutrition and lifestyle education intervention in patients with T2DM for a patient-approach medicine. At the same time, the level of nutritional knowledge of the patients was assessed and their problems in adopting appropriate nutritional approaches were recorded with the ultimate aim of improving them. Some small studies promote nutritional supplementation [21] or a diet that may affect postprandial glucose or other metabolic biomarkers [22–24] or alter the metabolic profile through intermittent fasting [25], but only a few have examined the impact of behavioral [26] and patient-centric therapeutic approaches [27] for diabetes, so original research could close the gap. This study aims to present multifaceted strategies compared with the clinical studies conducted in the last years, with diet playing a pivotal role in T2DM management, with a special focus not only on the Mediterranean diet but also on personalized dietary recommendations for patients and shedding light on the efficacy of a multidisciplinary team of health professionals.

2. Participants and Methods

2.1. Study Designed

The study was approved by the competent Bioethics Committee of the University of Thessaly (approval numbers 49162/13-10-2017 and 49161/13-10-2017), and it was in line with the Declaration of Helsinki. All volunteers signed a written informed consent prior to their participation in the study. The study period was from October 2017 to January 2018.

2.2. Participant Recruitment

Eighty-eight T2DM patients (forty males and forty-eight females, average age 51.4 years) were admitted to the Larissa General University Hospital, Greece, for 90.0 ± 3.3 days. Criteria for participation in the study were a diagnosis of diabetes established by hemoglobin A1C (HbA1c) or plasma glucose concentration. In addition, individuals with a recent diagnosis of T2DM (<3 months) and who were treatment-naive, 30–70 years of age, clinically and biochemically stable and without any acute metabolic complications of diabetes were also considered for the study. Individuals with history of recent alcohol use (<6 months), pregnant women and those in a severe comorbid state were excluded. The clinical samples for analysis were collected at the baseline (t = 0 months) and at the end (t = 3 months) of the study.

2.3. Data Collection and Measures

At baseline, patients filled out the questionnaire (demographic characteristics, personal information, complications of the disease), and data were recorded for biochemical indicators, including fasting glucose, blood glucose, HbA1c, total cholesterol, HDL cholesterol, LDL cholesterol and triglycerides, and also medications. Anthropometric data were also collected. Participants' weight was measured in light clothing and without shoes using a portable calibrated electronic weighing scale precision scale (TAN- ITA MC-780U Multi Frequency Segmental Body Composition Analyzer, Amsterdam, the Netherlands). Height was measured with portable measuring inflexible bars (Seca model 220, Seca, Hamburg, Germany). Waist (at umbilicus) and hip (at widest point) circumferences (WC and HC) were measured according to standard conditions using a measuring tape, and waist/hip circumference ratio (WHR) was calculated. Blood pressure was measured with a clever blood monitor (FYGB-869). All measurements were taken twice, and the average of the two values was reported, as suggested [28]. The follow-up was planned after 3 months and aimed to evaluate potential changes in the collected variables. Subjects recorded their physical activity using the International Physical Activity Questionnaire [29] and a validated diet recall Food Frequency Intake form [30].

Low-density lipoprotein cholesterol (LDL) was calculated according to the Friedewald formula [31]. Fasting glucose was calculated [32], and glycated hemoglobin (HbA1c) was measured by high-performance liquid chromatography (HPLC). All participants were evaluated for their changes in the collected variables.

2.4. Lifestyle and Nutritional Intervention

Each patient was allocated to one dietitian, a nutritionist, who was responsible for educating, measuring and adhering to the intervention. The intervention included one face-to-face meeting about nutrition under the direction of a dietician and the patient. During this visit, the patients were informed about the purpose of the study, answered the 24 h diet recall questionnaire [33] and were given an individualized nutritional plan (the energy requirements were calculated and personalized according to the preferences/needs of each patient) and detailed nutritional instructions, recommendations and advice, both for their diet and eating behavior. Patients were also educated regarding the choices of foods containing carbohydrates, with the aim of regulating sugar levels within normal limits as well as easily forming their daily diet. There were written specific nutritional recommendations: to have 5–6 meals per day, dividing the foods containing carbohydrates, to prefer foods rich in soluble fiber (legumes, fruits, vegetables, whole grains), to avoid the consumption of pure sugar and products containing it (sweets, cookies with sugar, cakes, jelly, ice creams with sugar, candies, sugared drinks, etc.), to reduce salt consumption and to avoid the consumption of saturated fat contained mainly in red meat, cold meats, egg yolks (up to 3 times/week) and butter. Emphasis was given to the consumption of vegetable fats, mainly olive oil, and to replace red meat with fish as much as possible. All participants' diets were evaluated to assess their compliance with the Mediterranean diet and standards recommended by the American Diabetes Association [34] and the Hellenic Diabetes Association [35]. Some dietary recommendations were given by the dietician, such as reducing the intake of calories, total fat to <30% of daily energy intake and saturated fat (including trans fatty acids) to <10% of daily energy intake and increasing fiber intake (15g to 30g/day) [36]. The patients were given a form with the food categories mentioned above and an individualized nutritional plan (55% carbohydrate, 15% protein, and 30% fat) that they could follow for weight loss, as this was a key target of the intervention for those who were overweight or obese. They also received instructions for exercise. More precisely, participants were instructed to follow a Mediterranean-type diet and to perform at least 150 min of moderate-intensity exercise a week. Adherence to recommendations from dietitians was reported after the three-month intervention. Finally, the patients could communicate, either through calling, messaging or visiting the dietician during the three-month intervention, for questions or for support for their efforts.

2.5. Statistical Analyses

Statistical analysis was performed with the Statistical Package for the Social Sciences (SPSS 21). A frequency analysis was performed for each of the variables in the questionnaires. The Kolmogorov–Smirnov test was used to test the variables for a normal distribution. Data are presented as mean ± standard deviation (SD) or as median value (interquartile range). A paired *t*-test was used to compare parametric variables before and after three months of the nutrition education intervention; the Wilcoxon test was used for non-parametric data; and the chi-square test was used for categorical variables. *p*-values < 0.05 (two-tailed) were considered statistically significant.

3. Results

3.1. Study Population

Individuals with a history of recent alcohol use (<3 months) or in a severe comorbid state were excluded. The study population consisted of 48 females (mean age 50.1 ± 9.1 years) and 40 males (mean age 51.4 ± 6.8 years). The average interval between baseline and the end of the follow-up was 90.0 ± 3.3 days, and all participants completed the study. The study was explained to all the individuals initially considered. Only 88 of the initial 100 patients gave informed written consent and so participated in the intervention at the Hospital of Thessaly.

3.2. Socio-Economic Data, Nutritional Knowledge and Perception of T2DM Treatment

Although, risk factors for people with T2DM [3] are well known. it is therefore necessary to select socio-economic data and clinical characteristics of the T2DM patients.

More than half of the patients were treating diabetes with antidiabetic medication only (56.8%), with 3.4% injecting insulin and only 9.1% taking diabetic medications and eating a balanced diet (Table 1).

The amount patients spent on food each month was also checked for correlation with dietary guidelines. Most (52% of the sample) spent EUR 200–300, 32% spent EUR 300–400, 11% spent less than EUR 200 and just under 5% spent more than EUR 400.

As part of the interview, patients were questioned about their educational level. The results showed that almost half of the patients had a high school degree (42.5%), while only 4.6% had an elementary school degree. The rest of the patients had a university degree (Table 1).

No comorbidities were recorded in 25 patients; 57.5% had a family history of diabetes. The frequency of comorbidities for the remaining 63 patients is recorded in Figure 1.

Table 1. Socio-economic data and clinical characteristics of the T2DM patients (N = 88).

		%
Sex	Men	45.5
	Women	54.5
Education	Primary Education	4.6
	Secondary Education	42.5
	Higher Education	52.9
Financial budget per month for food	EUR < 201	11.4
	EUR 201–300	52.3
	EUR 301–400	31.8
	EUR > 400	4.5
Treatment of T2DM	Exercise and insulin	1.1
	Exercise. and medication	1.1
	Exercise and diet	1.1
	Diet, exercise and medication	6.8
	Diet and medication	9.1
	Insulin	3.4
	Medication	56.8
	Medication and insulin	20.5

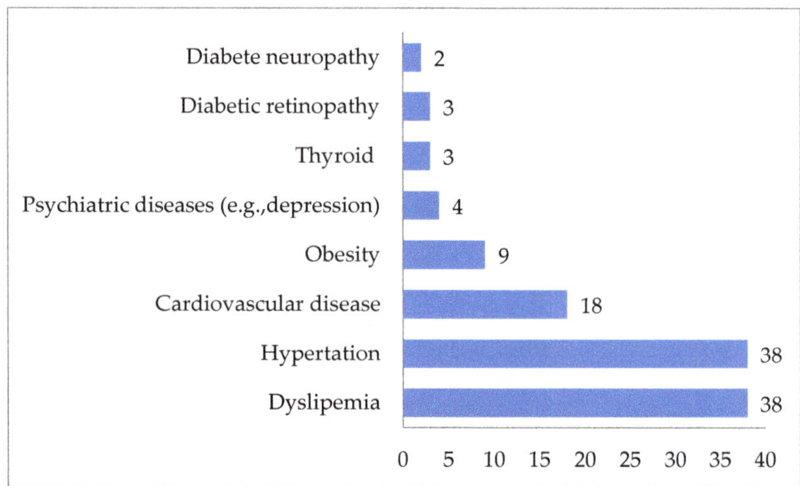

Figure 1. Comorbidity of the participants (%).

When asked about the frequency of blood glucose monitoring by the patients, 50% answered 1 time/day. After analyzing the patients' responses to the questionnaire, it was found that 90% of patients reported the frequency of their visits to the doctor for diabetes was every 3 months and 10% every year. A total of 89.8% of patients responded that they were not informed about the severity of the disease, while 10.2% of patients were informed. When asked if dietary instructions were given after the onset of the disease, the physician responded that it was the patient's responsibility. Almost half of the patients exercised every day. Finally, patients were asked if they smoked, and 25% of patients answered yes (Table 2).

Figure 2 shows the percentage of participants who adhered to the dietary guidelines prior to the intervention. Before the intervention, half the patients had not received clear dietary guidelines related to treating their diabetes. The others had been given some kind of dietary plan by the research dietician. For the majority of patients (56%), reducing carbohydrate intake was the most important dietary goal; 20% of patients tried to increase their intake of fruit and vegetables; 42% chose to include low-fat foods in their diet; 27% of patients did not follow any dietary guidelines. In terms of sodium consumption, only 5% restricted the consumption of processed foods, and 35% refrained from eating particularly salty foods.

Table 2. Therapeutic approaches and treatment of T2DM (%).

		N %
Frequency of blood glucose measurement	Occasionally	3.4
	Weekly	8.0
	1 time/day	50.0
	2–3 times/day	34.1
	>3 times/day	4.5
Frequency of doctor visits for type 2 diabetes mellitus	Once a year	13.6
	Once per 6 months	28.4
	Once per 3 months	35.2
	1 time/month	17.0
	2 times/month	5.7
Update on disease severity/consequences of non-regulation	No	89.8
	Yes	10.2
With the onset of the disease, were dietary instructions given (immediately)? If so, by whom?	No	11.4
	From the doctor	52.3
	By the dietitian with guidance from the attending physician	36.4
	From the dietitian during a visit on individual initiative	0.0
Reason for not visiting a dietitian	No answer	20.5
	Negligence	31.8
	The doctor is enough	36.4
	Cost	8.0
	No time available	3.4
Participation in physical activity	No	27.3
	Housework	6.8
	Outside work	2.3
	Walking	53.4
	Jogging	2.3
	Sport	8.0
Frequency of exercise	No answer	27.3
	1 time/week	2.3
	2 times/week	4.5
	3 times/week	10.2
	4 times/week	3.4
	5 times/week	5.7
	Every day	46.6
Smoking	No	75.0
	Yes	25.0

When asked about their knowledge of the proper diet for someone with diabetes, the largest percentage (57.5%) rated this as moderate, while 37.9% felt they were adequately informed. A small percentage of 3.4% felt that they had very good knowledge and 1.1% felt that they had no knowledge in this area.

3.3. Efficacy of the Nutrition Education Intervention

At the end of the dietary intervention period, changes were seen in all anthropometric and biochemical variables measured. Body weight and body mass index (BMI) decreased significantly, as did systolic blood pressure, fasting and mean glucose, total and LDL cholesterol and triglycerides (TGs) (Table 3). The most important outcome was the change in body composition (average body fat and vascular fat percentage averages) and waist circumference (Table 3).

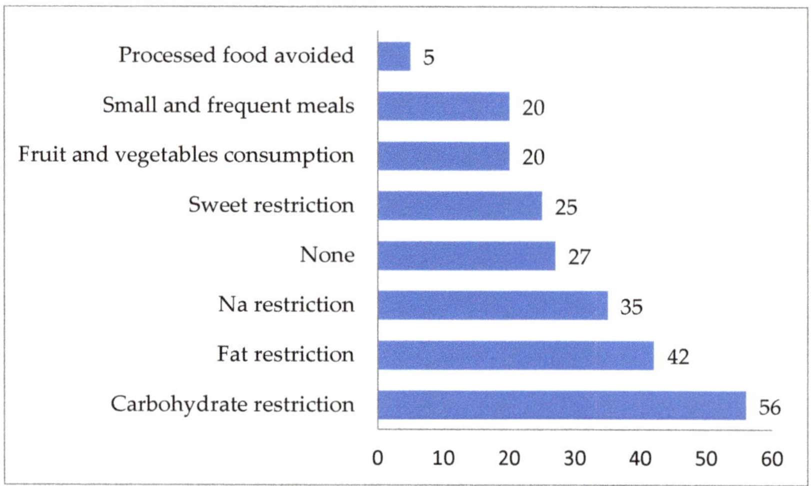

Figure 2. The participants who followed the nutritional guidelines before the intervention (%).

Table 3. Somatometric characteristics, biochemical variables and 24 h dietary recall results of the participants (N = 88) at baseline and after three months of the nutrition education intervention.

Characteristics *	At Baseline (t = 0 Months)	After Intervention (t = 3 Months)	*p*-Value
Body weight (Kg)	81.8 ± 17.2	77.8 ± 16.3	<0.001 *
BMI (kg/m^2)	28.8 ± 5.0	27.1 ± 5.4	<0.001 *
Waist circumference (cm)	107.4 ± 26.0	103.4 ± 25.5	<0.001 *
Hip circumference (cm)	96.8 ± 24.0	94.5 ± 23.6	<0.001 *
Waist–hip ratio	1.2 ± 0.1	1.1 ± 0.1	<0.001 *
HbA1c (%)	7.2 (6.4–8)	7.6 (6.7–7.7)	0.672 *
Postprandial blood glucose (mg/dL)	161.6 ± 36.0	131.9 ± 26.7	<0.001 *
Fasting glucose (mg/dL)	110.7 ± 19.5	99.5 ± 13.9	<0.001 *
Total cholesterol (mg/dL)	183.0 ± 41.1	165.8 ± 26.8	<0.001*
LDL-c (mg/dL)	96.2 ± 35.4	89.4 ± 20.0	0.002 *
HDL-c (mg/dL)	48.2 ± 15.5	47.1 ± 6.5	0.443 *
TG (mg/dL)	154.7 (150.7–154.7)	119.7 (69.7–120.7)	<0.001 **
SBP (mmHg)	14.2 ± 11.4	11.8 ± 0.9	0.056 *
DBP (mmHg)	8.8 ± 7.9	8.0 ± 0.4	0.342 *
Atherogenic index	4.1 ± 1.3	3.5 ± 0.5	<0.001 *
Uric acid (mg/dL)	6.5 ± 1.7	6.2 ± 1.4	0.0001 *
Body fat (%)	32.9 ± 7.8	29.2 ± 7.6	<0.001 *
Visceral fat (%)	11.8 ± 0.1	10.2 ± 0.1	<0.001 *
Muscle mass (%)	88.7 ± 4.3	83.1 ± 5.8	0.331 *
Daily calorie intake (kcal)	1659.7 ± 285.9	1631.4 ± 267.3	0.002 *

BW: body weight, BMI: body mass index, WC: waist circumference, HC: hip circumference, WHR: waist-to-hip ratio, HbA1c: glycated hemoglobin, BGP: blood glucose postprandial, FBG: fasting blood glucose, TC: total cholesterol, HDL: high-density lipoprotein cholesterol, LDL: low-density lipoprotein cholesterol, TG: triglycerides, SBP: systolic blood pressure, DBP: diastolic blood pressure, AI: atherogenic index, UA: uric acid, BF: body fat, VF: visceral fat, MM: muscle mass, DCI: daily calorie intake. Data are mean ± standard deviation (SD) or median value (interquartile range). * *p* values for the comparison with baseline by paired *t*-test. ** *p* values for the comparison with baseline by Wilcoxon test.

The atherosclerotic index and uric acid also decreased significantly at the end of the study period, while no significant changes occurred in HbA1c, HDL cholesterol, diastolic blood pressure and muscle mass (Table 3).

Based on the 24 h reminders at the beginning and end of the study, 73.4% of patients reduced their daily calorie intake and 5.9% increased their consumption of high-fiber foods by the third month, a large percentage (from 67.8% at the first visit to 59.8% in the third month) reduced their alcohol consumption, 5.9% opted for plant-based foods in place of meat, chicken and fish, while 5.2% made no changes to their dietary habits (Figure 3). Regarding patients' participation in physical activities, the percentage increased slightly by the third month, such as brisk walking, cycling or playing a friendly game of basketball or soccer (moderate to vigorous physical activity) (Figure 1). A significant percentage of patients skipped meals both at the beginning and at the end of the study (especially dinner (55.2%)) (Figure 4), although skipping breakfast and afternoon snacks was less frequent in the third month of the intervention and the differences were statistically significant ($p < 0.05$) (Figure 4). Lunch was the only meal that was not skipped by anyone.

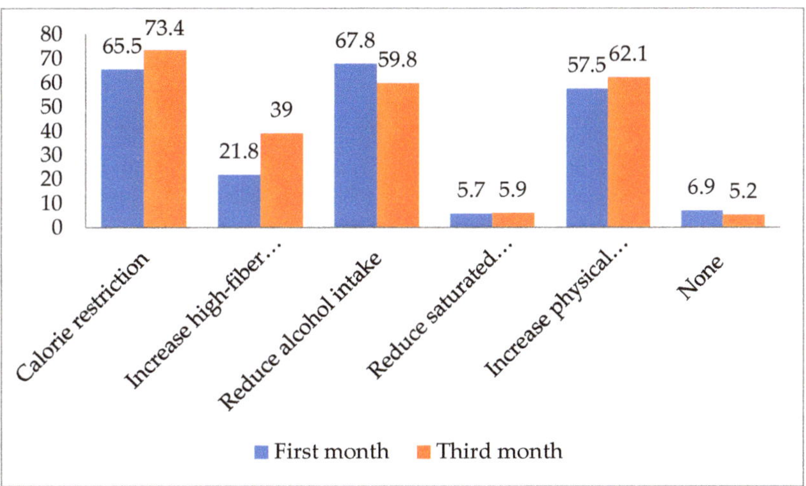

Figure 3. Percentage (%) of patients following the recommendations according to the ADA [22].

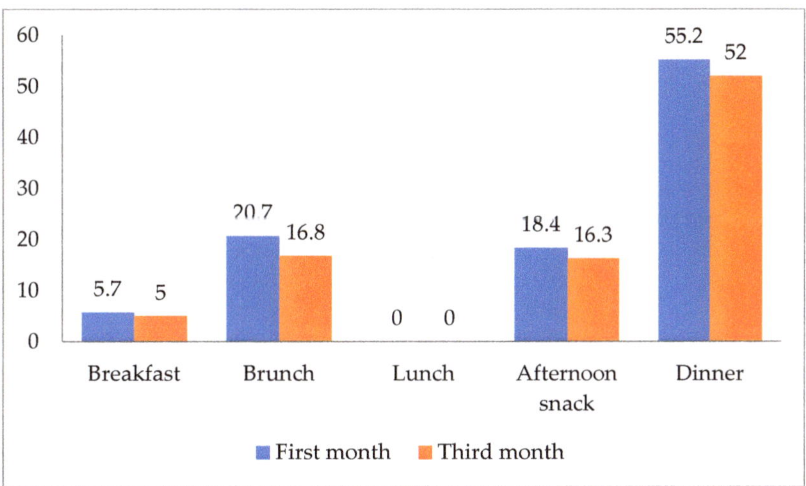

Figure 4. Percentage (%) of meal-skipping in participants.

Since dietitians should focus on individualized nutritional care to gain a holistic understanding of their patients, we recorded the frequency of patient communication with the dietitian during the three-month intervention and the results are shown in Figure 5.

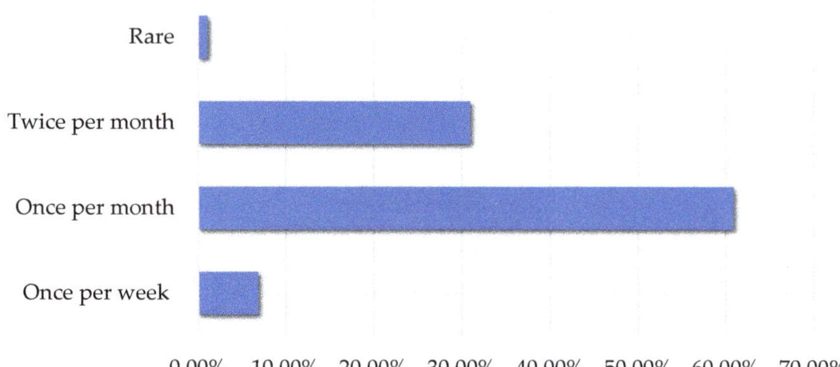

Figure 5. How often the patients communicated (either through calling, messaging or visiting) with the dietician during the three-month intervention.

4. Discussion

In this study, the nutritional and lifestyle recommendations given to diabetic patients appeared to cause significant changes in anthropometric characteristics and biochemical markers 3 months after a face-to-face meeting with a specialized dietitian. This is due to dietary instructions and changes in their physical activity (Tables 1 and 2). Main role in achieving and maintaining an appropriate body weight, preventing complications of the disease (Figure 1) and improving the overall clinical picture play also the appropriate dietary changes (Figure 2).

Other studies have also examined a similar research topic [37–39]. In most of them, a significant trend of a decrease in the recorded anthropometric characteristics and biochemical indices was observed [40]. In our study, body weight decreased from 81.8 to 77.8 kg after 3 months, and BMI also decreased from 28.8 to 27.1 kg/m^2. There was also a significant reduction in the average fasting blood glucose from 110.7 to 99.5 mg/dL and in the mean value of lipids (total cholesterol, LDL cholesterol, HDL cholesterol and triglycerides). The greatest change was observed in the mean total cholesterol and the mean triglyceride values. According to a recent study with a 3-month intervention to improve glucose control in individuals with type 2 diabetes through education, the HbA1c value was reduced by 1.1% [41]. In the ANODE study (2017), other secondary endpoints in the intention-to-treat analysis with web-based nutritional coaching were fasting blood glucose 1.46 mmol/L, total cholesterol 0.21 mg/dL, LDL-c 0.2 mg/dL, HDL-c 0.05 mg/dL, triglycerides 1.05 mg/dL and uric acid 45.45 μmol/L [42]. In our study, the changes in the other biomarkers were higher, with the exception of fasting blood glucose and uric acid, which were the same: fasting blood glucose 1.12 mmol/L, total cholesterol 17.2 mg/dL, LDL-c 6.8 mg/dL, HDL-c 1.1 mg/dL, triglycerides 35 mg/dL and uric acid 30 μmol/L. Other studies had positive results [41,42], as our study has. A possible explanation is the multi-layered strategies such as a personalized diet plan, written instructions with healthy food choices and personal contact with the dietician.

Some studies found varying effects of nutritional education [43–45]. For example, one study of 100 patients with type 2 diabetes found no significant effects on BMI or daily intake of fruit and vegetables, but there were positive changes in fasting blood glucose and HbA1c as well as weekly consumption of fruit and vegetables. Weekly moderate physical activity also improved [46]. Awareness about diabetes complications and consequent improvements in dietary knowledge, attitude and practices lead to better control of the disease [47]. Different medications may also have different effects [48] and it is important to provide appropriate motivation for physical activity [49]. Our study did not find any significant change in HbA1c values, although significant changes occurred in fasting blood glucose levels. The biggest change was found in those who followed the dietitian's nutritional advice.

Our results agree with other studies that both dietary guidelines and nutritional interventions contribute to the reduction in anthropometric characteristics and biochemical indicators of people with type 2 diabetes. Following specific dietary interventions also gives better results for weight loss than are found in people who follow more general dietary recommendations [50].

Our sample consisted of patients admitted to the hospital due to uncontrollable diabetes. It is possible that being admitted to the hospital and being given information about the possible complications that can occur if T2DM is not controlled contributed to their more positive attitude toward making changes in their behavior, both in terms of diet and physical activity (Figure 3). The intervention in the 88 adults with T2DM played an important role in their overall health.

Multiple supportive connections in community settings can also help people lose weight.

A study by Pellegrini and colleagues [47] aimed to evaluate the effectiveness of a technology-based system (TECH) on weight loss when used alone or in combination with a 6-month, in-person, behavioral weight loss intervention [47]. In another study of 212 patients [48], biomarkers such as fasting glucose, postprandial plasma glucose, triglyceride and HbA1c were lower at each follow-up meeting with the dietitian, which highlighted the improved patients' knowledge and behavior using the mobile interactive system as a helpful tool [49].

Nutrition is unquestionably important in the progression of T2DM, but most people rely more heavily on medication to manage their disease. In our study, the majority of the patients managed their diabetes by taking medicinal tablets (26.4%) or by combining antidiabetic drugs with appropriate nutrition/diet (27.6%). One review of 22 studies of overweight adults with type 2 diabetes [51] found that the most successful weight loss occurred with multicomponent interventions, including more intense physical activity and very low-calorie diets or low-calorie diets [51].

Programs delivered in primary care can produce meaningful weight loss. In our study, the average weight loss was 4 Kg. It is reported that a 5% to 10% weight loss is associated with health benefits, including lower systolic blood pressure along with reduced triglyceride and glucose levels, which may impact cardiac health [52]. In addition to weight control, the first goal is macronutrient and micronutrient intakes at the right amounts according to national dietary recommendations, i.e., a high-quality diet [34]. The proper dietary plan could be helpful for patients to follow the advice, as many do not realize that following the proper dietary rules prevents the occurrence of complications of the disease and improves the overall clinical picture of the patient [43]. In addition, more than half of the participants did not correctly identify alcohol as a factor that plays a role in the progression of the disease. Although abstaining from alcohol does not "reverse" diabetes per se, it significantly improves glycemic control and reduces the likelihood of complications [53]. Our study also showed that the consumption of breakfast is a factor in better metabolic biomarkers (Figure 4). This concurs with findings from a study of adults in families at high risk of T2DM in six European countries [54]. Another important finding of our study, as shown in the initial interviews of the patients with the dietitian, is that patients ignore the basic parameters of the disease. Although they know the importance of regulating blood glucose concentrations, they consider that reducing the calories they consume is enough to protect them from adverse situations. Observance of the frequency of meals by diabetic patients is of particular importance, but there was also a number of patients who skipped meals. Finally, the Mediterranean diet is considered a well-formed diet plan. However, with all dietary changes, we know that institutions and organizations need behavioral scientists to assist consumers in becoming more aware of healthy eating habits, nutritional labeling and checking overall health [27].

According to the Academy of Nutrition and Dietetics and the Endocrine Society, another factor that affects the success of a nutrition education program and may minimize the frequency of visits to a dietitian is the cost [55]. A potential solution could be for primary care to support the recommendations from the organizations through consultations with health professionals [56].

Other factors, such as emotional support as well as fast and effective access to care, have been found to be important [57]. A greater number of contacts between patients and providers led to greater weight loss (Figure 5). The dietitian in all the studies followed some strategies for weight loss (>1 kg/month) such as dietary plans with detailed diet information or advice on cooking and minimizing the frequency of eating out and some other support tools like a food exchange list. Some specifically list replacement foods organized by food groups and guidelines on healthy food choices or even healthy recipes [58].

After the nutritional intervention, a change was observed in all biochemical indices and anthropometric characteristics studied. Therefore, the adoption of a careful dietary pattern based on the Mediterranean diet can lead to significant and long-term changes in metabolic profile due to minimizing ultra-processed foods, which are high in saturated fats and added sugars. Such changes protect against cardiovascular complications [59,60].

Diet advice via individual sessions as treatment has occurred in many studies [61]. A single dietary counseling session [62], as in our study and others, did not have the same results as can occur in more intensified interventions such as weekly [63], fortnightly [64] or monthly sessions [20].

Dietary patterns such as the Mediterranean diet as well as physical activity are crucial for improving the metabolic disturbances seen in patients with diabetes [65]. The primary goal is to reduce saturated and trans fatty acids. Dietary plans documented to reduce body weight at least by 5% in overweight and obese people should also reduce total fat to <30% of daily energy intake, reduce saturated fat (including trans fatty acids) to <10% of daily energy intake, increase fiber intake (14g/1000 Kcal) [66], minimize refined and processed foods [67] and increase omega-3 fatty acid intake [68]. Increasing the daily consumption of natural foods rich in dietary antioxidants (tocopherols, carotenoids, vitamin C, flavonoids, polyphenols), trace elements and vitamins by increasing vegetable and fruit intake to 400 g/day should be encouraged [69]. Educational meetings with the dietitian in primary care will possibly help patients better understand type 2 diabetes and show how they can control it with easy-to-apply lifestyle changes. The interaction and exchanges between different specialists, a friendly atmosphere of meetings, weight loss tips, including the latest research-backed strategies, healthy eating plans, smart nutritional secrets, meal-planning help and lifestyle strategies to help manage diabetes could help patients control diabetes. Moreover, effective social support with assistance and encouragement from family members, adequate self-management skills and self-efficacy (confidence) may reduce the risk of developing diabetes complications [70].

In our study, despite recommendations from health professionals to adopt a healthier lifestyle, many patients avoided exercising, while a high percentage of patients smoked. These practices are anything but helpful in controlling the disease. This finding should concern and activate health professionals to mobilize change in the behavior and habits of patients. Exercise seems to have a positive association not only with the regulation of systolic and diastolic blood pressure [71] but also with glycemic control [72] and variability, insulin sensitivity, lipid profile, oxidative stress/antioxidative capacity and/or chronic inflammation [73]. Although there is limited evidence, stopping smoking has been shown to have benefits in reducing or slowing the risk of cardiovascular morbidity and mortality in people with diabetes [74].

The nutritional intervention adopted by our patients was based on the principles of the Mediterranean diet [69]. The Mediterranean diet is considered to be a model of healthy eating. Its beneficial actions are generally accepted both in the general population and in patients with T2DM [66]. It is a food pattern that satisfies all the conditions of a suitable diet for the regulation of the disease. It is therefore necessary to inform patients about the beneficial effects of this nutritional model in order to adopt it in their daily lives. Many people with diabetes initiate a conversation about diet with a health professional themselves [75]. The primary reason clinicians initiate a conversation about weight management is to avoid follow-on complications from T2DM [76]. A key strategic theme is to strengthen research-based understanding of T2DM and public health to improve the lives of young people and adults living with or at high risk of developing T2DM.

Therapeutic patient education for obesity or diabetes is a cost-effective intervention that improves patient outcomes [51]. The first step is to learn about the patient's circumstances and perspective, the second step is to help the patient identify their goals, the third step is to help the patient develop a plan (especially a dietary plan and oral health routine), the fourth step is to help the patient implement their plan and the fifth step is to review progress and adjust to changing circumstances. According to the most recently published WHO guidelines (2023), this review identified the factors that could be helpful for doctors, dietitians, nutritionists and other health professionals to promote health and prevent T2DM complications through education [27].

The main objective is to give health professionals better access to effective nutrition education for all people by identifying risk factors such as obesity, malnutrition, excessive sugar intake, weight, age, smoking, alcohol consumption and physical inactivity. A fundamentally different approach is then needed, one that emphasizes disease prevention and health management through a multidisciplinary, integrated and patient-centered approach to overall health [76].

The limitations of this study include the small sample size. The primary care physician was recommended and notified. Future research needs to be conducted using artificial intelligence and its applications to diabetes in order to detect and manage the disease and capture more data to be recorded for personalized healthcare with lower costs. A fully automated web-based program improves lifestyle habits and HbA1c in patients with type 2 diabetes and could be the next step for the researchers.

5. Conclusions

In conclusion, this study adds to the existing scientific evidence of the benefits of a nutrition education program for patients with type 2 diabetes mellitus by demonstrating relevant outcomes in both glycemic responses and dietary behaviors in a real-world setting. Overall, these data suggest that more frequent face-to-face contact with the dietitian may also be more helpful, and meal frequency and a daily breakfast followed by an afternoon snack of healthier foods may help to break down barriers and facilitate dietary self-management of diabetes. To achieve this, people with diabetes need education about their condition. They need to understand how to self-regulate their blood sugar not only in normal situations but also in stressful situations, such as intense physical activity, and when to seek early medical help. The earlier treatment begins, the better the prognosis.

Author Contributions: Conceptualization, O.G., A.B. and O.A.; methodology, O.G., A.B. and O.A.; software, M.D., A.V. and H.G.; validation, A.B., O.G., O.A. and M.D.; formal analysis, M.D.; investigation, M.D., A.V. and H.G.; resources, M.D.; data curation, O.G., A.B., O.A. and M.D.; writing—original draft preparation, M.D.; writing—review and editing, A.B., O.A., M.D. and O.G.; visualization, O.A. and A.B.; supervision, O.G.; project administration, O.G. All authors have read and agreed to the published version of the manuscript.

Funding: This research received no external funding.

Institutional Review Board Statement: The study was approved by the competent Bioethics Committee of the University of Thessaly (approval numbers 49162/13-10-2017 and 49161/13-10-2017), and it was in line with the Declaration of Helsinki.

Informed Consent Statement: Informed consent was obtained from all subjects involved in the study.

Data Availability Statement: The data presented in this study are available on request from the corresponding author. Data are unavailable due to privacy restrictions.

Acknowledgments: This work was supported by the Department of Endocrinology and Metabolic Diseases, Medical School, University of Thessaly, Larissa, Greece, and specifically by Bargiota Alexandra, Endocrinologist and Director of the Endocrinology Clinic of the University Hospital of Larissa, who granted us permission to enter the clinic and allowed us to talk with the patients. Additionally, Tsolaki Catherine, Clinical Dietitian and Head of the Dietetics Department of the University Hospital of Larisa, and Sotiriou Evangelia, Dietitian, for their cooperation in collecting data and conducting interviews with the patients.

Conflicts of Interest: The authors declare no conflicts of interest.

References

1. Saeedi, P.; Petersohn, I.; Salpea, P.; Malanda, B.; Karuranga, S.; Unwin, N.; Colagiuri, S.; Guariguata, L.; Motala, A.A.; Ogurtsova, K. Global and regional diabetes prevalence estimates for 2019 and projections for 2030 and 2045: Results from the International Diabetes Federation Diabetes Atlas. *Diabetes Res. Clin. Pract.* **2019**, *157*, 107843. [CrossRef]
2. Iacobini, C.; Vitale, M.; Pesce, C.; Pugliese, G.; Menini, S. Diabetic complications and oxidative stress: A 20-year voyage back in time and back to the future. *Antioxidants* **2021**, *10*, 727. [CrossRef] [PubMed]
3. Echouffo-Tcheugui, J.B.; Perreault, L.; Ji, L.; Dagogo-Jack, S. Diagnosis and Management of Prediabetes: A Review. *JAMA* **2023**, *329*, 1206–1216. [CrossRef] [PubMed]
4. Abdullah, A.; Peeters, A.; de Courten, M.; Stoelwinder, J. The magnitude of association between overweight and obesity and the risk of diabetes: A meta-analysis of prospective cohort studies. *Diabetes Res. Clin. Pract.* **2010**, *89*, 309–319. [CrossRef] [PubMed]
5. Field, A.E.; Coakley, E.H.; Must, A.; Spadano, J.L.; Laird, N.; Dietz, W.; Rimm, E.; Colditz, G.A. Impact of overweight on the risk of developing common chronic diseases during a 10-year period. *Arch. Intern. Med.* **2001**, *161*, 1581–1586. [CrossRef] [PubMed]
6. Taylor, R. Insulin resistance and type 2 diabetes. *Diabetes* **2010**, *61*, 778–779. [CrossRef]
7. Forouhi, N.G.; Misra, A.; Mohan, V.; Taylor, R.; Yancyet, W. Dietary and nutritional approaches for prevention and management of type 2 diabetes. *BMJ* **2018**, *361*, k2234. [CrossRef]
8. Wing, R.R. Looking back and forward from the Diabetes Prevention Program (DPP): A commentary on the importance of research aimed at intervention optimization. *Health Psychol.* **2021**, *40*, 1009–1016. [CrossRef]
9. Powers, M.A.; Bardsley, J.K.; Cypress, M.; Funnell, M.M.; Harms, D.; Hess-Fischl, A.; Hooks, B.; Isaacs, D.; Mandel, E.D.; Maryniuk, M.D.; et al. Diabetes self-management education and support in adults with type 2 diabetes: A consensus report of the American Diabetes Association, the Association of Diabetes Care & Education Specialists, the Academy of Nutrition and Dietetics, the American Academy of Family Physicians, the American Academy of PAs, the American Association of Nurse Practitioners, and the American Pharmacists Association. *Diabetes Care* **2020**, *43*, 1636–1649.
10. Haynes, A.; Kersbergen, I.; Sutin, A.; Daly, M.; Robinson, E. A systematic review of the relationship between weight status perceptions and weight loss attempts, strategies, behaviours and outcomes. *Obes. Rev.* **2018**, *19*, 347–363. [CrossRef]
11. Rawshani, A.; Rawshani, A.; Franzén, S.; Sattar, N.; Eliasson, B.; Svensson, A.M.; Zethelius, B.; Miftaraj, M.; McGuire, D.K.; Rosengren, A.; et al. Risk factors, Mortality, and Cardiovascular Outcomes in patients with type 2 diabetes. *N. Engl. J. Med.* **2018**, *379*, 633–644. [CrossRef]
12. Magkos, F.; Hjorth, M.F.; Astrup, A. Diet and exercise in the prevention and treatment of type 2 diabetes mellitus. *Nat. Rev. Endocrinol.* **2020**, *16*, 545–555. [CrossRef]
13. Han, H.H.; Cao, Y.; Feng, C.; Zheng, Y.; Dhana, K.; Zhu, S.; Shang, C.; Yuan, C.; Zong, G. Association of a Healthy Lifestyle with All-Cause and Cause-Specific Mortality Among Individuals with Type 2 Diabetes: A Prospective Study in UK Biobank. *Diabetes Care* **2022**, *45*, 319–329. [CrossRef] [PubMed]
14. Swift, D.L.; McGee, J.E.; Earnest, C.P.; Carlisle, E.; Nygard, M.; Johannsen, N.M. The Effects of Exercise and Physical Activity on Weight Loss and Maintenance. *Progress. Cardiovasc. Dis.* **2018**, *61*, 206–213. [CrossRef] [PubMed]
15. Meng, Y.; Bai, H.; Wang, S.; Li, Z.; Wang, Q.; Chen, L. Efficacy of low carbohydrate diet for type 2 diabetes mellitus management: A systematic review and meta-analysis of randomized controlled trials. *Diabetes Res. Clin. Pract.* **2017**, *131*, 124–131. [CrossRef] [PubMed]
16. McMacken, M.; Shah, S. A plant-based diet for the prevention and treatment of type 2 diabetes. *J. Geriatr. Cardiol.* **2017**, *14*, 342–354. [CrossRef]
17. Hall, K.D.; Kahan, S. Maintenance of lost weight and long-term management of obesity. *Med. Clin.* **2018**, *102*, 183–197. [CrossRef] [PubMed]
18. Bailey, R.R. Goal setting and action planning for health behavior change. *Am. J. Lifestyle Med.* **2019**, *13*, 615–618. [CrossRef] [PubMed]
19. Sheeran, P.; Klein, W.M.; Rothman, A.J. Health behavior change: Moving from observation to intervention. *Annu. Rev. Psychol.* **2017**, *68*, 573–600. [CrossRef]
20. Newson, L.; Parody, F.H. Investigating the experiences of low-carbohydrate diets for people living with Type 2 Diabetes: A thematic analysis. *PLoS ONE* **2022**, *17*, e0273422. [CrossRef]

21. Behrouz, V.; Dastkhosh, A.; Sohrab, G. Overview of dietary supplements on patients with type 2 diabetes. *Diabetes Metab. Syndr. Clin. Res. Rev.* **2020**, *14*, 325–334. [CrossRef] [PubMed]
22. Papakonstantinou, E.; Oikonomou, C.; Nychas, G.; Dimitriadis, G.D. Effects of Diet, Lifestyle, Chrononutrition and Alternative Dietary Interventions on Postprandial Glycemia and Insulin Resistance. *Nutrients* **2022**, *14*, 823. [CrossRef]
23. Vlachos, D.; Malisova, S.; Lindberg, F.A.; Karaniki, G. Glycemic Index (GI) or Glycemic Load (GL) and Dietary Interventions for Optimizing Postprandial Hyperglycemia in Patients with T2 Diabetes: A Review. *Nutrients* **2020**, *12*, 1561. [CrossRef]
24. Esposito, K.; Maiorino, M.I.; Di Palo, C.; Giugliano, D. Adherence to a Mediterranean diet and glycaemic control in Type 2 diabetes mellitus. *Diabet. Med.* **2009**, *26*, 900–907. [CrossRef]
25. Rajpal, A.; Ismail-Beigi, F. Intermittent fasting and 'metabolic switch': Effects on metabolic syndrome, prediabetes and type 2 diabetes. *Diabetes Obes. Metab.* **2020**, *22*, 1496–1510. [CrossRef]
26. Salvia, M.G.; Quatromoni, P.A. Behavioral Approaches to Nutrition and Eating Patterns for Managing Type 2 Diabetes: A Review. *Am. J. Med. Open* **2023**, *9*, 100034. [CrossRef]
27. WHO. *Technical Advisory Group on Diabetes: Hybrid Meeting, 30 November–1 December 2022*; World Health Organization: Geneva, Switzerland, 2023.
28. WHO Multicentre Growth Reference Study Group; de Onis, M. Reliability of anthropometric measurements in the WHO Multicentre Growth Reference Study. *Acta Paediatr.* **2006**, *95*, 38–46.
29. Craig, C.L.; Marshall, A.L.; Sjöström, M.; Bauman, A.E.; Booth, M.L.; Ainsworth, B.E.; Pratt, M.; Ekelund, U.; Yngve, A.; Sallis, J.F. International physical activity questionnaire: 12-country reliability and validity. *Med. Sci. Sports Exerc.* **2003**, *35*, 1381–1395. [CrossRef]
30. Katsouyanni, K.; Rimm, E.B.; Gnardellis, C.; Trichopoulos, D.; Polychronopoulos, E.; Trichopoulou, A. Reproducibility and relative validity of an extensive semi-quantitative food frequency questionnaire using dietary records and biochemical markers among Greek schoolteachers. *Int. J. Epidemiol.* **1997**, *26*, 118–127. [CrossRef]
31. Friedewald, W.T.; Levy, R.I.; Fredrickson, D.S. Estimation of the concentration of low-density lipoprotein cholesterol in plasma, without use of the preparative ultracentrifuge. *Clin. Chem.* **1972**, *18*, 499–502. [CrossRef] [PubMed]
32. Matthews, D.R.; Hosker, J.P.; Rudenski, A.S.; Naylor, B.; Treacher, D.F.; Turner, R.C. Homeostasis model assessment: Insulin resistance and β-cell function from fasting plasma glucose and insulin concentrations in man. *Diabetologia* **1985**, *28*, 412–419. [CrossRef] [PubMed]
33. Blanton, C.A.; Moshfegh, A.J.; Baer, D.J.; Kretsch, M.J. The USDA Automated Multiple-Pass Method accurately estimates group total energy and nutrient intake. *J. Nutr.* **2006**, *136*, 2594–2599. [CrossRef]
34. ElSayed, N.A.; Aleppo, G.; Aroda, V.R.; Bannuru, R.R.; Brown, F.M.; Bruemmer, D.; Collins, B.S.; Hilliard, M.E.; Isaacs, D.; Johnson, E.L. 5. Facilitating Positive Health Behaviors and Well-being to Improve Health Outcomes: Standards of Care in Diabetes—2023. *Diabetes Care* **2023**, *46* (Suppl. S1), S68–S96. [CrossRef] [PubMed]
35. Bimpas, N.G.; Auyeung, V.; Tentolouris, A.; Tzeravini, H.; Eleutheriadou, I.; Tentolouris, N. Adoption of and adherence to the Hellenic Diabetes Association guidelines for the management of subjects with type 2 diabetes mellitus by Greek physicians. *Hormones* **2021**, *20*, 347–358. [CrossRef]
36. Reynolds, A.N.; Akerman, A.P.; Mann, J. Dietary fibre and whole grains in diabetes management: Systematic review and meta-analyses. *PLoS Med.* **2020**, *17*, e1003053. [CrossRef]
37. Group, L.A.R. The Look AHEAD study: A description of the lifestyle intervention and the evidence supporting it. *Obesity* **2006**, *14*, 737–752.
38. Sharma, R.; Prajapati, P. Diet and lifestyle guidelines for diabetes: Evidence based ayurvedic perspective. *Rom. J. Diabetes Nutr. Metab. Dis.* **2014**, *21*, 335–346. [CrossRef]
39. Ajala, O.; English, P.; Pinkney, J. Systematic review and meta-analysis of different dietary approaches to the management of type 2 diabetes. *Am. J. Clin. Nutr.* **2013**, *97*, 505–516. [CrossRef]
40. ElSayed, N.A.; Aleppo, G.; Aroda, V.R.; Bannuru, R.R.; Brown, F.M.; Bruemmer, D.; Collins, B.S.; Hilliard, M.E.; Isaacs, D.; Johnson, E.L. 3. Prevention or Delay of Type 2 Diabetes and Associated Comorbidities: Standards of Care in Diabetes—2023. *Diabetes Care* **2023**, *46* (Suppl. S1), S41–S48. [CrossRef]
41. Chaib, A.; Zarrouq, B.; El Amine Ragala, M.; Lyoussi, B.; Giesy, J.P.; Aboul-Soud, M.A.M.; Halim, K. Effects of nutrition education on Metabolic profiles of patients with type 2 diabetes mellitus to improve glycated hemoglobin and body mass index. *J. King Saud. Univ. —Sci.* **2023**, *35*, 1018–3647. [CrossRef]
42. Hansel, B.; Giral, P.; Gambotti, L.; Lafourcade, A.; Peres, G.; Filipecki, C.; Kadouch, D.; Hartemann, A.; Oppert, J.; Bruckert, E.; et al. A fully automated web-based program improves lifestyle habits and HbA1c in patients with type 2 diabetes and abdominal obesity: Randomized trial of patient e-coaching nutritional support (the ANODE study). *J. Med. Internet Res.* **2017**, *19*, e360. [CrossRef]
43. Maheri, A.; Asnaashari, M.; Joveini, H.; Tol, A.; Firouzian, A.A.; Rohban, A. The impact of educational intervention on physical activity, nutrition and laboratory parameters in type II diabetic patients. *Electron. Physician* **2017**, *9*, 4207. [CrossRef]
44. Sami, W.; Ansari, T.; Butt, N.S.; Hamid, M.R.A. Effect of diet on type 2 diabetes mellitus: A review. *Int. J. Health Sci. (Qassim)* **2017**, *11*, 65–71. [PubMed]

45. Vaccaro, O.; Masulli, M.; Bonora, E.; Del Prato, S.; Giorda, C.B.; Maggioni, A.P.; Mocarelli, P.; Nicolucci, A.; Rivellese, A.A.; Squatrito, S.; et al. Addition of either pioglitazone or a sulfonylurea in type 2 diabetic patients inadequately controlled with metformin alone: Impact on cardiovascular events. A randomized controlled trial. *Nutr. Metab. Cardiovasc. Dis.* **2012**, *22*, 997–1006. [CrossRef] [PubMed]
46. Duclos, M.; Oppert, J.-M.; Verges, B.; Coliche, V.; Gautier, J.-F.; Guezennec, Y.; Reach, G.; Strauch, G. Physical activity and type 2 diabetes. Recommandations of the SFD (Francophone Diabetes Society) diabetes and physical activity working group. *Diabetes Metab.* **2013**, *39*, 205–216. [CrossRef] [PubMed]
47. Pellegrini, C.A.; Verba, S.D.; Otto, A.D.; Helsel, D.L.; Davis, K.K.; Jakicic, J.M. The Comparison of a Technology-Based System and an In-Person Behavioral Weight Loss Intervention. *Obesity* **2012**, *20*, 356–363. [CrossRef]
48. Hu, Y.; Wen, X.; Wang, F.; Yang, D.; Liu, S.; Li, P.; Xu, J. Effect of telemedicine intervention on hypoglycaemia in diabetes patients: A systematic review and meta-analysis of randomised controlled trials. *J. Telemed. Telecare* **2019**, *25*, 402–413. [CrossRef] [PubMed]
49. Guo, S.H.-M. Assessing quality of glycemic control: Hypo- and hyperglycemia, and glycemic variability using mobile self-monitoring of blood glucose system. *Health Inform. J.* **2019**, *26*, 287–297. [CrossRef] [PubMed]
50. ElSayed, N.A.; Aleppo, G.; VR Aroda, V.R.; RR Bannuruet, R.R. 8. Obesity and Weight Management for the Prevention and Treatment of Type 2 Diabetes: Standards of Care in Diabetes—2023. *Diabetes Care* **2023**, *46* (Suppl. S1), S128–S139. [CrossRef]
51. Norris, S.L.; Zhang, X.; Avenell, A.; Gregg, E.; Bowman, B.; Serdula, M.; Brown, T.J.; Schmid, C.H.; Lau, J. Long-term effectiveness of lifestyle and behavioral weight loss interventions in adults with type 2 diabetes: A meta-analysis. *Am. J. Med.* **2004**, *117*, 762–774. [CrossRef]
52. Ryan, D.H.; Yockey, S.R. Weight Loss and Improvement in Comorbidity: Differences at 5%, 10%, 15%, and Over. *Curr. Obes. Rep.* **2017**, *6*, 187–194. [CrossRef]
53. Marathe, P.H.; Gao, H.X.; Close, K.L. American Diabetes Association Standards of Medical Care in Diabetes 2017. *J. Diabetes* **2017**, *9*, 320–324. [CrossRef]
54. Apergi, K.; Karatzi, K.; Reppas, K.; Karaglani, E.; Usheva, N.; Giménez-Legarre, N.; Moreno, L.A.; Dimova, R.; Antal, E.; Jemina, K. Association of breakfast consumption frequency with fasting glucose and insulin sensitivity/b cells function (HOMA-IR) in adults from high-risk families for type 2 diabetes in Europe: The Feel4Diabetes Study. *Eur. J. Clin. Nutr.* **2022**, *76*, 1600–1610. [CrossRef]
55. Sun, Y.; You, W.; Almeida, F.; Estabrooks, P.; Davy, B. The Effectiveness and Cost of Lifestyle Interventions Including Nutrition Education for Diabetes Prevention: A Systematic Review and Meta-Analysis. *J. Acad. Nutr. Diet.* **2017**, *117*, 404–421.e36. [CrossRef]
56. Rosenfeld, R.M.; Kelly, J.H.; Agarwal, M.; Aspry, K.; Barnett, T.; Davis, B.C.; Fields, D.; Gaillard, T.; Gulati, M.; Guthrie, G.E. Dietary interventions to treat type 2 diabetes in adults with a goal of remission: An expert consensus statement from the American College of Lifestyle Medicine. *Am. J. Lifestyle Med.* **2022**, *16*, 342–362. [CrossRef]
57. Flint, S.W.; Leaver, M.; Griffiths, A.; Kaykanloo, M. Disparate healthcare experiences of people living with overweight or obesity in England. *eClinicalMedicine* **2021**, *41*, 101140. [CrossRef]
58. Early, K.B.; Stanley, K. Position of the Academy of Nutrition and Dietetics: The Role of Medical Nutrition Therapy and Registered Dietitian Nutritionists in the Prevention and Treatment of Prediabetes and Type 2 Diabetes. *J. Acad. Nutr. Diet.* **2018**, *118*, 343–353. [CrossRef]
59. Acosta-Navarro, J.; Antoniazzi, L.; Oki, A.M.; Bonfim, M.C.; Hong, V.; Acosta-Cardenas, P.; Strunz, C.; Brunoro, E.; Miname, M.H.; Salgado Filho, W. Reduced subclinical carotid vascular disease and arterial stiffness in vegetarian men: The CARVOS Study. *Int. J. Cardiol.* **2017**, *230*, 562–566. [CrossRef]
60. Afshin, A.; Sur, P.J.; Fay, K.A.; Cornaby, L.; Ferrara, G.; Salama, J.S.; Mullany, E.C.; Abate, K.H.; Abbafati, C.; Abebe, Z. Health effects of dietary risks in 195 countries, 1990–2017: A systematic analysis for the Global Burden of Disease Study 2017. *Lancet* **2019**, *393*, 1958–1972. [CrossRef]
61. Cradock, K.A.; ÓLaighin, G.; Finucane, F.M.; Gainforth, H.L.; Quinlan, L.R.; Ginis, K.M. Behaviour change techniques targeting both diet and physical activity in type 2 diabetes: A systematic review and meta-analysis. *Int. J. Behav. Nutr. Phys. Act.* **2017**, *14*, 18. [CrossRef]
62. BANERJEE, M.; MACDOUGALL, M.; LAKHDAR, A.F. Impact of a single one-to-one education session on glycemic control in patients with diabetes. *J. Diabetes* **2012**, *4*, 186–190. [CrossRef] [PubMed]
63. Napoleone, J.M.; Miller, R.G.; Devaraj, S.M.; Rockette-Wagner, B.; Arena, V.C.; Venditti, E.M.; Kramer, K.; Strotmeyer, E.S.; Kriska, A.M. Impact of Maintenance Session Attendance and Early Weight Loss Goal Achievement on Weight Loss Success in a Community-Based Diabetes Prevention Program Intervention. *Sci. Diabetes Self-Manag. Care* **2021**, *47*, 279–289. [CrossRef]
64. West, D.S.; DiLillo, V.; Bursac, Z.; Gore, S.A.; Greene, P.G. Motivational Interviewing Improves Weight Loss in Women with Type 2 Diabetes. *Diabetes Care* **2007**, *30*, 1081–1087. [CrossRef]
65. Castro-Barquero, S.S.; Ruiz-León, A.M.; Sierra-Pérez, M.; Estruch, R.; Casas, R. Dietary strategies for metabolic syndrome: A comprehensive review. *Nutrients* **2020**, *12*, 2983. [CrossRef]
66. American Diabetes Association Professional Practice Committee. 5. Facilitating behavior change and well-being to improve health outcomes: Standards of Medical Care in Diabetes—2022. *Diabetes Care* **2022**, *45* (Suppl. S1), S60–S82. [CrossRef]
67. American Diabetes Association Professional Practice Committee. 3. Prevention or delay of type 2 diabetes and associated comorbidities: Standards of Medical Care in Diabetes—2022. *Diabetes Care* **2022**, *45* (Suppl. S1), S39–S45. [CrossRef]

68. Chew, E.Y. Dietary Intake of Omega-3 Fatty Acids from Fish and Risk of Diabetic Retinopathy. *JAMA* **2017**, *317*, 2226–2227. [CrossRef]
69. Tosti, V.; Bertozzi, B.; Fontana, L. Health benefits of the Mediterranean diet: Metabolic and molecular mechanisms. *J. Gerontol. Ser. A* **2018**, *73*, 318–326. [CrossRef]
70. Adu, M.D.; Malabu, U.H.; Malau-Aduli, A.E.; Malau-Aduli, B.S. Enablers and barriers to effective diabetes self-management: A multi-national investigation. *PLoS ONE* **2019**, *14*, e0217771. [CrossRef]
71. Heberle, I.; de Barcelos, G.T.; Silveira, L.M.P.; Costa, R.R.; Gerage, A.M.; Delevatti, R.S. Effects of aerobic training with and without progression on blood pressure in patients with type 2 diabetes: A systematic review with meta-analyses and meta-regressions. *Diabetes Res. Clin. Pract.* **2021**, *171*, 108581. [CrossRef]
72. Savikj, M.; Zierath, J.R. Train like an athlete: Applying exercise interventions to manage type 2 diabetes. *Diabetologia* **2020**, *63*, 1491–1499. [CrossRef] [PubMed]
73. Meuffels, F.M.; Isenmann, E.; Strube, M.; Lesch, A.; Oberste, M.; Brinkmann, C. Exercise Interventions Combined with Dietary Supplements in Type 2 Diabetes Mellitus Patients—A Systematic Review of Relevant Health Outcomes. *Front. Nutr.* **2022**, *9*, 817724. [CrossRef]
74. Campagna, D.; Alamo, A.; Di Pino, A.; Russo, C.; Calogero, A.; Purrello, F.; Polosa, R. Smoking and diabetes: Dangerous liaisons and confusing relationships. *Diabetol. Metab. Syndr.* **2019**, *11*, 85. [CrossRef]
75. ElSayed, N.A.; Aleppo, G.; Aroda, V.R.; Bannuru, R.R.; Brown, F.M.; Bruemmer, D.; Collins, B.S.; Hilliard, M.E.; Isaacs, D.; Johnson, E.L. 1. Improving Care and Promoting Health in Populations: Standards of Care in Diabetes—2023. *Diabetes Care* **2023**, *46* (Suppl. S1), S10–S18.76. [CrossRef] [PubMed]
76. American Diabetes Association. Standards of Care in Diabetes—2023 Abridged for Primary Care Providers. *Clin. Diabetes* **2023**, *41*, 4–31. [CrossRef] [PubMed]

Article

The Mulberry Juice Fermented by *Lactiplantibacillus plantarum* O21: The Functional Ingredient in the Formulations of Fruity Jellies Based on Different Gelling Agents

Aleksandra Szydłowska [1,*], Dorota Zielińska [1], Barbara Sionek [1] and Danuta Kołożyn-Krajewska [1,2]

[1] Department of Food Gastronomy and Food Hygiene, Institute of Human Nutrition Sciences, Warsaw University of Life Sciences (WULS), Nowoursynowska St. 159C, 02-776 Warsaw, Poland; dorota_zielinska@sggw.edu.pl (D.Z.); barbara_sionek@sggw.edu.pl (B.S.); danuta_kolozyn_krajewska@sggw.edu.pl (D.K.-K.)

[2] Department of Dietetics and Food Studies, Faculty of Science and Technology, Jan Dlugosz University in Czestochowa, Al. Armii Krajowej 13/15, 42-200 Częstochowa, Poland

* Correspondence: aleksandra_szydlowska@sggw.edu.pl

Abstract: This study aimed to investigate the effects of adding probiotics, prebiotics, and different types of jelly agents on a few key quality attributes of potentially functional mulberry jellies throughout a 10-day storage period at 4 °C. Mullbery juice was separately fermented at 37 °C for 24 h using *Lactiplantibacillus plantarum* O21; it was a favorable matrix for the proliferation of probiotics. Lactic acid fermentation positively affected the total anthocyanin concentration of investigated products. Also, antioxidant capacities of mulberry juices were improved by *L. plantarum* O21 fermentation. The results showed that the applied prebiotic–inulin addition and agar–agar addition, as a gelling agent in recipes of potentially functional mulberry jellies, were proved to be beneficial technological solutions, both in fresh and stored products, and obtained an appropriate, high number of LAB bacteria, good sensory quality, and beneficial antioxidant properties.

Keywords: mulberry juice; probiotics; lactic acid fermentation; antioxidant activity; gelatin; agar–agar; sensory properties

Citation: Szydłowska, A.; Zielińska, D.; Sionek, B.; Kołożyn-Krajewska, D. The Mulberry Juice Fermented by *Lactiplantibacillus plantarum* O21: The Functional Ingredient in the Formulations of Fruity Jellies Based on Different Gelling Agents. *Appl. Sci.* **2023**, *13*, 12780. https://doi.org/10.3390/app132312780

Academic Editors: Theodoros Varzakas and Maria Antoniadou

Received: 2 November 2023
Revised: 21 November 2023
Accepted: 27 November 2023
Published: 28 November 2023

1. Introduction

Customers' appreciation of the relationship between food and wellness has resulted in a significant shift in eating habits and lifestyle changes. The rise in consumer knowledge has been one of the main forces behind the development of functional food products that may satisfy both basic nutritional needs and offer health advantages. The term "functional food" refers to natural and industrially processed foods, which "when regularly consumed within a diverse diet at efficacious levels have potentially positive effects on health beyond basic nutrition" [1]. One of the functional food segments is the so-called "probiotic food".

It is generally known that lactic acid bacteria (LAB) fermentation has a considerable potential to enhance the functional, nutritional, and sensory qualities of both plant-based and animal feeds [2,3]. The definition of probiotic bacteria is "live microorganisms that, when administered in adequate amounts, confer a health benefit on the host". This definition captures the essence of probiotics (microbial, viable, and health-promoting) while embracing a wide range of microorganisms and uses [4]. As of right now, probiotics for human usage are limited to lactic acid bacteria that have been isolated from the gastrointestinal system.

However, the International Scientific Association of Probiotics and Prebiotics (ISAPP) in a position statement [5] agreed with the definition that was offered [4]. The term "probiotic" in this context does not refer only to traditional probiotics. Innovation will surely lead to the isolation of promising probiotics from novel sources with intriguing new health benefits and hitherto undiscovered functions [6]. Moreover, some authors report that strains considered probiotics can be extracted from fermented products of animal origin [7–9] and plant origin [10–13]. Traditional fermented products are a rich source of microorganisms, some of which may have probiotic properties [14]. On the other hand, prebiotics set an example of food ingredients, which are not digested by endogenous enzymes of the human gastrointestinal tract, but they end up intact in the colon, where they ferment, providing food for probiotic microorganisms. In addition, prebiotics are broken down by sucrose bacteria present in the lower gastrointestinal tract and have the ability to stimulate their growth. Inulin and oligofructose are the most effective and most commonly used prebiotics [15].

Growth can be noted in in the popularity of using plant diets among consumers in the world, due to the growing level of knowledge regarding its health-promoting aspects and the availability of products on the food market that enables the incorporation of such a diet [16,17]. It has recently appeared that there is a modern tendency in the development of vegan probiotic food products. The failure to comply with the criterion of "plant origin" in the case of the probiotic bacterial strain used may deprive the final product of the status of vegan food [18]. Considering the biochemical composition of fruits and vegetables, it has been discovered that they are excellent starting materials for the fermentation of probiotics. In addition to their phytochemical content, which promotes health, fruits and vegetables also provide a number of benefits (such as being high in sugar and nutrients) for the growth and survival of probiotics [19]. Fruits and vegetables have a limited shelf life, thus it is critical to perform probiotic fermentation to add value and extend the shelf life of processed foods [20]. The black mulberry (*Morus nigra* L.) belongs to the *Moraceae* family. It is a multipurpose plant that can be used for a variety of purposes including fuel, fodder, fiber, and fruit. It is abundant in bioactive substances, such as bioflavonoids and non-anthocyanin, which are responsible for its medicinal properties and has earned it a "superfood" status in European countries [21–23]. Recently, the mulberry has gained prominence because of its phytochemical makeup and positive health impacts on humans, including its immunomodulatory, antidiabetic, antioxidant, and anticancer qualities [24]. This plant has also been reported to be used in Chinese conventional medicine for fever treatment, diabetes, obesity, blood pressure, urinal disorders, liver damage prevention, atherosclerosis, inflammation, and body joint strengthening, among other things [25,26]. On the other hand, Bong et al. 2019 [27] reported that LAB-biotransformed mulberry fruit extract showed antibacterial action by inhibiting *S. Typhimurium* growth and biofilm formation, and showed an anti-inflammatory effect in the bacterially infected intestinal epithelial cells.

Fruits of the black mulberry (*Morus nigra* L.) have a distinct flavor with juicy and acidic qualities and high staining activity, and are therefore attractive for usage in the food processing industry. Mulberry-based products such as fruit juice [28], liquor [29] ice cream, jam or muffins have been developed and manufactured [30]. Mulberry fruits, due to their mouth-watering taste, are a plant used both fresh and added to food products such as yoghurt, vegetables or muesli [31,32]; they also have a culinary use [33]. By the relatively high water activity (A_w) and low acidity, mulberry fruits are difficult to preserve. Due to this, there is a great need to use technologies to extend the mulberry fruit's shelf life and improve its nutritional, organoleptic, and health benefits in order to increase the commercial production for benefits to both the economy and public health. In recent years, researchers have demonstrated that mulberry fruit can be also a healthy food matrix for probiotic delivery [34–37].

In light of this, the objective of this study was to estimate the possibility of developing potentially functional mulberry jellies based on a probiotic bacterial strain isolated from traditional plant food and impact of the type of jelly agent used for the selected quality properties of these products during 10 days of refrigerated storage.

2. Materials and Methods

2.1. Materials

Mulberry (*Morus nigra* L.) pasteurized juice (Eka Medica Co., Kozy, Poland), inulin (Frutafit® Tex, Roosendaal, The Netherlands), agar–agar (Agnex, Białystok, Poland), and Sucrose (Diamant, Gostyń, Poland) were purchased at the local market in the city (Warsaw), Poland. Distilled water was used in the preparation of the products.

2.2. Preparing of Bacterial Cultures

In this study, one bacterial strain with probiotic properties—*Lactiplantibacillus plantarum* O21 (GenBank accession KM 186159)—was applied. It was obtained from a pure culture maintained in the collection of the Department of Food Gastronomy and Food Hygiene, Institute of Human Nutrition Sciences at Warsaw University of Life Sciences. Some probiotic properties of the used strain were previously described [38].

The bacterial strains were cultured by a two-fold passage in modified Vegitone MRS broth (Sigma–Aldrich Co., Darmstadt, Germany) with 2% (*v/v*) of inoculum after being kept at −80 °C in 20% (*m/w*) glycerol.

For 5 min, the bacterial culture was centrifuged at 10,000× *g*. After obtaining the cell pellets, they were cleaned and resuspended to their original volume in 10 mL of sterile 0.85% saline solution. The ready bacterial culture was then added to 400 mL of mulberry juice that contained 5% *v/v* sucrose.

In the prepared starting culture, there were roughly 9.75 log CFU (colony forming units) mL^{-1} of probiotic bacteria. A volume of mulberry juice containing 2% (*v/v*) of the bacterial strain was added (Table 1).

Table 1. Mulberry jelly formulations.

Formulation	Ingredient [% *w/v*]						
	Inulin	Agar–Agar	Gelatin	Mulberry Juice	Distilled Water	Inoculum	Sucrose
G 0P	0	0	2	70	21.0	2	5
G I-1	1	0	2	70	20.0	2	5
G I-3	3	0	2	70	18.0	2	5
A 0P	0	2	0	70	21.0	2	5
A I-1	1	2	0	70	20.0	2	5
A I-3	3	2	0	70	18.0	2	5

Explanatory notes: G—product based on gelatin; A—product based on agar–agar; 0P—no addition of prebiotics; I-1—with 1% *w/v* of inulin addition; I-3—with 3% *w/v* of inulin addition.

Every determination was made three times. The results that were collected were presented as mean ± standard deviation (SD).

2.3. Preparing of Jelly Formulations

For the study, a total of 6 formulations of jellies were produced: 3 samples with a gelatin addition and 3 samples with an agar–agar addition (I-1 with 1% of inulin addition; I-3 with 3% of inulin addition; 0P—no addition of prebiotics). All manufactured samples are shown in Table 1. All mixtures contained mulberry juice (70% *v/v*), sucrose (5% *w/v*), 1% (*w/v*) or 3% (*w/v*) of a prebiotic addition or without a prebiotic addition, and 2% (*v/v*) of an inoculum addition, respectively. It was fermented using the potentially probiotic bacterial strain *L. plantarum* O21 strain, at 37 °C for 24 h. Gelatin and agar–agar were used as the gelling agents.

The fermentation process was carried out in sterile, glass vessels of a 500 mL capacity. The fermented juices were then combined with the other recipe ingredients; gelatin (2% *w/v*) or agar–agar (2% *w/v*) prepared with distilled water, with a final volume of 400 mL.

The samples were prepared as follows:

Gelatin: gelatin was thoroughly dissolved in a suitable portion of hot water, cooled and added to fermented juice. Then, 200 mL were placed in plastic vessels and cooled. After 3 h, it was placed at 4 °C.

Agar–agar version: a large portion of agar–agar was poured into a suitable portion of cold water and left until it swelled (approximately, 10 min.). Then, it was transferred to a small pot and heated until boiling (90–100 °C), dissolving the agar–agar. Then, it was cooled and other ingredients were added. Finally, the mixtures were divided into 200 mL pieces, each of which were put into a plastic container.

The product remained at room temperature (22 °C to concentration; after 3 h, it was placed in the above-refrigerated conditions, at 4 °C).

The investigated samples were stored at refrigerated conditions for 24 h. Next, the particular determinations were performed.

All determinations were performed in triplicate. The obtained results were expressed as a mean ± standard deviation (SD).

2.4. Microbiological Analysis

The TEMPO® System, an automated quality indicator system (used by BioMérieux, Mercy Etoile, France), performed the analyses. The TEMPO® System was used to calculate the amount of LAB bacteria present in investigated samples (log CFU g^{-1}). The most probable number (MPN) algorithm was used to create this system. The samples were diluted by 1/400 in a single vial. The Tempo Filler moved the inoculated medium into the Tempo card.

The cards were incubated at 37 °C for 48 h before the Tempo LAB test was able to obtain performance levels that were comparable to the NF ISO15214: 1998 standard [39]. The data were automatically processed by the software system that determines which of the wells tested positive. The volume of wells and sample dilution used to count the positive wells allowed for an automatic conversion of the results to log CFU g^{-1}.

All determinations were performed in triplicate. The obtained results were expressed as a mean ± standard deviation (SD).

2.5. Acidity Analysis (pH)

Using a pH meter, the studied jellies' pH values were determined three times. (Elmetron, CP551, Zabrze, Poland). A total of 20 g of the product sample was utilized for each measurement. The acquired results were interpreted with 0.001 accuracy.

All determinations were performed in triplicate. The obtained results were expressed as a mean ± standard deviation (SD).

2.6. Total Anthocyanin Concentration

The total anthocyanin concentration (TAC) was determined by a pH differential method, which was adapted from the methods described by Tchabo et al., 2017 [40]. Two buffer solutions were prepared: CH_3COONa (0.4 mol/L) at pH 4.5; KCl (0.025 mol/L) at pH 1. Two sets of trials were prepared. A total of 100 µL was measured. All determinations were performed in triplicate. The obtained results were expressed as a mean ± standard deviation (SD).

2.7. Extract Preparation

The extraction of gels was conducted before the total anthocyanin content and antioxidant activity were assessed. Then, 2 g of gels with 20 mL of acidified methanol (the HCl: methanol ratio was 1:99) were mixed and homogenized. After the homogenized mixture was filtered and allowed to stand for 24 h, the extracted mixture was utilized for the aforementioned analyses.

Cyanidin-3-glucoside chloride was used to express the results because it is the most common anthocyanin in plants. A 2 mL buffer with pH = 1 (KCl/HCl) was added to the sample, and the absorbance of the solution at 520 nm and 700 nm was measured by using the spectrophotometer Genesys TM 20 (Thermo Scientific Co., Waltham, MA, USA). The 2 mL buffer with a pH (4.5) $CH_3COONa/CH_3 COOH$ = was added to the second portion of the sample.

The TAC was expressed as milligram equivalent of cyanidin 3-gucoside per mL of juice. It was calculated according to the following Equation (1):

$$TAC = [(A_1 - A_2) - (A_3 - A_4)] \times \frac{MW \times DF \times 10^2}{\varepsilon \times L} \tag{1}$$

where A_1 stands for the absorbance at 520 nm with pH 1.0; A_2 stands for the absorbance at 700 nm with pH 1.0; A_3 stands for the absorbance at 520 nm with pH 4.5; A_4 stands for the absorbance at 700 nm with pH 4.5; *MW*—molecular weight of cyanidin-3-glucoside (449.2 g/mol); *DF*—dilution factor (100); L—path length (1 cm); ε—molar extinction coefficient for cyanidin-3-glucoside (26.900 L/mol·cm).

All determinations were performed in triplicate. The obtained results were expressed as a mean $\pm$ standard deviation (SD).

2.8. Determination of Antioxidant Activity

2.8.1. DPPH Radical Scavenging Assay

The antioxidant activity was determined by updated existing methods for measuring Alothman, Bhat, and Karim (2009) [41] and Brand–Wiliams et al. (1995) [42]. 2-diphenyl-1-picrylhydrazyl (DPPH), a synthetic radical, was used in the procedure (Sigma–Aldrich, Darmstadt, Germany). A total of 0.012 g of DPPH (M = 394.32 g/mol), produced as a DPPH solution, was dissolved in 100 cm^3 of ethanol. Crushed bar samples weighing 25 g were added to 100 mL of ethanol to create extracts. The solution was shaken. Then, after 20 h, it was filtered.

The solution was kept in a dark place. By mixing up to 0.2 cm^3 of DPPH solution with 0.8 cm^3 of ethanol, the absorbance A0 was calculated. A total of 0.2 cm^3 of DPPH solution, 0.6 cm^3 of ethanol, and 0.2 cm^3 of the test extract were all present in the test sample. The absorbance was measured at 517 nm using a spectrophotometer GenesysTM 20 (Thermo Scientific Co., Waltham, MA, USA). A measurement of absorbance (A) was made 5 and 30 min. after the process started. The solution's mean absorbance value (AR) was computed after three repetitions of each measurement. When the DPPH solution was lowered by test samples of bars, the violet color vanished.

The spectrophotometer at wavelength = 517 nm captured the reduction in absorbance. Ethanol was used to calibrate the utilized machinery. Applying the subsequent Equation (2), the *inhibition %* of the DPPH radical discoloration was determined:

$$\% \ Inhibition = 100 \ (A_0 - Ar)/A_0 \tag{2}$$

where A_0 is the absorbance of the control; Ar is the absorbance of the extract.

All determinations were performed in triplicate. The obtained results were expressed as a mean $\pm$ standard deviation (SD).

2.8.2. ABTS Radical Scavenging Assay

Following Re et al., 1999 [43], the radical scavenging assay for ABTS (2,2′-azino-bis 3-ethylbenzothiazoline-6-sulphonic acid, Sigma Aldrich, Darmstadt, Germany) was carried out. After 6 min of incubation with the extracts, fluctuations in the concentration of the ABTS•+ radical cations were assessed using a spectrophotometer, the GenesysTM 20 (Thermo Scientific Co., Waltham, MA, USA). The decrease in the absorbance of the solution at a 734 nm wavelength indicates that the antioxidative characteristics of these extracts have lowered the levels of ABTS•+. Equation (2) is used to calculate the extracts' capacity to thwart the oxidation reaction based on the formula.

All determinations were performed in triplicate. The obtained results were expressed as a mean $\pm$ standard deviation (SD).

2.9. Color Measurement

The Konica Minolta CM-2300d spectrophotometer (Konica Minolta Business Technologies, Inc., Osaka, Japan) was used to measure the CIE color parameters (L*, a* and b*) according to the method described by Fazaeli et al., 2013 [44]. A white reference tile was used to calibrate the spectrophotometer.

- The relative color difference index (ΔE) and the index of the hue angle (H^0) were calculated according to the following Equation (3):

$$\Delta E = \left[\left(L_0 - L\right)^2 + \left(a_0 - a\right)^2 + \left(b_0 - b\right)^2 \right]^{0.5} \tag{3}$$

where:

Parameter L*—lightness coefficient (L* = 0 indicates black and L* = 100 indicates white [dimensionless value]);
Parameter a*—red color coefficient (dimensionless value);
Parameter b*—yellow color coefficient (dimensionless value);
L*0, a*0, b*0—color coefficients related to for unfermented juice (dimensionless value).

- The index of hue angle (H^0), Equation (4):

$$H^0 = \tan^{-1}(b*/a*) \tag{4}$$

- Chroma (C), Equation (5):

$$C = (a^2 + b^2)^{1/2} \tag{5}$$

All determinations were performed in triplicate. The obtained results were expressed as a mean ± standard deviation (SD).

2.10. Sensory Analysis

The quantitative descriptive analysis (QDA) following the ISO procedure [45] was performed to evaluate the sensory characteristics of the investigated jellies. The sensory estimation of the products was carried out, 1 day after the manufacturing process and 10 days after the refrigerated storage. Ten panelists of WULS's Department of Food Gastronomy and Food Hygiene took part in the sensory evaluation. The panelists received instruction in the fundamentals of sensory evaluation methodology [46]. Experts in an age range from 22 to 55 who possessed strong expertise in sensory assessment techniques, such as estimating the profiles of frozen delicacies such as fruit jellies, were included.

The chosen sensory attributes of mulberry jellies fermented by *Lactiplantibacillus plantarum* O21 were as follows: color, compactness, smoothness, watery, sweet odor, bitter odour, acid odor, other odor, sweet taste, bitter taste, acid flavor, and other flavors as well as overall quality.

The panelists' task was to mark the intensity of each of the quality attributes and conduct their assessment on an appropriate scale (linear graphical scale of 0 (low)–10 (high) conventional units (c.u.).

All determinations were performed in triplicate. The obtained results were expressed as a mean ± standard deviation (SD).

2.11. Statistical Analysis

All measurements were performed in triplicates. A one-way analysis of variance (ANOVA) test was followed by Fisher NIR test, with the overall significance level set to 0.05.

Pearson's correlation was used to identify the correlation of TAC and different color values (L* a* b* H^0 C ΔE) in investigated samples.

Principal component analysis (PCA) is a statistical method of reducing the dimensionalities of high-dimensional datasets in a variety of research areas [47]. In this study, the PCA method was used to estimate the changes in the sensory quality of potentially functional mulberry jellies during storage and to interpret the overall sensory quality, color evaluation, antioxidant activity, and the total anthocyanin content results of investigated products.

All statistical analyses were performed using STATISTICA 13.3 PL software (StatSoft, Kraków, Poland).

3. Results

3.1. Microbiological Analysis

At the first stage of the study, during the fermentation process and as a result of the metabolic activity of *Lactiplantibacillus plantarum* O21, a decrease in pH values in each sample was noted. The pH of fermented mulberry juice dropped from the initial 4.8 to 3.94 (juice with 3% of inulin addition), and then to 4.09 (control sample, without prebiotic addition), respectively.

The results of the study indicate that the highest significant count of LAB bacteria ($p < 0.05$) among all fermented samples of fruit juice was observed in juice with a 3% of inulin addition at 9.35 log CFU g^{-1}, respectively. On the other hand, in the case of the control sample, the lowest level of the count of bacteria was noted at the level of 9.05 log CFU g^{-1}, respectively. These determinations were the initial stages necessary to select fermentation conditions and to determine additive levels. However, they were not critical to the aim of this study.

The results of changes in the pH value and the count of LAB bacteria in potentially functional mulberry jellies fermented by *Lactiplantibacillus plantarum* O21, based on different gelling agents during 10 days of storage at 4 °C, are shown in Figure 1a,b.

As a result of the fermentation process and the addition of fermented mulberry juice into the jelly recipes, the pH decrement was accompanied by culture growth in the investigated products (Figure 1a). As illustrated in Figure 1, the statistically significant highest count of LAB bacteria was obtained in sample AI-3 (with 3% *w/v* of inulin addition; 2% *w/v* agar–agar; 7.5 log CFUg^{-1}), respectively. In addition, this results in a pH value that is noticeably lower (3.6), respectively (Figure 1a), when compared to other investigated products. A similar trend in the pH value and bacterial count was also observed in the stored products (Figure 1b).

In summary, regardless of the recipe composition and storage time, each of the evaluated products had a sufficient amount of LAB probiotic bacteria in their concentration (higher than 6.5 log CFU g^{-1}, respectively) at a minimum level of 6 log CFU g^{-1}.

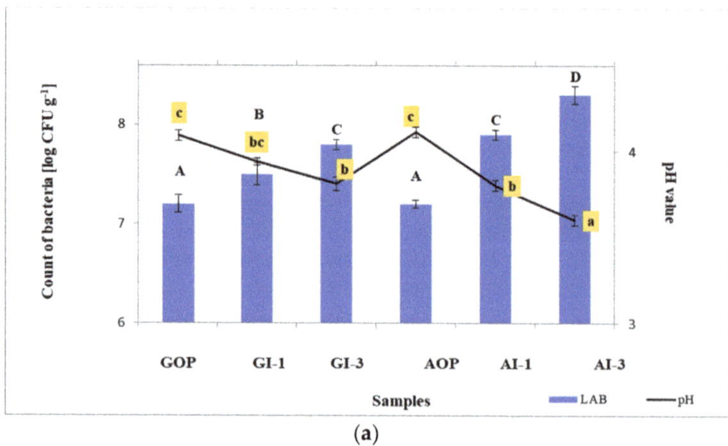

(a)

Figure 1. *Cont.*

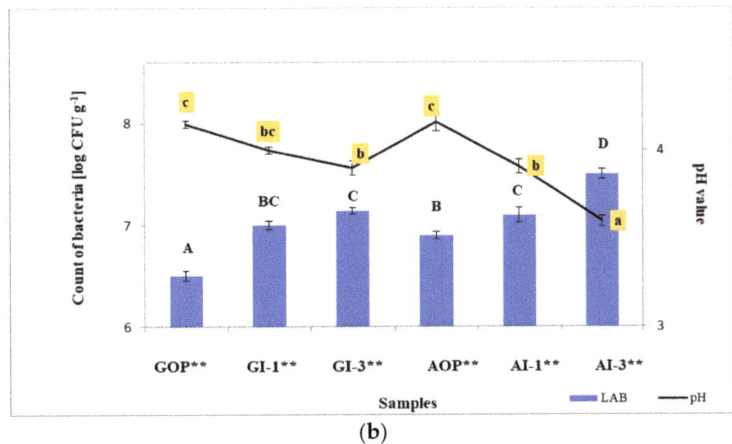

(b)

Figure 1. Changes in pH value and the count of LAB bacteria in investigated mulberry jellies fermented by *Lactiplantibacillus plantarum* O21, based on different gelling agents during 10 days of storage at 4 °C ((**a**)—fresh products; (**b**)—stored products). ** Explanatory notes: The presented samples are coded according to Section 2, Table 1. The results are expressed as the mean ± standard deviation ($n = 3$). Mean values denoted by the same letters (upper case letters—in regards to the LAB count, and lower case letters—in regards to the pH value) do not differ significantly ($p > 0.05$).

3.2. Color Measurement and Total Anthocyanin Concentration

The changes in total anthocyanin content and color values (L* a* b* H⁰ C ΔE) in the investigated samples of fruity jellies are shown in Table 2. The effect of fermentation on color properties of investigated samples was examined along with the following color features (L*, a* and b*) and the total color difference (ΔE).

Table 2. Changes in total anthocyanin concentration and color values (L* a* b* H⁰ C ΔE) in investigated mulberry jellies fermented by *Lactiplantibacillus plantarum* O21, based on different gelling agents during 10 days of storage at 4 °C.

Sample Code	Total Anthocyanin Concentration (TAC) (mg/mL)	L*	a*	b*	H⁰	C*	ΔE
UMJ	0.90 ± 0.24	10.5 ± 0.10	31.50 ± 0.13	9.80 ± 0.16	17.28 ± 0.16	32.90 ± 0.11	-
FMJ	1.70 ± 0.12	10.00 ± 0.14	31.82 ± 0.10	9.45 ± 0.17	16.54 ± 0.13	33.19 ± 0.12	0.72 ± 0.16
GOP	1.20 ± 0.55 [Ab]	9.50 ± 0.10 [Ca]	32.30 ± 1.14 [Aa]	9.10 ± 0.15 [Eb]	15.75 ± 0.09 [Fb]	33.5 ± 0.90 [Ab]	1.46 ± 0.85 [Aa]
GOP **	0.95 ± 0.03 [aba]	10.45 ± 0.14 [bb]	31.80 ± 0.14 [ab]	8.15 ± 0.10 [aa]	14.57 ± 0.08 [aa]	32.82 ± 0.09 [Aa]	2.35 ± 0.15 [cdh]
GI1	1.40 ± 0.41 [Bb]	9.31 ± 0.06 [Ba]	32.80 ± 0.09 [Fb]	8.80 ± 0.12 [Cb]	15.11 ± 0.13 [Da]	33.96 ± 0.90 [Fab]	1.83 ± 0.08 [Ca]
GI1 **	1.25 ± 0.21 [ca]	9.71 ± 0.11 [cb]	32.40 ± 0.09 [bca]	9.20 ± 0.09 [da]	15.80 ± 0.09 [db]	33.68 ± 0.90 [ba]	1.86 ± 0.32 [ba]
GI3	1.55 ± 0.11 [Cab]	9.26 ± 0.23 [Aa]	33.58 ± 0.19 [Db]	8.65 ± 1.23 [Ba]	14.57 ± 0.11 [Ca]	32.26 ± 0.90 [Aa]	2.35 ± 0.10 [Da]
GI3 **	1.50 ± 0.25 [ea]	9.75 ± 0.22 [cdb]	32.80 ± 0.0 [da]	8.95 ± 0.19 [ca]	15.21 ± 0.09 [bb]	34.00 ± 0.09 [cb]	2.37 ± 0.12 [cda]
AOP	1.25 ± 0.06 [ABb]	9.35 ± 0.21 [Ba]	33.21 ± 0.09 [Ca]	8.95 ± 0.28 [Da]	15.59 ± 0.09 [Ea]	33.32 ± 0.22 [Ba]	1.55 ± 0.10 [Bb]
AOP **	0.90 ± 0.18 [aa]	10.22 ± 0.21 [ebc]	32.60 ± 0.18 [cb]	9.55 ± 0.21 [eab]	16.17 ± 0.12 [eb]	33.94 ± 0.11 [bb]	1.14 ± 0.11 [aa]
AI1	1.40 ± 0.09 [Bb]	9.25 ± 0.26 [Aa]	33.50 ± 0.19 [Db]	8.60 ± 0.61 [Ba]	14.41 ± 0.10 [Ba]	34.60 ± 0.03 [Da]	2.65 ± 0.15 [Eb]
AI1 **	1.15 ± 0.19 [ba]	9.55 ± 0.09 [bb]	33.35 ± 0.08 [ea]	8.95 ± 0.13 [cab]	15.37 ± 0.11 [cb]	34.50 ± 0.09 [da]	2.22 ± 0.09 [ca]

Table 2. *Cont.*

Sample Code	Total Anthocyanin Concentration (TAC) (mg/mL)	L*	a*	b*	H^0	C*	ΔE
AI3	1.63 ± 0.03 [Db]	9.31 ± 0.15 [Ba]	33.60 ± 0.11 [DEb]	8.50 ± 0.10 [Aa]	14.20 ± 0.11 [Aa]	34.65 ± 0.11 [DEb]	2.74 ± 0.07 [Fb]
AI3 **	1.45 ± 0.06 [da]	9.36 ± 0.09 [aab]	33.40 ± 0.0 5 [fa]	8.70 ± 0.21 [bb]	14.57 ± 0.13 [ab]	34.51 ± 0.05 [da]	2.47 ± 0.11 [ea]

Explanatory notes: The presented samples are coded according to Section 2, Table 1. **—product stored during 10 days storage at 4 °C. The results are expressed as the mean ± standard deviation ($n = 3$). Values denoted by different letters differ significantly ($p < 0.05$). UMJ—unfermented mulberry juice; FMJ—fermented mulberry juice; TAC—Total anthocyanin concentration; L*—Lightness; a*—Redness; b*—Yellowness; C*—Chroma; H^0—Hue angle; ΔE—Relative color difference index. Values denoted by different capital letters in the fresh products, differ significantly ($p < 0.05$). Values are denoted by different lowercase letters in the stored products, significantly ($p < 0.05$). Values are denoted by different color lowercase letters in the same fresh and stored batch products, significantly ($p < 0.05$).

Results posted in Table 2 allow for an estimation of the impact of the fermentation process on color properties of investigated products. The recorded color parameters in the case of fruity jellies based on fermented mulberry juice, testify to the greater saturation of the red component—the value of the parameter a* (+31.8–+33.6) and lower yellow saturation (parameter value b* (+8.15–+9.55) (Table 2).

Due to the process of fermentation, there was an increase in the content of the total anthocyanin concentration (TAC) in mulberry juice, from 0.90 mg/mL to 1.70 mg/mL, respectively. Despite a slight decrease in the TAC content following the combination of all the recipe's ingredients, in fresh jelly samples, significantly higher ($p < 0.05$) (1.20–1.55 mg/mL) TAC compared to the sample of unfermented juice (slightly higher values were obtained for agar-concentrated samples—AI 1 and AI 3). Regardless of the gelling agent used, the inulin addition statistically and significantly influenced ($p < 0.05$) the increase in TAC content in fresh products, which, at the same time, corresponded to the increased number of bacteria *Lactiplantibacillus plantarum* O21 in these samples (Figure 1a). As a result of the refrigerated storage, TAC content significantly decreased ($p < 0.05$), regardless of the product type (prebiotic addition; used gelling agent). Also note that the used gelling agent and prebiotic addition affect TAC content after the refrigerated storage. In products based on gelatin, with the addition of a prebiotic (regardless of the added level), a smaller loss of TAC content (Table 2) was observed as compared to agar–agar-based tests.

In the finished goods, a slight decrease in L * was observed in fermented mulberry juice ($p < 0.05$), which indicated the darkening of the color after combining all recipes. The values of the L* and b* parameter decreased during juice fermentation and during the production of potentially functional jellies. An increase in the color parameter—a* was also observed in finished products compared to the fermented intermediate—mulberry juice (Table 2).

There was an increase in the value of the ΔE parameter as a result of combining all components and obtaining mulberry jelly tests. Fresh products based on both agar and gelatin (GO1, GI3, AI1, and AI3), with the addition of a prebiotic, displayed noticeably greater ΔE values ($p < 0.05$) in relation to control tests (G0P and A0P). However, significantly higher values of the ΔE parameter were recorded for the jelly agar-agar-based tests. As a result of the refrigerated storage, there was a slight but statistically significant ($p < 0.05$) reduction in these values in the case of the agar–agar product-based and the control sample, G0P (Table 2).

Results shown in Table 2 are verified by a noteworthy inverse relationship between parameters L* and TAC (R2 = −0.763; $p < 0.05$), and b* and TAC (R2 = −0.353; $p < 0.01$), as well as between H^0 and TAC (R2 = −0.649; $p < 0.05$) (Table 3). Nevertheless, the parameters b* and L are positively related to the color value H^0, with coefficients (R2 = +0.896; $p < 0.05$; R2 = +0.412; $p < 0.01$), respectively.

Table 3. Pearson's correlation coefficients of total anthocyanin concentration and color attributes of investigated mulberry jellies fermented by *Lactiplantibacillus plantarum* O21, including refrigerated storage.

	TAC	L*	a*	b*	H^0	C*
TAC	1					
L*	−0.763 *	1				
a*	0.282	−0.121	1			
b*	−0.358 **	0.232	0.353	1		
H^0	−0.649 *	0.412 **	0.017	0.896 *	1	
C*	0.216	−0.206	0.192	0.136	−0.097	1

Explanatory notes: TAC—Total anthocyanin concentration; L*—Lightness; a*—Redness; b*—Yellowness; C*—Chroma; H^0—Hue angle; ΔE—Relative color difference index. * Correlation is significant at $p < 0.05$. ** Correlation is significant at $p < 0.01$.

3.3. Determination of Antioxidant Activity

The changes in antioxidant activity of the investigated functional mulberry jellies fermented by *Lactiplantibacillus plantarum* O21 are presented in Figure 1a,b.

The obtained results illustrate the positive effect that lactic acid fermentation had on the scavenging activities of the DPPH and ABTS radical. Regardless of the gelling agent utilized, the lactic fermentation procedure produced fermented mulberry jellies with antioxidant activity that was roughly two to three times higher than unfermented mulberry juice. Regardless of the level used, fresh and inulin samples after storage had a higher antioxidant activity value on the DPPH and ABTS assay.

The DPPH activity of fresh fruity jellies, referred to as % DPPH inhibition, ranged from 55.0% (sample G0P) to 81.0% in the sample GI3, respectively (Figure 2a). In stored products, regardless of the gelling agent used, these values have changed significantly ($p < 0.05$) and have decreased. In both gelatin and agar–agar-based trials, the trend in assessing fresh products in the antioxidant capacity detected by the DPPH and ABTS assay continued. For stored samples with a 1% and 3% inulin addition, statistically significant higher values of antioxidant activity ($p < 0.05$) were reported.

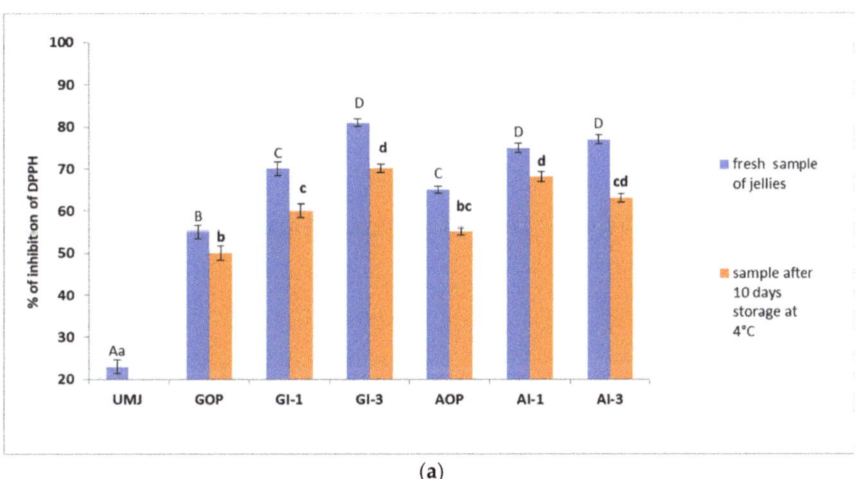

(a)

Figure 2. *Cont.*

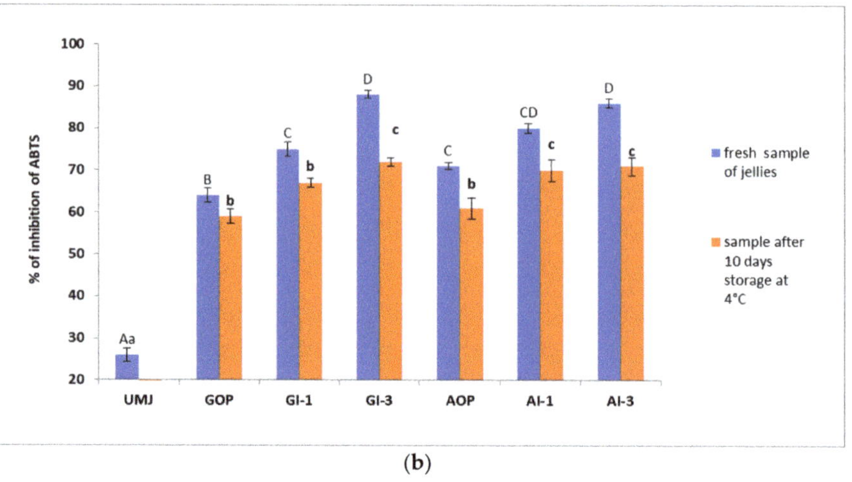

(b)

Figure 2. Changes in antioxidant activity of functional mulberry jellies fermented by *Lactiplantibacillus plantarum* O21, based on different gelling agents stored at 4 °C during 10 days: (**a**) % inhibition of DPPH; (**b**) % inhibition of ABTS. Explanatory notes: Data are expressed as mean values ± standard deviations. Refer to Table 1 for identification of test samples. Values denoted by different capital letters in the fresh samples differ significantly ($p < 0.05$); $n = 3$. Values denoted by different lowercase letters in the samples after 10 days of storage at 4 °C differ significantly ($p < 0.05$); $n = 3$.

When comparing the results of the DPPH assay to the results of the ABTS assay, the antioxidant capacity of the examined unfermented mulberry juice, and fresh and preserved functional jellies, was slightly greater (Figure 2a,b).

3.4. Sensory Characteristics of Functional Mulberry Jellies

The changes in sensory profiles of investigated mulberry jellies fermented by *Lactiplantibacillus plantarum* O21 are shown in Figure 3a,b.

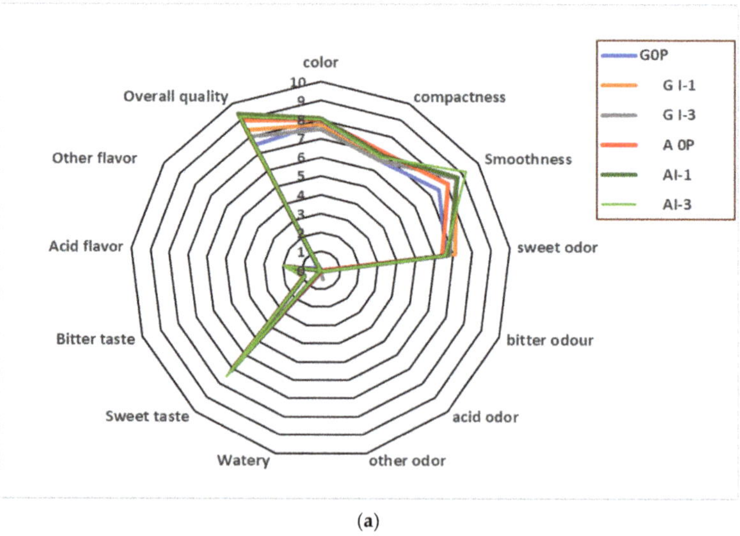

(a)

Figure 3. *Cont.*

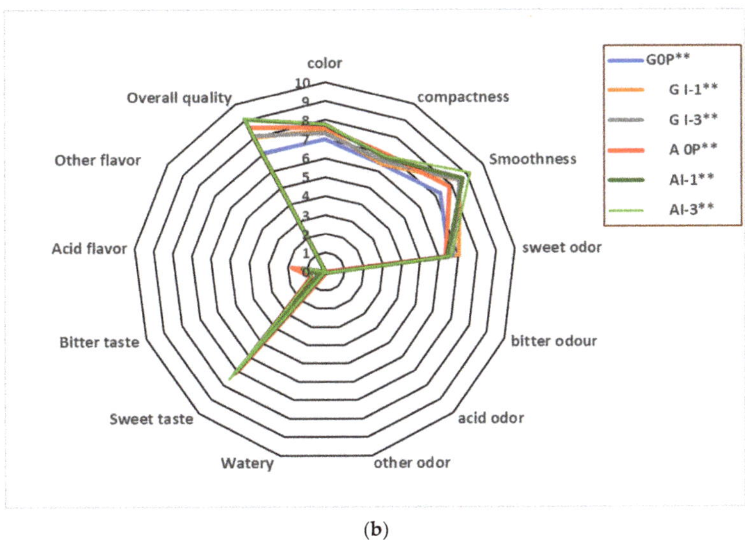

(b)

Figure 3. Sensory profile of functional mulberry jellies fermented by *Lactiplantibacillus plantarum* O21, based on different gelling agents: (**a**) fresh mulberry jellies; (**b**) products stored at 4 °C during 10 days. ** Explanatory notes: The presented samples are coded according to Section 2, Table 1.

The highest value was the smoothness (8.60–9.23 c.u.), and intensity of the sweet taste sensation (6.0–7.5 c.u.) of functional jellies with the addition of inulin (I-1; AI-3) (Section 2, Table 1), regardless of the level of the additive and gelling agent used. In contrast, in the control tests of the G (G0P) (Section 2, Table 1), a gelatin-based series, the highest level of intensity of the consistency distinguishing feature, i.e., "watery", was recorded at 0.1–0.15, respectively (Figure 3a,b).

Gelatin-based samples (G0P; GI-1, GI-3) (Section 2, Table 1) obtained slightly higher color distinguishing values (Figure 3a). However, these differences were not statistically significant ($p > 0.01$).

In agar–agar-based trials (A0P; AI-1; AI-3) (Section 2, Table 1), the highest intensity of the sour taste sensation was noted. All treatments of investigated products were characterized by a good overall quality (meaning higher than 7.5 c.u.). All of the samples in the A series (agar–agar-based) and products that were gelatin-based with the prebiotic addition (GI-1, GI-3) (Section 2, Table 1) showed the highest level of overall quality intensity (8.0—9.2 c.u.), which can be attributed to the high intensity of positive sensory factors such as sweet taste and smoothness (Figure 3a).

During storage, the value of the assessed sensory quality characteristics of innovative mulberry jellies has not changed significantly ($p > 0.01$), and therefore the sensory overall quality has not significantly deteriorated either. However, the overall quality of stored products was rated at 7.1–9.2 c.u., respectively (Figure 3b).

The results of the sensory evaluation of potentially functional mulberry jellies fermented by *Lactiplantibacillus plantarum* O21 during refrigerated storage were also elaborated by the PCA method. The PCA graph of the selected sensory attributes in investigated products is shown in Figure 4.

The analysis of the study's collected results employed the sum of the first two principal components. The first two main components accounted for 42.2% of the difference between the samples, according to a multivariate analysis of the data. The investigated products were grouped into four clusters with their substance gel and inulin addition (Figure 4).

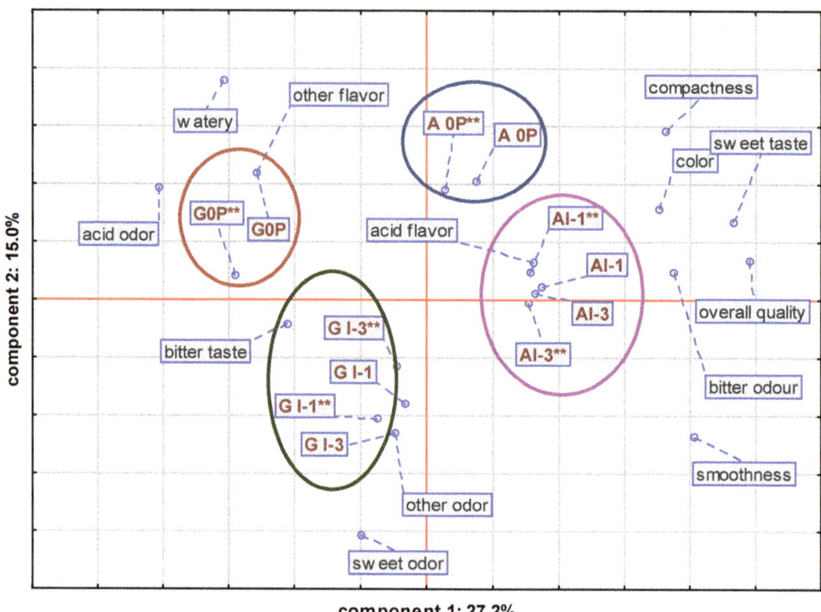

Figure 4. Principal Component Analysis (PCA) graph of the selected sensory attributes and overall quality in investigated samples of mulberry jellies fermented by *Lactiplantibacillus plantarum* O21, based on different gelling agents. ** Explanatory notes: The presented samples are coded according to Section 2, Table 1.

The first component explained 27.2% of the total variability and was mainly related to "color", "compactness", "smoothness", "bitter odor", "sweet taste", "acid flavor", and "overall quality". However, the second component explained 15.0% of the total variability and was mainly related to for such distinguishing features as "watery", "other flavor", or "acid flavor" (Figure 4).

Analyzing the arrangement of tested samples in the space of the plot, it can be concluded that the gelling agents differentiated the products located on the right side (A series; Section 2, Table 1) and the left side (G series; (Section 2, Table 1) of the plot (Figure 4).

The first homogeneous group, in terms of quality, were fresh and stored jelly samples based on gelatin (G 0P; G0P**) (Section 2, Table 1). These samples were characterized by a high intensity of *"watery"* and low intensity of "acid odor" (Figure 4).

On the other hand, the second group consisted of fresh and stored jellies that are agar–agar based (A 0P; A0P**) (Section 2, Table 1). These samples received the highest marks among the products evaluated, in the case of the "compactness" distinguishing feature.

The third cluster is constituted of fresh and stored samples of agar–agar-based jellies (Section 2, Table 1). This region of the plot's properties shows that the goods under investigation share similar qualities, including the strongest overall quality intensity, sweetest flavor, and strongest color notes.

The fourth indicated cluster was fresh and stored gelatin-based products, seria G (Section 2, Table 1). These goods are distinguished by their strong *"sweet odor"*. Also in the GI3 sample, the highest value for the "other odor" distinguishing feature was recorded, which was described by the evaluators as "meaty".

4. Discussion

New carriers introduce beneficial bacteria into the diet, which makes us enrich and rebuild our microbiome. Bioactive ingredients contained in food support the survival and growth of microorganisms in the intestines. Currently, there is a high demand for this type of food.

The probiotic features and health benefits conferred are known to be LAB-strain-specific [48]. Probiotic food products must have a high concentration of microorganism cells [$\geq 10^6$ CFU mL^{-1}], or between 10^8 and 10^{11} CFU per day, to have the required positive impact [49]. In our study, all investigated, fresh, and stored samples of developed mulberry jellies contained an appropriate and high number of LAB bacteria *Lactiplantibacillus plantarum* O21, which was higher than 6.5 log CFU g^{-1} of product. The addition of inulin significantly affected $p < 0.05$ (Figure 1a,b) to increase this number. This is in accordance with Farinha et al., 2015 [50], who reports that to improve probiotic survival, prebiotic ingredients can be added to food preparations containing probiotic microbiota. Moreover, our results are in agreement with the report of other authors, who confirmed that the addition of agar–agar into fermented coconut jelly can improve probiotics' survivability, and phenolic and antioxidant compounds during in vitro digestion [51].

Plants are valuable raw materials for the fermentation of food, due to them being rich in bioactive ingredients that can increase the survival rate of probiotics. On the other hand, probiotic bacteria can affect the higher content of polyphenols and antioxidant activity.

An increase in the antioxidant activity of plant-based foods as a result of lactic fermentation occurs due to an increase in the release of antioxidant compounds, mainly due to an increase in phenolic compounds and flavonoids via microbial hydrolysis [52]. Changes in the level of antioxidant activity compounds' production during lactic acid fermentation may be at different levels, which depends on used microorganisms, cultivation medium, pH, temperature, inhibitors, stimulators, or the atmosphere. Rodríguez et al., 2009 noted that *L. plantarum* was able to break down several phenolic compounds found in food, producing molecules that affect the aroma of the meal as well as compounds with higher antioxidant activity [53]. In turn, it was also found that a prebiotic addition can affect the antioxidant activity of plant-based food. For example, Michalska et al., 2019 [54] suggested that inulin is a better protection agent for anthocyanins than maltodextrin, except for vacuum drying at 90 °C, which probably causes inulin to form 5-(Hydroxymethyl)furfural (HMF), thus limiting its ability to protect anthocyanins. The authors suggest that the higher pace of the process meant a stronger anthocyanin degradation. In our study, regardless of the gelling agent used, significantly higher ($p < 0.05$) antioxidant activity values were observed in samples with the addition of a prebiotic assessed by two methods, DPPH and ABTS, while maintaining this trend after refrigerated storage.

Phenolic compounds are closely related to the sensory quality of food, through the impact on such distinguishing features as astringency, color, and flavor [55]. The color of mulberry fruit-based products is mainly dependent on the level of anthocyanins [56]. In all fruity jellies based on fermented juice, a slight increase in (b) and a decrease in (L) and (b) was observed compared to unfermented mulberry juice [57] (Table 2). Our results were in agreement with Kwaw et al. 2016 [34], who also demonstrated that an effect of lactic fermentation on mulberry juices' quality includes a color assessment. The reason for such changes may be the increase in the monomeric anthocyanin concentration in attempts based on fermented juice. As reported by Boranbayeva et al. 2014 [58], antioxidant activity in black mulberry juice is correlated with total monomeric anthocyanins.

It is well known that variations in the products' sensory profiles are caused by the fermentation technique used to produce fermented goods, in addition to the ingredients in the recipe [59]. According to Tkacz et al. 2021 [59], in assessing the sensory quality of novel functional products, the differentiating factor was the intensity of the sour taste sensation. In our study, in agar–agar-based trials, especially in (AI-1; AI-3) (Section 2, Table 1) with the addition of inulin, the highest intensity of the sour taste sensation was observed, which at the same time was correlated with a significantly ($p < 0.05$) higher number of bacteria *Lactiplantibacillus plantarum* O21 in these products (Figure 1a,b). Inulin as a prebiotic is used by probiotic bacteria as a source of carbon. Because probiotics selectively break down inulin, they can increase their population in the environment. A negative consequence of this process from the point of view of product quality is the overproduction of organic acids by LAB bacteria, mainly lactic acid, and their accumulation in food. The metabolic activity of bacteria can be reduced, for example, by storing food in refrigerated conditions.

Because of its technological qualities, inulin is also employed in food technology as a low-calorie sweetener, fat substitute, and texture adjuster [60]. In the present study, the highest values of the texture distinguishing feature—smoothness and the intensity of the sweet taste sensation—was noted in the case of mulberry jellies with the addition of inulin (GO-1, GI-3; AI-1; AI-3) (Section 2, Table 1), regardless of the level of this additive used and the gelling agent used. Moreover, the results showed that agar–agar-based samples of jellies were best evaluated in terms of the texture-distinguishing feature, "compactness" (Figure 3a,b). Domínguez-Murillo and Urías-Silvas, 2023, suggested that the addition of 1% of agar to samples of coconut jellies presented the maximum hardness ($p < 0.05$), due to more bonding points and intermolecular interactions [51]. It can be concluded that inulin can be a good alternative for sucrose, which is in line with current food design trends. In contrast, the addition of agar is both texture-forming and protective against probiotic bacteria at the same time. The above properties of these additives are conducive to their use in the production of jellies.

5. Conclusions

This study demonstrated the use of a half-product in the form of mulberry juice fermented with the *Lactiplantibacillus plantarum* O21 strain with probiotic properties, including such additives as: the gelling agent (gelatin; agar–agar), prebiotic (inulin), impact on the color parameters, TAC content, antioxidant activities, and sensory quality of final products, such as functional jellies.

Results generated in this study suggest that the agar–agar-based products with the addition of inulin were characterized by high parameters in the field of microbiological assessment (significantly higher numbers of LAB), antioxidant activity values, and good notes in sensory evaluation—"overall sensory quality". However, the use of gelatin and a higher 3% level of an inulin supplement as the gelling agent means that the sensory evaluation of this product variant could taste a bit "meaty".

It should be further noted that the functional mulberry jellies in the agar–agar-based version can be beneficial alternatives for a wide range of consumers, for example: people with milk protein allergy, lactose intolerance, and vegans, as well children. In conclusion, our research will continue in order to refine production technology, as well as optimize parameters to obtain a symbiotic product.

Author Contributions: A.S., Conceptualization, Investigation, Supervision, Methodology, Software, Validation, Formal analysis, Investigation, Writing—review and editing, Visualization, Writing—original draft, and Data Curation. D.Z., Writing—review and editing. B.S., Investigation. D.K.-K., Formal analysis. All authors have read and agreed to the published version of the manuscript.

Funding: This research received no external funding.

Institutional Review Board Statement: This study was conducted in accordance with the Declaration of Helsinki, and approved by the Rector's Committee for the Ethics of Scientific Research Involving Humans at WULS-SGGW (Resolution No. 28/RKE/2023/U of 6 July 2023).

Informed Consent Statement: Informed consent was obtained from all subjects involved in the study.

Data Availability Statement: Data is contained within the article.

Conflicts of Interest: The authors declare no conflict of interest.

References

1. Granato, D.; Barba, F.J.; Kovačević, D.B.; Lorenzo, J.M.; Cruz, A.G.; Putnik, P. Functional Foods: Product Development, Technological Trends, Efficacy Testing, and Safety. *Annu. Rev. Food Sci. Technol.* **2020**, *11*, 93–118. [CrossRef] [PubMed]
2. Kumar, B.V.; Vijayendra, S.V.N.; Reddy, O.V.S. Trends in dairy and non-dairy probiotic products - a review. *J. Food Sci. Technol.* **2015**, *52*, 6112–6124. [CrossRef] [PubMed]

3. Zhang, Y.; Hu, P.; Lou, L.; Zhan, J.; Fan, M.; Li, D.; Liao, Q. Antioxidant activities of lactic acid bacteria for quality improvement of fermented sausage. *J. Food Sci.* **2017**, *82*, 2960–2967. [CrossRef] [PubMed]

4. Hill, C.; Guarner, F.; Reid, G.; Gibson, G.R.; Merenstein, D.J.; Pot, B.; Morelli, L.; Canani, R.B.; Flint, H.J.; Salminen, S.; et al. The international scientific association for probiotics and prebiotics consensus statement on the scope and appropriate use of the term probiotic. *Nat. Rev. Gastroenterol. Hepatol.* **2014**, *11*, 506–514. [CrossRef] [PubMed]

5. ISAPP. *Minimum Criteria for Probiotics*; International Scientific Association for Probiotics and Prebiotics: Sacramento, CA, USA, 2018.

6. Johansen, E. Future access and improvement of industrial lactic acid bacteria cultures. *Microb. Cell Factories* **2017**, *16*, 230. [CrossRef] [PubMed]

7. Łepecka, A.; Szymański, P.; Okoń, A.; Zielińska, D. Antioxidant activity of environmental lactic acid bacteria strains isolated from organic raw fermented meat products. *LWT* **2023**, *174*, 114440. [CrossRef]

8. Javed, G.A.; Arshad, N.; Munir, A.; Khan, S.Y.; Rasheed, S.; Hussain, I. Signature probiotic and pharmacological attributes of lactic acid bacteria isolated from human breast milk. *Int. Dairy J.* **2022**, *127*, 105297. [CrossRef]

9. Falfán-Cortés, R.N.; Mora-Peñaflor, N.; Gómez-Aldapa, C.A.; Rangel-Vargas, E.; Acevedo-Sandoval, O.A.; Franco-Fernández, M.J.; Castro-Rosas, J. Characterization and Evaluation of the Probiotic Potential In Vitro and In Situ of Lacticaseibacillus paracasei Isolated from Tenate Cheese. *J. Food Prot.* **2022**, *85*, 112–121. [CrossRef]

10. Trindade, D.P.d.A.; Barbosa, J.P.; Martins, E.M.F.; Tette, P.A.S. Isolation and identification of lactic acid bacteria in fruit processing residues from the Brazilian Cerrado and its probiotic potential. *Food Biosci.* **2022**, *48*, 101739. [CrossRef]

11. Sampaio, K.B.; de Albuquerque, T.M.R.; Rodrigues, N.P.A.; de Oliveira, M.E.G.; de Souza, E.L. Selection of Lactic Acid Bacteria with In Vitro Probiotic-Related Characteristics from the Cactus *Pilosocereus gounellei* (A. Weber ex. K. Schum.) Bly. ex Rowl. *Foods* **2021**, *10*, 2960. [CrossRef]

12. Argyri, A.A.; Zoumpopoulou, G.; Karatzas, K.A.G.; Tsakalidou, E.; Nychas, G.-J.E.; Panagou, E.Z.; Tassou, C.C. Selection of potential probiotic lactic acid bacteria from fermented olives by in vitro tests. *Food Microbiol.* **2013**, *33*, 282–291. [CrossRef] [PubMed]

13. Gupta, A.; Sharma, N. Probiotic Potential of Lactic Acid Bacteria Ch-2 Isolated from Chuli Characterization of Potential Probiotic Lactic Acid Bacteria- Pediococcus acidilactici Ch-2 Isolated from Chuli- A Traditional Apricot Product of Himalayan Region for the Production of Novel Bioactive Compounds with Special Therapeutic Properties. *J. Food Microbiol. Saf. Hyg.* **2017**, *2*, 1–11. [CrossRef]

14. Lactic Acid Bacteria Isolated from Food Can Have Probiotic Properties. Available online: https://foodfakty.pl/bakterie-mlekowe-wyizolowane-z-zywnosci-moga-miec-wlasciwosci-probiotyczne (accessed on 20 January 2023).

15. Markowiak, P.; Śliżewska, K. Effects of probiotics, prebiotics, and synbiotics on human health. *Nutrients* **2017**, *9*, 1021. [CrossRef] [PubMed]

16. Wu, J.Y. Everyone vegetarian, world enriching. *Open J. Philos.* **2014**, *4*, 160–165. [CrossRef]

17. Kessler, C.S.; Holler, S.; Joy, S.; Dhruva, A.; Michalsen, A.; Dobos, G.; Cramer, H. Personality profiles, values and empathy: Differences between lacto-Ovo-vegetarians and vegans. *Forsch. Komplementärmedizin* **2016**, *23*, 95–102. [CrossRef]

18. Pimentel, T.C.; da Costa, W.K.A.; Barão, C.E.; Rosset, M.; Magnani, M. Vegan probiotic products: A modern tendency or the newest challenge in functional foods. *Food Res. Int.* **2021**, *140*, 110033. [CrossRef]

19. Tamang, J.P.; Tamang, B.; Schillinger, U.; Guigas, C.; Holzapfel, W.H. Functional properties of lactic acid bacteria isolated from ethnic fermented vegetables of the Himalayas. *Int. J. Food Microbiol.* **2009**, *135*, 28–33. [CrossRef]

20. Panghal, A.; Janghu, S.; Virkar, K.; Gat, Y.; Kumar, V.; Chhikara, N. Potential non-dairy probiotic products—A healthy approach. *Food Biosci.* **2018**, *2*, 80–89. [CrossRef]

21. Ercisli, S.; Orhan, E. Some physico-chemical characteristics of black mulberry (*Morus nigra* L.) genotypes from Northeast Anatolia region of Turkey. *Sci. Hortic.* **2008**, *116*, 41–46. [CrossRef]

22. Hojjatpanah, G.; Fazaeli, M.; Emam-Djomeh, Z. Effects of heating method and conditions on the quality attributes of black mulberry (*Morus nigra*) juice concentrate. *Int. J. Food Sci. Technol.* **2007**, *46*, 956–962. [CrossRef]

23. Krishna, H.; Singh, D.; Singh, R.S.; Kumar, L.; Sharma, B.D.; Saroj, P.L. Morphological and antioxidant characteristics of mulberry (*Morus* spp.) genotypes. *J. Saudi Soc. Agric. Sci.* **2018**, *19*, 136–145. [CrossRef]

24. Memete, A.R.; Timar, A.V.; Vuscan, A.N.; Miere, F.; Venter, A.C.; Vicas, S.I. Phytochemical Composition of Different Botanical Parts of *Morus* Species, Health Benefits and Application in Food Industry. *Plants* **2022**, *11*, 152. [CrossRef] [PubMed]

25. Wang, Y.; Xiang, L.; Wang, C.; Tang, C.; He, X. Antidiabetic and antioxidant effects and phytochemicals of mulberry fruit (*Morus alba* L.) polyphenol enhanced extract. *PLoS ONE* **2013**, *8*, e71144. [CrossRef] [PubMed]

26. Bhattacharjya, D.; Sadat, A.; Dam, P.; Buccini, D.F.; Mondal, R.; Biswas, T.; Biswas, K.; Sarkar, H.; Bhuimali, A.; Kati, A.; et al. Current concepts and prospects of mulberry fruits for nutraceutical and medicinal benefits. *Curr. Opin. Food Sci.* **2021**, *40*, 121–135. [CrossRef]

27. Kim, B.S.; Kim, H.; Kang, S.-S. In vitro anti-bacterial and anti-inflammatory activities of lactic acid bacteria-biotransformed mulberry (*Morus alba* Linnaeus) fruit extract against Salmonella Typhimurium. *Food Control.* **2019**, *106*, 106758. [CrossRef]

28. Wang, F.; Du, B.-L.; Cui, Z.-W.; Xu, L.-P.; Li, C.-Y. Effects of high hydrostatic pressure and thermal processing on bioactive compounds, antioxidant activity, and volatile profile of mulberry juice. *Food Sci. Technol. Int.* **2017**, *23*, 119–127. [CrossRef] [PubMed]

29. Tang, C.; Wu, J.; Luo, G.; Wu, F.; Yang, Q.; Xiao, G. Wine-making experiment using mulberry fruits from different fruit mulberry varieties. *Acta Sericologica Sin.* **2008**, *34*, 24–27.
30. Lee, J.; Choi, S. Quality characteristics of muffins added with mulberry concentrate. *Korean J. Culin. Res.* **2011**, *17*, 285–294. [CrossRef]
31. Jan, B.; Parveen, R.; Zahiruddin, S.; Khan, M.U.; Mohapatra, S.; Ahmad, S. Nutritional constituents of mulberry and their potential applications in food and pharmaceuticals: A review. *Saudi J. Biol. Sci.* **2021**, *28*, 3909–3921. [CrossRef]
32. Zhang, H.; Ma, Z.F.; Luo, X.; Li, X. Effects of Mulberry Fruit (*Morus alba* L.) Consumption on Health Outcomes: A Mini-Review. *Antioxidants* **2018**, *7*, 69. [CrossRef]
33. Kobus-Cisowska, J.; Gramza-Michalowska, A.; Kmiecik, D.; Flaczyk, E.; Korczak, J. Mulberry fruit as an antioxidant component in muesli. *Agric. Sci.* **2013**, *4*, 130–135. [CrossRef]
34. Kwaw, E.; Ma, Y.; Tchabo, W.; Apaliya, M.T.; Wu, M.; Sackey, A.S.; Xiao, L.; Tahir, H.E. Effect of lactobacillus strains on phenolic profile, color attributes and antioxidant activities of lactic-acid-fermented mulberry juice. *Food Chem.* **2018**, *250*, 148–154. [CrossRef] [PubMed]
35. Chaudhary, A.; Sharma, V.; Saharan, B.S. Probiotic Potential of Noni and Mulberry Juice Fermented with Lactic Acid Bacteria. *Asian J. Dairy Food Res.* **2019**, *38*, 114–120. [CrossRef]
36. Chuah, H.Q.; Tang, P.L.; Ang, N.J.; Tan, H.Y. Submerged fermentation improves bioactivity of mulberry fruits and leaves. *Chin. Herb. Med.* **2021**, *13*, 565–572. [CrossRef] [PubMed]
37. Lv, M.; Aihaiti, A.; Liu, X.; Tuerhong, N.; Yang, J.; Chen, K.; Wang, L. Development of Probiotic-Fermented Black Mulberry (*Morus nigra* L.) Juice and Its Antioxidant Activity in C2C12 Cells. *Fermentation* **2022**, *8*, 697. [CrossRef]
38. Zielińska, D.; Rzepkowska, A.; Radawska, A.; Zieliński, K. In Vitro Screening of Selected Probiotic Properties of Lactobacillus Strains Isolated from Traditional Fermented Cabbage and Cucumber. *Curr. Microbiol.* **2015**, *70*, 183–194. [CrossRef] [PubMed]
39. *NF ISO 15214*; Microbiology of Food and Animal Feeding Stuffs. Horizontal Method for the Enumeration of Mesophilic Lactic Acid Bacteria. Colony Count Technique at 30 °C. ISO: Geneva, Switzerland, 1998.
40. Tchabo, W.; Ma, E.; Kwaw, Y.; Zhang, X.; Li, H.; Afoakwah, N.A. Effects of ultrasound, high pressure, and manosonication processes on phenolic profile and antioxidant properties of a sulfur dioxide-free mulberry (*Morus nigra*) wine. *Food Bioprocess Technol.* **2017**, *10*, 1210–1223. [CrossRef]
41. Alothman, M.; Bhat, R.; Karim, A. UV radiation-induced changes of antioxidant capacity of fresh-cut tropical fruits. *Innov. Food Sci. Emerg. Technol.* **2009**, *10*, 512–516. [CrossRef]
42. Brand-Williams, W.; Cuvelier, M.E.; Berset, C. Use of a free radical method to evaluate antioxidant activity. *LWT Food Sci. Technol.* **1995**, *28*, 25–30. [CrossRef]
43. Re, R.; Pellegrini, N.; Proteggente, A.; Pannala, A.; Yang, M.; Rice-Evans, C. Antioxidant activity applying an improved ABTS radical cation decolorization assay. *Free Radic. Biol. Med.* **1999**, *26*, 1231–1237. [CrossRef]
44. Fazaeli, M.; Hojjatpanah, G.; Emam-Djomeh, Z. Effects of Heating Method and Conditions on the Evaporation Rate and Quality Attributes of Black Mulberry (*Morus nigra*) Juice Concentrate. *J. Food Sci. Technol.* **2013**, *50*, 35–43. [CrossRef] [PubMed]
45. Sensory Analysis—Methodology—General Guidance for Establishing a Sensory Profile. Available online: https://www.iso.org/standard/58042.html (accessed on 22 March 2023).
46. *ISO 8586-2:1994*; Sensory Analysis. General Guidance for the Selection, Training and Monitoring of Assessors—Part 2: Experts. ISO: Geneva, Switzerland, 1994.
47. Kitao, A. Principal Component Analysis and Related Methods for Investigating the Dynamics of Biological Macromolecules. *Multidiscip. Sci. J.* **2022**, *5*, 298–317. [CrossRef]
48. Jampaphaeng, K.; Cocolin, L.; Maneerat, S. Selection and evaluation of functional characteristics of autochthonous lactic acid bacteria isolated from traditional fermented stinky bean (Sataw-Dong). *Ann. Microbiol.* **2016**, *67*, 25–36. [CrossRef]
49. Uriot, O.; Denis SJuniua, M.; Roussel, Y.; Dary-Mourot, A.; Blanquet-Diot, S. *Streptococcus thermophilus*: From yogurt starter to a new promising probiotic candidate? *J. Funct. Food.* **2017**, *37*, 74–89. [CrossRef]
50. Farinha, L.R.L.; Sabo, S.S.; Porto, M.C.; Souza, E.C.; Oliveira, M.N.; Oliveira, R.P.S. Influence of prebiotic ingredients on the growth kinetics and bacteriocin production of *Lactococcus lactis*. *Chem. Eng. J.* **2015**, *43*, 313–318.
51. Domínguez-Murillo, A.; Urías-Silvas, J. Fermented coconut jelly as a probiotic vehicle, physicochemical and microbiology characterisation during an in vitro digestion. *Int. J. Food Sci. Technol.* **2022**, *58*, 45–52. [CrossRef]
52. Hur, S.J.; Lee, S.Y.; Kim, Y.-C.; Choi, I.; Kim, G.-B. Effect of fermentation on the antioxidant activity in plant-based foods. *Food Chem.* **2014**, *160*, 346–356. [CrossRef]
53. Rodríguez, H.; Curiel, J.A.; Landete, J.M.; De las Rivas, B.; de Felipe, F.L.; Gómez-Cordovés, C.; Mancheño, J.M.; Muñoz, R. Food phenolics and lactic acid bacteria. *Int. J. Food Microbiol.* **2009**, *132*, 79–90. [CrossRef]
54. Michalska, A.; Wojdyło, A.; Brzezowska, J.; Majerska, J.; Ciska, E. The Influence of Inulin on the Retention of Polyphenolic Compounds during the Drying of Blackcurrant Juice. *Molecules* **2019**, *24*, 4167. [CrossRef]
55. Natić, M.M.; Dabić, D.C.; Papetti, A.; Fotirić Akšić, M.; Ognjanov, V.; Ljubojević, M.; Tešić, Ž. Analysis and characterisation of phytochemicals in mulberry (*Morus alba* L.) fruits grown in Vojvodina, North Serbia. *Food Chem.* **2015**, *171*, 128–136. [CrossRef]
56. Tchabo, W.; Ma, Y.; Kwaw, E.; Zhang, H.; Xiao, L.; Apaliya, M.T. Statistical interpretation of chromatic indicators in correlation to phytochemical profile of a sulphur dioxide-free mulberry (*Morus nigra*) wine submitted to non-thermal maturation processes. *Food Chem.* **2018**, *239*, 470–477. [CrossRef] [PubMed]

57. Chen, L.; Xin, X.L.; Yuan, Q.P.; Su, D.H.; Liu, W. Phytochemical properties and antioxidant capacities of various colored berries. *J. Sci. Food Agric.* **2014**, *94*, 180–188. [CrossRef] [PubMed]
58. Boranbayeva, T.; Karadeniz, F.; Yılmaz, E. Effect of Storage on Anthocyanin Degradation in Black Mulberry Juice and Concentrates. *Food Bioprocess Technol.* **2014**, *7*, 1894–1902. [CrossRef]
59. Tkacz, K.; Wojdyło, A.; Turkiewicz, I.P.; Nowicka, P. Anti-diabetic, anti-cholinesterase, and antioxidant potential, chemical composition and sensory evaluation of novel sea buckthorn-based smoothies. *Food Chem.* **2021**, *338*, 128105. [CrossRef]
60. Canbulat, Z.; Ozcan, T. Effects of short-chain and long-chain inulin on the quality of probiotic yogurt containing Lactobacillus rhamnosus. *J. Food Process. Preserv.* **2015**, *39*, 1251–1260. [CrossRef]

Article

Body Weight Loss Efficiency in Overweight and Obese Adults in the Ketogenic Reduction Diet Program—Case Study

Gordana Markovikj [1], Vesna Knights [1,*] and Jasenka Gajdoš Kljusurić [2]

[1] Faculty of Technology and Technical Sciences-Veles, University St. Kliment Ohridski-Bitola, Dimitar Vlahov bb, 1400 Veles, North Macedonia; gordana-goca@hotmail.com

[2] Faculty of Food Technology and Biotechnology, University of Zagreb, Pierottijeva 6, 10000 Zagreb, Croatia; jasenka.gajdos@pbf.unizg.hr

* Correspondence: vesna.knights@uklo.edu.mk

Featured Application: The appropriate application of data analysis techniques provides relevant data that can be used in designing strategies and programs for the general population, such as the effectiveness of weight loss programs and key similarities or differences in the eating habits of similar and/or different population groups.

Abstract: Obesity stands out as an ongoing pandemic today, and it is crucial to recognize the basic factors that influence it in the observed group and to intervene through lifestyle changes. Therefore, in this work, the k ketogenic diet (E = 6280 $\pm$ 210 kJ) was used in a weight loss program for two regionally different groups (including 200 participants) from southeastern European countries (Republic of North Macedonia (n = 100) and Kosovo (n = 100)). The applied data analysis revealed similarities and differences in (ii) the consumption of certain food groups (e.g., 0.5–1 kg Nuts/week; in region 1 is consumed by 11.3% of participants while in region 2 by 37.8%, respectively) and (ii) anthropometric indicators of excess body mass (body mass index and waist-to-hip ratio). Nutritional intervention with a ketogenic diet also reduces the intake of sweet and salty snacks that are rich in carbohydrates. The average expected time to reach the target body mass was 112 days, and the results of the progress of all participants were presented after 120 days. The results show regional differences, especially in women; in group 1, 73.91% achieved a body mass index in the healthy range (<25 kg/m^2), while in group 2, the success rate was 81.69%. Understanding the different eating habits in the mentioned regions is key here, and it was shown that in region 2, over 40% of the participants consume 500–1000 g of seeds per week. The above indicates that the results of this study and regional differences can be considered when designing strategies and intervention programs in the lifestyle of overweight and obese people in similar environments. The study also shows that the ketogenic diet is one of the useful dietary intervention approaches used to change eating habits that will show results relatively quickly.

Keywords: overweight; obesity; ketogenic diet; weight loss; regional differences; applied data analysis

Citation: Markovikj, G.; Knights, V.; Gajdoš Kljusurić, J. Body Weight Loss Efficiency in Overweight and Obese Adults in the Ketogenic Reduction Diet Program—Case Study. *Appl. Sci.* **2023**, *13*, 10704. https://doi.org/10.3390/app131910704

Academic Editors: Theodoros Varzakas and Maria Antoniadou

Received: 1 September 2023
Revised: 22 September 2023
Accepted: 25 September 2023
Published: 26 September 2023

1. Introduction

In the human population, obesity has become a pandemic of the modern era [1]. Numerous diseases and health complications are associated with obesity and excess body weight and are manifested in cardiovascular diseases, various types of cancer, hypertension, type 2 diabetes, polycystic ovary syndrome (PCOS), and many others [2,3]. But less often considered is the social isolation and mental state [4] often accompanied with anxiety and depression [5], which Berry [6] describes as follows: "Imprisoned in every fat man, a thin man is wildly signaling to be let out". Research shows that one of the important steps in "releasing the thin man" and solution to "globesity" [7] is the establishment of a balanced diet, adequate physical activity, and changing behavior and lifestyle [2–8].

Therefore, a number of strategies have been launched at the world level, in the design of appropriate educational programs and body weight regulation programs, because this implies an improvement in the general condition of the population [9,10], with the aim to start at childhood because prevention is better than to cure [11].

The parameter used to classify moderately active people into groups of malnourished, normally nourished, obese, or overweight people is the body mass index (BMI) as a ratio of body mass to the square of body height (kg/m^2). Values of the BMI for normal nourished individual is in the range of 18.5–25 kg/m^2, while the classification of "overweighed" starts with a BMI > 25 kg/m^2, and obesity with a BMI $\geq$ 30 kg/m^2 [7].

Flegal et al. [12] investigated the changes in body height, body mass, and BMI in the period from 1999 to 2016, analyzing the data available in the sources as (i) the National Health and Nutrition Examination Survey (NHANES) [13], (ii) National Health Interview Survey (NHIS) [14], and (iii) Behavioral Risk Factor Surveillance System (BRFSS) [15]. The mean height of the female and male population remained the same (1.76 m for men and 1.62 m for women), while the body mass values have shown a proportional increase within the observed years (1999 to 2016), and thus BMI increased proportionally. Unfortunately, from the average BMI in the population in 1999 (men: 27.6 kg/m^2; women 28.1 kg/m^2), the values for men and women increased to 29.1 kg/m^2 and 29.6 kg/m^2, respectively [12].

Individuals who decide to reduce their body mass, which will consequently result in a reduction in their BMI, often want to see some results hastily, otherwise they rapidly return to their old food patterns and/or behavior, wherefore the yo-yo effect occurs, meaning that the person then feels even worse [16]. Therefore, the perception of effective weight loss is especially necessary for people with a BMI $\geq$ 30 kg/m^2 in order to remain stable after restrictive energy intake.

Forehand recognition of the factors that influence an individual's increased body weight (eating habits) helps nutritionists and medical doctors design appropriate body weight regulation programs in which the ketogenic diet pattern is often used, which may modify lipid profile and the inflammatory state of the body [17] and can help with weight loss, visceral adiposity, and appetite control [18]. Drabińska et al. [17] highlighted the presence of different ketogenic diets depending on the amount of carbohydrates, but the majority of them are characterized by the minimization of energy and carbohydrate intake [19]. In the standard ketogenic diet, there is a preferred share of carbohydrates under 5% in the daily energy intake, while the share of fats is dominant ($\approx$60%), and the remaining part is made up of proteins (35%) [20]. The limited intake of carbohydrates, with a dominant share of fats and a moderate intake of protein, results in the use of ketone bodies from fat metabolism. Because of the limited intake of carbohydrates, the primary way of producing energy from glucose is utilizing fat as its primary source of energy, i.e., nutritional ketosis [17,18]. The ketogenic diet with a reduced daily energy intake can be beneficial for weight loss [17–19]; however, one should not ignore the regional differences (due to, for example, body composition, eating habits, and preferences) that have proven to be significant in weight loss programs as well [20].

Therefore, the aim of our study was to investigate the (i) similarities and/or differences in the food habits of the participants from two southeastern European countries, (ii) the relationship of the anthropometric indicators related to unhealthy nourishment (body mass index and waist-to-hip ratio) with the consumption of certain foods and health problems, as well as the (iii) effectiveness of the ketogenic diet pattern on two regionally different groups in the same program of body weight reduction.

2. Materials and Methods

This study included 200 overweight and obese adults (% of them are females) who participated in a cross-sectional case study. The study was conducted in two countries in southern Europe, the Republic of North Macedonia (Group 1: 100 participants, 31% male and 69% female) and Kosovo (Group 2: also 100 participants, 13% male and 87% female). All participants were enrolled in the body weight reduction program and signed consent for their data to be used for scientific purposes, considering and respecting all GDPR principles. None of the participants dropped out of the program. The study was conducted in accordance with ethical permission (code 10-168/1 approved on 28 April 2022).

The data collected for each participant were based on their (i) medical condition and family history related to overweight/obesity, (ii) dietary habits related to the intake of foods with high (snacks (salty/sweet), seeds, nuts, sweet and carbonated drinks) and low energy densities (fruits and vegetable), before the weight loss program, and (iii) anthropometric data (body height, body mass, chest, abdomen, waist, hips, biceps, and thigh circumferences) before and after 120 days of being enrolled in the program. The participants' anthropometric data were collected following the recommendation [20]. For each participant, a targeted body mass was set (which corresponds to a body mass index in the range of 24–25 kg/m^2). Basic data of the participants at the beginning of the program are listed in Table 1.

Table 1. Basic data presented as average value with corresponding range (given in the brackets) for the participants of two groups collected during the first medical visit.

Basic Information	Group 1		Group 2	
	Male (n = 31)	Female (n = 69)	Male (n = 13)	Female (n = 87)
Age (years)	35.3 (18–569)	37.6 (18–68)	41.8 (18–57)	38 (18–67)
Body mass (kg)	103.3 (60.3–237)	78.8 (50.7–152.5)	108.4 (85.1–157.6)	91.3 (62.5–161.2)
Body height (m)	1.76 (1.62–1.92)	1.64 (1.47–1.84)	1.74 (1.55–1.85)	1.65 (1.52–1.83)
Body mass index (kg/m^2)	38.3 (26.3–64.3)	33.6 (25.4–63.5)	35 (28–49)	32.8 (25.1–54.5)
Consumed meals per day (No.)	2.5 (1–4)	2.1 (1–4)	2.5 (2–4)	2.6 (1–5)
Waist Circumference (cm)	110.7 (80–164)	100.9 (75–162)	105.4 (77–138)	101.3 (66–166)
Waist-to-hip ratio	0.97 (0.91–1.05)	0.95 (0.89–1.05)	0.94 (0.85–1)	0.95 (0.85–1.03)

Group 1: participants from North Macedonia; Group 2: participants from Kosovo.

The food intake of both groups during the weight loss program based on the ketogenic diet was controlled and based on the average daily energy and nutrient intake presented in Table 2.

Table 2. Average energy and nutrient content of the controlled ketogenic diet.

Controlled Dietary Parameters	Content
Energy, kJ	6280 ± 210
Proteins [#], g	136 ± 5.6
Fats [#], g	95 ± 4.3
Carbohydrates [#], g	360 ± 1.9
Vitamins	
Thiamine (vit. B$_1$), mg	1.4 ± 0.02
Riboflavin (vit. B$_2$), mg	1.6 ± 0.03
Niacin (vit. B$_3$), mg NE	18 ± 0.8
Pantothenic acid (vit. B$_5$), mg	6 ± 0.3
Vitamin B6, mg	2 ± 0.08
Biotin (vit. B$_7$), μg	150 ± 6.2
Folic acid (vit. B$_9$), μg	200 ± 10.4
Vitamin B12, μg	1 ± 0.2
Vitamin C, mg	60 ± 2.7
Vitamin D, μg	5 ± 0.25

Table 2. *Cont.*

Controlled Dietary Parameters	Content
Vitamin E, mg	7.4 ± 0.4
Vitamin K, µg	30 ± 1.1
Vitamin A, µg RE	800 ± 32.7
Minerals	
Calcium, mg	360 ± 17.8
Chromium, µg	25 ± 0.9
Cupper, mg	1 ± 0.01
Iodine, µg	100 ± 4.2
Iron, mg	8 ± 2.1
Magnesium, mg	360 ± 15.9
Manganese, mg	1 ± 0.03
Molybdenum, µg	25 ± 1.1
Potassium, mg	1200 ± 48.3
Selenium, µg	25 ± 1.3
Sodium, mg	1200 ± 44.2
Zinc, mg	10 ± 0.9
polyunsaturated fats	
Omega-3, mg	1000 ± 30

[#] share of proteins/fats/carbohydrates (35:60:5) followed the recommendations [21].

To determine the differences in the mean values of consumed certain foods between the observed groups, an analysis of variance (ANOVA) was performed, while a principal components analysis was performed to determine the factors of similarity or difference (dietary habits vs. anthropometric parameters and personal data) in the mentioned regions. In order to graphically highlight the prevalence of certain diseases (which are more common in overweight and obese people) in the observed regions and according to gender, a heat map was used, where higher and/or lower values were highlighted in different colors. With the aim of clearer data analysis (changes in WHR, BMI, and BM in the weight loss program) and potential deviation and characterized symmetry of data distribution (depending on region, age, gender, and family prevalence of obesity/overweight), a box plot was used.

3. Results

3.1. Determining the Level of Nutrition and Eating Habits before Body Weight Reduction Program

In order to assess the condition of each individual included in the program, it is important to determine the degree of overweight and adjust the approach depending on whether the participant is overweight or obese. For each participant, the expected time to reach the target body mass was calculated according to the Wishnofsky equation [8]. The average expected duration of the adapted ketogenic diet was 112 days (ranging from 60 to 380 days). For the majority of the male population, the time frame for reaching the targeted body weight was 180 to 380 days.

3.1.1. Consumption of Different Food Groups

In order to determine the intake of certain foods during the day, the respondents provided information on quantitative intake according to the presented qualitative models [22], and the results are shown in Table 3. The daily consumption of drinks (carbonated drinks and water) is presented in the Appendix A (Table A1).

The results show significant differences for the consumed amounts for the participants from the same group but related to gender, as well as between groups.

Female participants have a strong preference for sweets, in both regions, and their consumption between 101 and 200 g (37.7% from Group 1) dominates, while almost a third of respondents from Group 2 (29.9%) consume more than 200 g of sweet food per day, which, according to the interview data, included chocolate, cakes, biscuits, and candies.

Table 3. Daily consumed amounts of foods from different food groups presented as frequencies for each gender from the observed groups.

Amount of Daily Consumed Foods	Frequency (% per Day or per Week #)			
	Group 1		Group 2	
	Male	Female	Male	Female
Sweets				
0 g	9.7 [A,a]	5.8 [B,a]	0.0 [A,b]	0.0 [A,b]
<50 g	38.7 [A,a]	26.1 [B,a]	25.0 [A,b]	18.2 [B,b]
50–100 g	16.1 [A,a]	23.2 [A,a]	8.3 [A,b]	28.6 [B,a]
101–200 g	16.1 [A,a]	37.7 [B,a]	0.0 [A,b]	23.4 [B,b]
>200 g	19.4 [A,a]	7.2 [B,a]	66.7 [A,b]	29.9 [B,b]
Chips				
0 g	35.5 [A,a]	18.8 [B,a]	25.0 [A,b]	29.9 [A,a]
<50 g	25.8 [A,a]	33.3 [B,a]	16.7 [A,b]	26.0 [B,b]
50–100 g	19.4 [A,a]	30.4 [B,a]	41.7 [A,b]	35.1 [B,a]
101–200 g	9.7 [A,a]	13.0 [A,a]	8.3 [A,a]	6.5 [A,b]
>200 g	9.7 [A,a]	4.3 [A,a]	8.3 [A,a]	2.6 [B,a]
Vegetables				
0 g	12.9 [A,a]	10.1 [A,a]	0.0 [A,b]	4.0 [B,a]
<500 g	87.1 [A,a]	88.4 [A,a]	0.0 [A,b]	12.0 [B,b]
500–1000 g	0.0 [A,a]	1.4 [A,a]	91.7 [A,b]	70.7 [B,b]
1001–1500 g	0.0 [A,a]	0.0 [A,a]	0.0 [A,a]	4.0 [B,b]
>1500 g	0.0 [A,a]	0.0 [A,a]	8.3 [A,b]	9.3 [A,b]
Fruits				
0 g	19.4 [A,a]	17.4 [A,a]	0.0 [A,b]	2.6 [A,b]
<500 g	54.8 [A,a]	49.3 [A,a]	36.4 [A,b]	46.1 [B,a]
500–1000 g	16.1 [A,a]	18.8 [A,a]	36.4 [A,b]	31.6 [A,b]
1001–1500 g	6.5 [A,a]	8.7 [A,a]	18.2 [A,b]	11.8 [A,a]
>1500 g	3.2 [A,a]	5.8 [A,a]	9.1 [A,b]	7.9 [A,a]
Nuts #				
0 g	22.6 [A,a]	26.1 [A,a]	45.5 [A,b]	16.2 [B,b]
<500 g	32.3 [A,a]	31.9 [A,a]	9.1 [A,b]	29.7 [B,a]
500–1000 g	9.7 [A,a]	13.0 [A,a]	36.4 [A,b]	39.2 [A,b]
1001–1500 g	22.6 [A,a]	26.1 [A,a]	0.0 [A,b]	9.5 [B,b]
200–500 g	12.9 [A,a]	2.9 [B,a]	9.1 [A,a]	5.4 [A,a]
>500 g	0.0 [A,a]	0.0 [A,a]	9.1 [A,b]	2.7 [B,b]
Seeds #				
0.0 g	54.8 [A,a]	56.5 [A,a]	33.3 [A,a]	24.7 [A,a]
<500 g	22.6 [A,a]	21.7 [A,a]	25.0 [A,a]	23.3 [A,a]
500–1000 g	6.5 [A,a]	4.3 [A,a]	41.7 [A,a]	43.8 [A,a]
1001–1500 g	16.1 [A,a]	17.4 [A,a]	0.0 [A,a]	6.8 [A,a]
>1500 g	0.0 [A,a]	0.0 [A,a]	0.0 [A,a]	1.4 [A,a]

Consumption per week; different capital letters in the same line: significant differences ($p < 0.05$) by gender (within the same regional group); different lowercase letters: significant differences ($p < 0.05$) for the same gender (different groups).

For respondents who singled out high values of fruit intake, it is important to emphasize that according to the interview, it is clear that it was about candied and dried fruit, while in the case of vegetables, fried vegetables and stuffed vegetables were often mentioned. Southeast Europe abounds in vegetables that are stuffed with cheese and/or meat and vegetables that are consumed as a salad, but are fried or baked in oil (e.g., roasted peppers in oil with garlic), the energy value of which is significantly higher compared to raw vegetables prepared for a salad or cooked. Comparing the nutrition facts of 100 g of raw green papers [23] with 30 kcal and 0% fat with the same pepper but stuffed with cheese [24], the energy values are almost seven times higher (200 kcal, 67.5% of fat).

Specific foods eaten during the day are nuts and seeds, for which the amounts are different regarding regional differences and different food habits. A study conducted in

Serbia gave an overview of the chemical composition and nutritional characteristics of nuts and seeds as food daily consumed in this region [25], indicating the nutrition richness and corresponding composition of this food, but also how it is high in energy.

3.1.2. Body Mass Index as the Indicator of Level of Nourishment

The first indicator of the need for body weight reduction, in the average population, is the body mass index (BMI). It is calculated as the ratio of body mass (in kg) to the square of body height (in meters). According to the classification according to the body mass index [26], obese people have a body mass index in the range of 25–29.9 kg/m^2, and the obesity category is divided into three classes: I obesity class (BMI 30–34.9 kg/m^2) and extreme obesity, which is defined as BMI > 35 kg/m^2, where the II class of obesity is in the BMI range of 35–39.9, and the third (obesity class III) denotes BMI values > 40 kg/m^2. In the first group (Region 1) the overweight group is dominant (30%), while in Group 2 (Region) the obesity group, class I, is the dominant one, followed by the overweight group (25%), as presented in Figure 1.

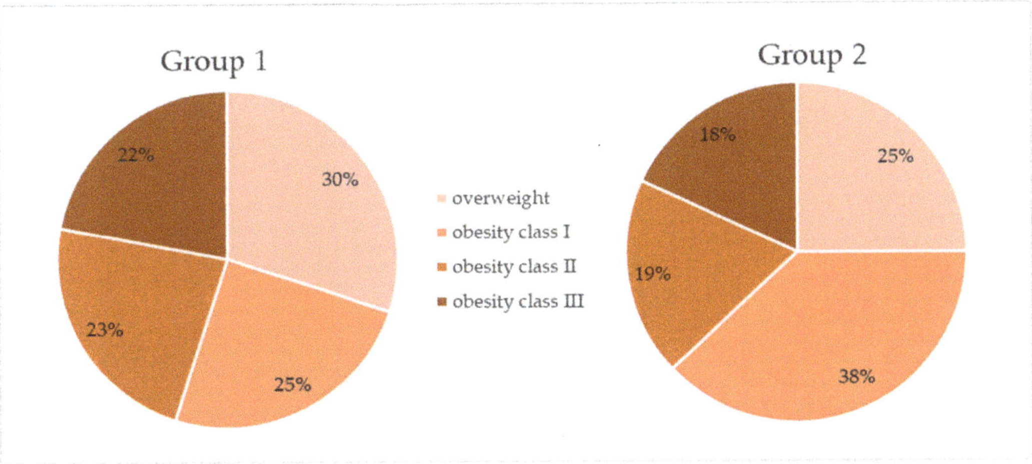

Figure 1. Share of two group participants in the overweight and obesity classes.

During the first medical examination, answers were collected through an interview, and one of the important issues was the incidence of overweight/obesity in the family (Table A2) as well as diagnosed health problems, for which the prevalence in the examined group is presented using a heat map (Figure 2) because this enables the observation of potential grouping in the observed groups [27] and at the same time highlights the differences in values (dark green—the lowest proportion and gradation to dark red—the highest value). Helicobacter pylori was not present in the group of respondents, while hypertension was extremely high in the male population in Group 2. Thyroid problems were dominant in the female population and steatosis in the male population. Dyslipidemia is a health problem equally appearing in the obese/overweight participants with a slightly higher incidence in the female population in Group 2.

3.2. Relating Overweight/Obesity with Eating Habits

In order to determine what needs to be done in particular for people with excess body weight and different classes of obesity, a qualitative analysis and a quantitative analysis were carried out. Principal components analysis served as a qualitative tool for the entire set of respondents (Figure 3). This made it possible to determine the grouping of the observed parameters according to similarity/difference [27,28], but also to determine which of the observed parameters describe the connection between obesity and eating habits, with a share of variations greater than 85%, in the observed set.

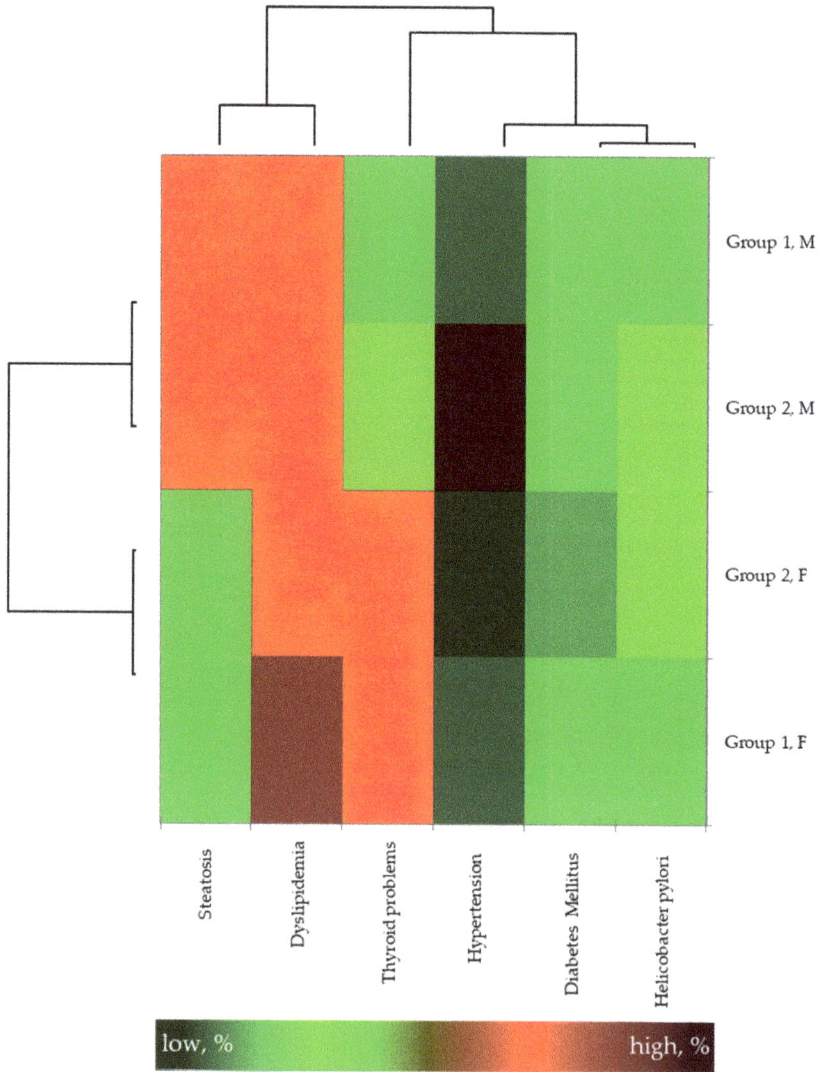

Figure 2. Prevalence (%) of diagnosed diseases in overweight and obese subjects.

Higher consumption of sweets is related with age under 30 years, as well those aged over 50 years. The consumption of a larger amount of vegetables was specified by the female population, while carbonated drinks were consumed more by the male population. The qualitative variable regions are located on opposite sides of the coordinate system, which is an indication that there are a number of differences in their dietary patterns.

The investigated food groups are also associated with the anthropometric indicators, such as body mass, body mass index, and waist-to-hip ratio, by use of the same qualitative multivariate tool as presented in the previous figure (Figure 3), principal component analysis. Such an approach greatly helps in understanding the observed problem [29], especially when the share that describes the set of all variations in the observed set is as high as it is in the case outlined in Figure 4, where the first and second main components describe almost 96.7% of all variations. As expected, the variables "Body mass" and "Body mass index" are positioned in the same coordinate quadrant. This was expected because body mass is in the numerator of the BMI calculation equation. What should definitely be highlighted in this figure are the variables opposite the first quadrant, namely the variables positioned in the third quadrant: the number of meals and the consumption of vegetables. Namely, the variables in the opposite quadrants should be inversely proportional [23], which would lead to the conclusion that an increase in the number of meals and the amount of vegetables consumed reduces BM and BMI. However, the explanation is the following, which is that the number of meals was not singled out as a significant variable, and the proportion of vegetables has already been commented on, and its consumption should be taken with a grain of salt, as well as the "form" of vegetables consumed.

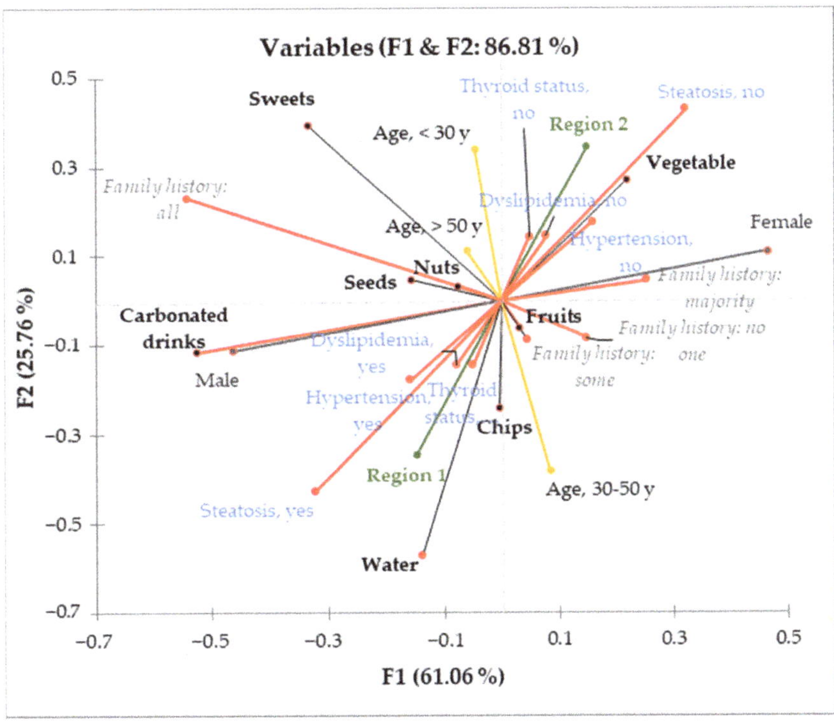

Figure 3. Principal component analysis indicating relationship of eating habits (bolded text) with the incidence of some healthy issues (blue text), according to (i) belonging to one of the two observed regions (green), (ii) gender, (iii) prevalence of overweight/obesity in the family (gray text), and (iv) age (yellow lines).

3.3. Successful Implementation of the Body Weight Reduction Program

For all participants (regardless of whether the target body mass was achieved), after 4 months, a comparison of parameters that are indicators of the success of body mass reduction, and which directly results in the reduction in parameters such as WHR and BMI, was carried out, as shown in Figure 5. In order to show the range, average, and potential of respondents who are the so-called outliers according to the observed parameter [8], the Box–Whisker diagram was used. The effectiveness of the ketogenic diet after 120 days in the program is presented in Table 4.

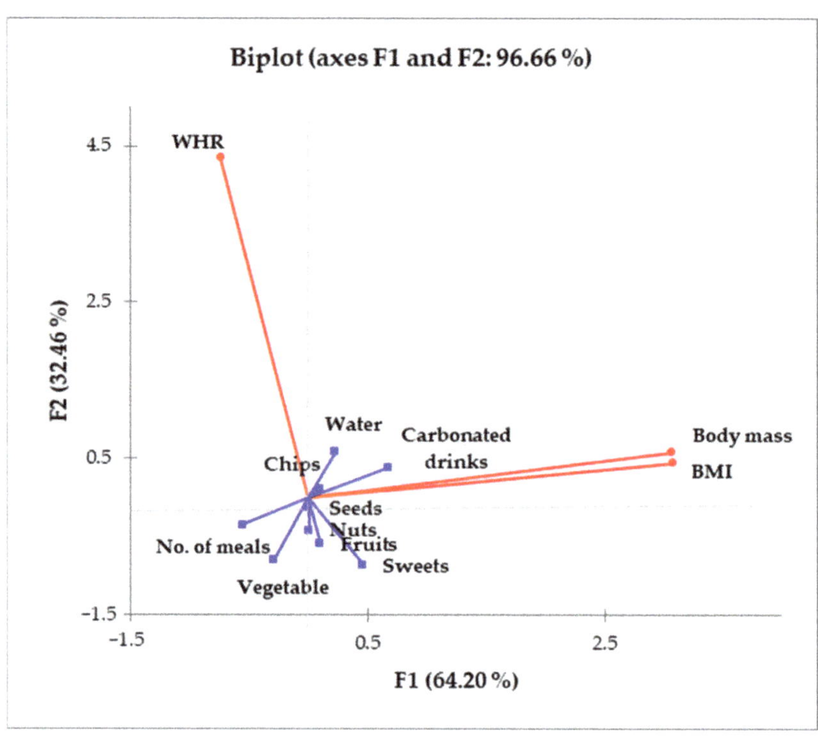

Figure 4. Principal component analysis indicating relationship of numbers of meals per day and consumption of observed foods and drinks (blue lines, bolded text) with the anthropometric indicators (red lines) as waist hip ration (WHR), body mass and body mass index (BMI).

Table 4. Proportion of successfully reduced anthropometric indicators (BMI < 25 kg/m^2 and WHR < 0.85 and 0.9 for women and men, respectively) with corresponding mean values and standard deviations according to gender and regions (Group 1 and 2).

Participants		BMI		WHR	
		in the Healthy Range (%)	Average ± SD (kg/m^2)	in the Healthy Range (%)	Average ± SD (kg/m^2)
Group 1	Male	35.71	27.76 ± 3.44	35.71	0.92 ± 0.05
	Female	73.91	25.41 ± 4.35	69.57	0.84 ± 0.06
Group 2	Male	30.77	27.9 ± 3.37	46.15	0.88 ± 0.09
	Female	81.69	25.14 ± 3.06	98.59	0.82 ± 0.06

The female population was more effective considering the proportion of those participants who achieved the healthy range of BMI (73.9% from Region 1; 81.7% from Region 2) and WHR (over 69% in Region 1 and over 98% in Region 2). Table A3 presents the progress in the body weight reduction program, and this is presented in Table 1 (at the beginning of the program) according to groups and gender. Only age and body height are omitted because they did not change during the weight loss program.

Significant differences in the BM, BMI, and WHR are related with the familiar prevalence of overweight/obesity and age (only for the WHR).

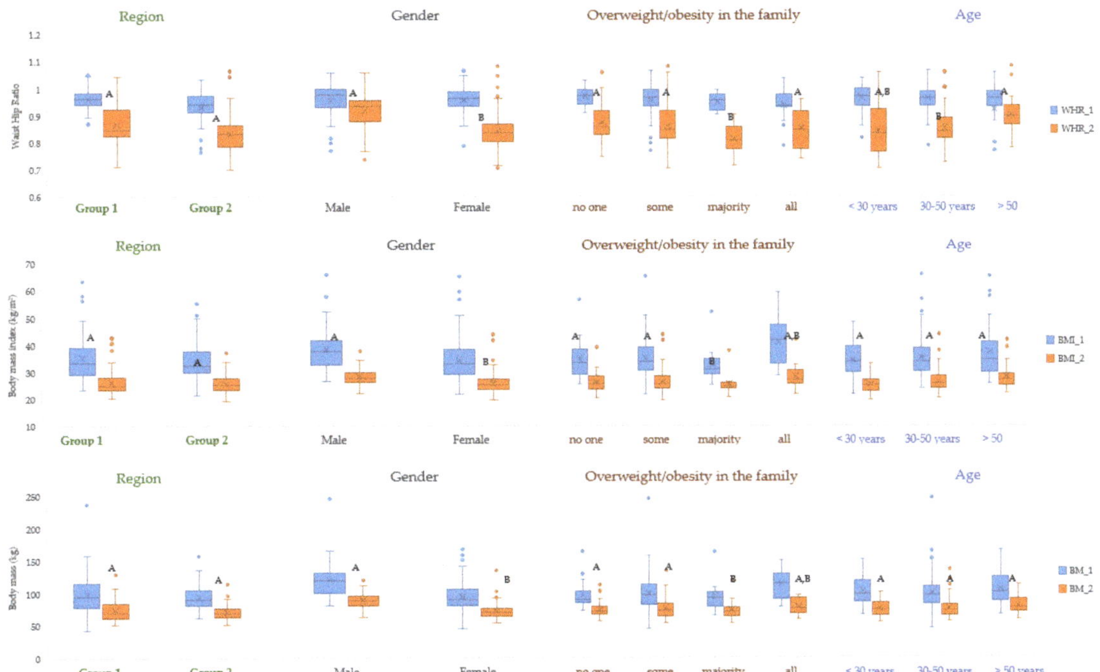

Figure 5. Box plot success in reducing WHR, body mass index, and body weight loss after participating in a weight loss program. Different capital letters within the observed groups in (i) a certain region, (ii) according to gender, (iii) prevalence of overweight/obesity in the family, and (iv) age indicate significant differences. Compared are the body mass indexes at the beginning (BMI_1, blue) and at the end (BMI_2, orange) of the weight loss program.

4. Discussion

Our finding indicated regional differences in the daily and weekly consumption of foods from all observed food categories (Sweets, Chips, Vegetables, Fruits, and Nuts and Seeds, but not in the category "Drinks"). However, gender, as a qualitative variable, showed that the female population consumed at least 101 g of these (44.9% from Group 1 and 53.3%, from Group 2, respectively). Causes of globesity are related with overfeeding [30], ultra-processed food [31], and fast food [32], but also with the gut microbiota [33,34], inheritance and genetics [35], etc., indicating its multiplex etiology [36]. But when obesity is related with food intake, the causes are high energy and low nutrient density [37], with a high proportion of salty snacks and sweets [38] and an extremely low proportion of fruits and vegetables [39]. As Nour and coworkers presented [40], the consumption of more than four servings per day reduces the risk of weight gain but also the inverse association with waist circumference in women (≈ -0.4 cm per daily vegetable serving), while another study has related vegetable intake with a higher intake of fibers, relating it with weight loss among Full Plate Living program participants [41].

A high share of the obese participants (73% obese and 27% overweight) in the total set of 200 participants (100 per region/group) is also related with the estimation of overweight/obesity, where a "few pounds more" is mostly taken seriously when obesity is more likely than overweight, especially in the male population [42].

Unfortunately, the findings of our study also confirmed the relation of health problems such as steatosis, dyslipidemia, thyroid problems, hypertension, diabetes, and Helicobacter pylori with high values of WHR and BMI (over the values related with normal health). A study which investigated hypertension (HTA) in obese patients showed that it is present in 58% of male and 49% of female patients [43]; luckily, in our group, only eight male participants were diagnosed with HTA, but eight females were also diagnosed too. In the group of our respondents, dyslipidemia dominated (14% of respondents), followed by a thyroid problem (11%), which was exclusively present in the female population, followed by steatosis and HTA (8% of patients) and diabetes (2%). Although the study of Baradaran at al. [44] showed that participants infected with H. pylori have a higher risk of obesity (OR of 1.01), in our study groups, no one reported this health problem during the first interview.

The final results of this study present significant weight loss after four months in the program (or earlier for those for whom the expected time frame was under the average of 112 days for achieving the targeted body mass). Although the proportions of BMI and WHR in the healthy range are in favor of the female population (BMI = 77.8%; WHR = 84.08%), the male population initially had more kilos to lose, and the time span in which the target body mass was expected to be reached was 6 to 12 months for most of them. This corresponds to the share of almost a third (31.8%) who achieved the desired body mass in four months. What should definitely be pointed out is that the proportion of the male population who, through the weight loss program, reduced their waist and hip circumferences to a value of less than 0.9 also achieved the expected body mass index that corresponds to a healthy BMI, which is 25 kg/m^2. Similar findings were the outcomes of a study which is among one of the few investigating gender differences in weight loss and weight loss maintenance, where the male population lost a smaller amount compared to females in the same time frame [45].

Research investigating optimal dietary strategies for weight loss and body weight maintenance [46] highlighted the ketogenic diet as effective in suppressing hunger (during the diet), and some therapeutic effects were found in patients with type 2 diabetes, as well as in cardiovascular and neurological diseases and polycystic ovary syndrome. However, the aforementioned research by Kim [46], as well as other studies, state that additional studies are needed to confirm the effectiveness and safety of the ketogenic diet [47,48]. A safe diet after the weight loss program which can help maintain the targeted body weight would be a plant-based diet [49], as well as the Mediterranean diet [50]. The ketogenic diet was suggested as one of the options for losing weight in obese people [17], but there are certain risks, such as kidney disease [51], osteoporosis, and liver disease, while a high share of fats increases the daily intake of saturated fats, which can lead to heart disease [47]. The study of Gomez-Arbelaz et al. [52] showed that 120 days of ketogenic diet will not influence the values of the plasma bicarbonate level and acid-base status or blood pH, indicating "that this diet can be considered as a safe nutritional solution for obese individuals" [53].

Several key takeaways which emerge from this study are:

Regional and Gender Differences: This research highlighted substantial differences in food consumption patterns between the two studied regions, as well as gender-based variations. These variations underscore the importance of tailoring interventions to specific populations and considering cultural and regional factors in weight loss programs.

Health Implications: The prevalence of health problems, such as hypertension, dyslipidemia, and thyroid issues, was found to be closely linked to higher body mass index (BMI) and waist-to-hip ratio (WHR) values. This underscores the critical role of weight management in reducing the risk of various health complications.

Eating Habits and Weight Loss: This study demonstrated that dietary habits have a significant impact on weight loss outcomes. Participants who adopted healthier eating habits and reduced their consumption of high-calorie, low-nutrient foods achieved more successful weight loss results.

Ketogenic Diet Efficacy: The ketogenic diet approach was shown to be effective in facilitating weight loss and improving anthropometric indicators, especially for females. The program's success suggests that dietary interventions, such as the ketogenic diet, can play a crucial role in achieving relatively rapid weight loss results.

Importance of Individualization: This study emphasized the importance of tailoring weight loss interventions to individual needs, including factors like age, gender, family history, and personal preferences. This individualized approach enhances the likelihood of sustainable weight loss and improved overall well-being.

Consideration of Lifestyle and Cultural Factors: Lifestyle changes, including dietary modifications, play a crucial role in addressing the obesity epidemic. Understanding cultural and regional differences in eating habits can help design more effective and targeted intervention programs.

5. Conclusions

This case study investigated the effectiveness of a ketogenic diet program for weight loss among overweight and obese adults from two distinct regions. The study revealed noteworthy findings that shed light on the complexities of weight management and its relation to dietary habits, health issues, gender, and regional variations. Several key takeaways which emerge from this study are (i) the confirmed regional (cultural) and gender differences; (i) health implications related to BMI over 30 kg/m^2; and (iii) the effectiveness of the ketogenic diet for weight reduction.

In essence, this case study underscores the multifaceted nature of weight management and the importance of comprehensive strategies that encompass dietary changes, lifestyle modifications, and tailored approaches. These findings contribute valuable insights for healthcare professionals, policymakers, and individuals seeking effective ways to combat obesity and promote healthier living.

Author Contributions: Conceptualization, V.K. and J.G.K.; methodology, J.G.K..; validation, G.M., V.K. and J.G.K.; formal analysis, J.G.K.; investigation, G.M.; resources, G.M.; data curation, V.K. and J.G.K.; writing—original draft preparation, G.M., V.K. and J.G.K.; writing—review and editing, V.K. and J.G.K.; visualization, V.K. and J.G.K.; supervision, V.K. and J.G.K.; project administration, G.M.; funding acquisition, G.M. All authors have read and agreed to the published version of the manuscript.

Funding: This research received no external funding.

Institutional Review Board Statement: The study was conducted in accordance with the Declaration of Helsinki and approved by the Institutional Review Board of Универзитет „Св. Климент Охридски"—Битола, Технолошко—технички факултет Велес (protocol code 10-168/1 approved on 28 April 2022).

Informed Consent Statement: Informed consent was obtained from all subjects involved in the study.

Data Availability Statement: The data that support the findings of this study are available from the corresponding author upon reasonable request.

Acknowledgments: We would like to thank our participants for providing their written consent to the use of their data for processing and publication in this paper.

Conflicts of Interest: The authors declare no conflict of interest.

Appendix A

Table A1. Frequency of consumed foods from different food groups presented as average and the ranges for each gender from the observed groups.

Drink Consumed per Day	Frequency (% per Day)			
	Group 1		Group 2	
	Male	Female	Male	Female
Carbonated drinks 0 L	25.8 [A,a]	39.1 [B,a]	30.8 [A,a]	32.9 [A,a]

Table A1. *Cont.*

| | Frequency (% per Day) | | | |
| | Group 1 | | Group 2 | |
Drink Consumed per Day	**Male**	**Female**	**Male**	**Female**
<0.5 L	12.9 [A,a]	29.0 [B,a]	38.5 [A,b]	17.1 [B,b]
0.5–1 L	16.1 [A,a]	20.3 [A,a]	23.1 [A,b]	25.7 [A,a]
1–2 L	38.7 [A,a]	7.2 [B,a]	23.1 [A,b]	17.1 [A,b]
>2 L	6.5 [A,a]	4.3 [A,a]	15.4 [A,b]	7.1 [B,a]
Water				
0 L	74.2 [A,a]	56.5 [B,a]	75.0 [A,a]	88.5 [B,b]
<0.5 L	19.4 [A,a]	33.3 [B,a]	0.5 [A,b]	2.3 [A,b]
0.5–1 L	6.5 [A,a]	10.1 [A,a]	16.7 [A,b]	3.4 [A,b]
1–2 L	0.0 [A,a]	0.0 [A,a]	0.0 [A,a]	2.3 [A,b]
>2 L	0.0 [A,a]	0.0 [A,a]	8.3 [A,b]	3.4 [A,b]

Different capital letters in the same line: significant differences ($p < 0.05$) by gender (within the same regional group); different lowercase letters: significant differences ($p < 0.05$) for the same gender (different groups).

Table A2. Frequency of incidence of overweight/obesity in the family for investigated groups and gender.

| **Family Incidence of Overweight/Obesity** | Group 1 | | Group 2 | | Total | |
	Male	**Female**	**Male**	**Female**	**Male**	**Female**
No one	16.1 [A,a]	17.2 [A,a]	7.7 [A,b]	22.1 [A,a]	11.9	19.7
Some	77.4 [A,a]	73.4 [A,a]	46.2 [A,b]	65.1 [A,b]	61.8	69.3
Majority	3.2 [A,a]	1.6 [A,a]	23.1 [A,b]	5.8 [B,a]	13.2	3.7
All	3.2 [A,a]	7.8 [A,a]	23.1 [A,b]	7.0 [B,a]	13.2	7.4

Different capital letters in the same line: significant differences ($p < 0.05$) by gender (within the same regional group); different lowercase letters: significant differences ($p < 0.05$) for the same gender (different groups).

Table A3. Average values with corresponding range (given in the brackets) of the anthropometric parameters for the participants of two groups after 120 days in the weight loss program based on the ketogenic diet.

| **Basic Information** | Group 1 | | Group 2 | |
	Male ($n = 31$)	**Female ($n = 69$)**	**Male ($n = 13$)**	**Female ($n = 87$)**
Body mass (kg)	85.4 (60.3–107.7)	68.1 (50.7–129.1)	84.5 (68.0–114.8)	68.5 (51.1–89.6)
Body mass index (kg/m^2)	27.8 (21.6–38.2)	25.4 (20.0–42.6)	27.9 (23.4–37.1)	25.1 (19.2–33.7)
Consumed meals per day (No.)	2.6 (1–5)	2.8 (1–6)	2.5 (2–4)	2.5 (1–5)
Waist Circumference (cm)	86.7 (72–114)	80.2 (71–120)	84.3 (66–102)	79.1 (58–84)
Waist-to-hip ratio	0.92 (0.82–1.04)	0.84 (0.71–0.93)	0.88 (0.71–1)	0.82 (0.70–0.91)

Group 1: participants from Republic of North Macedonia; Group 2: participants from Kosovo.

References

1. Meldrum, D.R.; Morris, M.A.; Gambone, J.C. Obesity pandemic: Causes, consequences, and solutions-but do we have the will? *Fertil. Steril.* **2017**, *107*, 833–839. [CrossRef]
2. Ahluwalia, M.K. Chrononutrition—When We Eat Is of the Essence in Tackling obesity. *Nutrients* **2022**, *14*, 5080. [CrossRef]
3. Di Rosa, C.; Lattanzi, G.; Taylor, S.F.; Manfrini, S.; Khazrai, Y.M. Very low calorie ketogenic diets in overweight and obesity treatment: Effects on anthropometric parameters, body composition, satiety, lipid profile and microbiota. *Obes. Res. Clin. Pract.* **2020**, *14*, 491–503. [CrossRef]

4. Tham, K.W.; Abdul Ghani, R.; Cua, S.C.; Deerochanawong, C.; Fojas, M.; Hocking, S.; Lee, J.; Nam, T.Q.; Pathan, F.; Saboo, B.; et al. Obesity in South and Southeast Asia-A new consensus on care and management. *Obes. Rev.* **2023**, *24*, e13520. [CrossRef]
5. Zavala, G.A.; Kolovos, S.; Chiarotto, A.; Chiarotto, A.; Bosmans, J.E.; Campos-Ponce, M.; Rosado, J.L.; Garcia, O.P. Association between obesity and depressive symptoms in Mexican population. *Soc. Psychiatry Psychiatr. Epidemiol.* **2018**, *53*, 639–646. [CrossRef]
6. Berry, E.M. The Obesity Pandemic-Whose Responsibility? No Blame, No Shame, Not More of the Same. *Front. Nutr.* **2020**, *7*, 2. [CrossRef]
7. Vasileva, L.V.; Marchev, A.S.; Georgiev, M.I. Causes and solutions to "globesity": The new FA(S)T alarming global epidemic. *Food Chem. Toxicol.* **2018**, *121*, 173–193. [CrossRef]
8. Markovikj, G.; Knights, V.; Kljusurić, J.G. Ketogenic Diet Applied in Weight Reduction of Overweight and Obese Individuals with Progress Prediction by Use of the Modified Wishnofsky Equation. *Nutrients* **2023**, *15*, 927. [CrossRef]
9. Frank, J.W. Controlling the obesity pandemic: Geoffrey Rose revisited. *Can. J. Public Health* **2022**, *113*, 736–742. [CrossRef]
10. Abarca-Gomez, L.; NCD Risk Factor Collaboration. Worldwide trends in body-mass index, underweight, overweight, and obesity from 1975 to 2016: A pooled analysis of 2416 population-based measurement studies in 128·9 million children, adolescents, and adults. *Lancet* **2017**, *390*, 2627–2642. [CrossRef]
11. Pandita, A.; Sharma, D.; Pandita, D.; Pawar, S.; Tariq, M.; Kaul, A. Childhood obesity: Prevention is better than cure. *Diabetes Metab. Syndr. Obes.* **2016**, *9*, 83–89. [CrossRef]
12. Flegal, K.M.; Ogden, C.L.; Fryar, C.; Afful, J.; Klein, R.; Huang, D. Comparisons of Self-Reported and Measured Height and Weight, BMI, and Obesity Prevalence from National Surveys: 1999–2016. *Obesity* **2019**, *27*, 1711–1719. [CrossRef] [PubMed]
13. CDC. National Health and Nutrition Examination Survey 1999–2016 Survey Content Brochure. Available online: https://www.cdc.gov/nchs/nhanes/index.htm (accessed on 7 June 2023).
14. CDC. National Center for Health Statistics- National Health Intervju Survey. Available online: https://www.cdc.gov/nchs/nhis/index.htm (accessed on 7 June 2023).
15. CDC. Behavioral Risk Factor Surveillance System. Available online: https://www.cdc.gov/brfss/index.html (accessed on 7 June 2023).
16. Contreras, R.E.; Schriever, S.C.; Pfluger, P.T. Physiological and Epigenetic Features of Yoyo Dieting and Weight Control. *Front. Genet.* **2019**, *10*, 1015. [CrossRef]
17. Drabińska, N.; Wiczkowski, W.; Piskuła, M.K. Recent advances in the application of a ketogenic diet for obesity management. *Trends Food Sci. Technol.* **2021**, *110*, 28–38. [CrossRef]
18. Dowis, K.; Banga, S. The Potential Health Benefits of the Ketogenic Diet: A Narrative Review. *Nutrients* **2021**, *13*, 1654. [CrossRef] [PubMed]
19. Knights, V.; Kolak, M.; Markovikj, G.; Gajdoš Kljusurić, J. Modeling and Optimization with Artificial Intelligence in Nutrition. *Appl. Sci.* **2023**, *13*, 7835. [CrossRef]
20. Papadaki, A.; Linardakis, M.; Plada, M.; Larsen, T.M.; van Baak, M.A.; Lindroos, A.K.; Pfeiffer, A.F.; Martinez, J.A.; Handjieva-Darlenska, T.; Kunešová, M.; et al. A multicentre weight loss study using a low-calorie diet over 8 weeks: Regional differences in efficacy across eight European cities. *Swiss Med. Wkly.* **2013**, *143*, w13721. [CrossRef]
21. Sreenivas, S. Keto Diet for Beginners—Nourish byWebMD. Available online: https://www.webmd.com/diet/keto-diet-forbeginners (accessed on 4 July 2022).
22. Maltarić, M.; Ruščić, P.; Kolak, M.; Bender, D.V.; Kolarić, B.; Ćorić, T.; Hoejskov, P.S.; Bošnir, J.; Kljusurić, J.G. Adherence to the Mediterranean Diet Related to the Health Related and Well-Being Outcomes of European Mature Adults and Elderly, with an Additional Reference to Croatia. *Int. J. Environ. Res. Public Health* **2023**, *20*, 4893. [CrossRef]
23. Fatsecret. Green Peppers—Nutrition Facts. Available online: https://www.fatsecret.com/calories-nutrition/usda/green-peppers (accessed on 30 August 2023).
24. Fatsecret. Hot & Sweet Peppers Stuffed with Cheese—Nutrition Facts. Available online: https://www.eatthismuch.com/food/nutrition/hot-sweet-peppers-stuffed-with-cheese,2508247/ (accessed on 30 August 2023).
25. Dodevska, M.; Kukic Markovic, J.; Sofrenic, I.; Tesevic, V.; Jankovic, M.; Djordjevic, B.; Ivanovic, N.D. Similarities and differences in the nutritional composition of nuts and seeds in Serbia. *Front. Nutr.* **2022**, *9*, 1003125. [CrossRef] [PubMed]
26. Weir, C.B.; Jan, A. BMI Classification Percentile and Cut Off Points. [Updated 2023 Jun 26]. Available online: https://www.ncbi.nlm.nih.gov/books/NBK541070/ (accessed on 16 August 2023).
27. Maraschim, J.; Honicky, M.; Moreno, Y.M.F.; Hinnig, P.d.F.; Cardoso, S.M.; Back, I.d.C.; Vieira, F.G.K. Consumption and Breakfast Patterns in Children and Adolescents with Congenital Heart Disease. *Int. J. Environ. Res. Public Health* **2023**, *20*, 5146. [CrossRef] [PubMed]
28. Addas, A. Understanding the Relationship between Urban Biophysical Composition and Land Surface Temperature in a Hot Desert Megacity (Saudi Arabia). *Int. J. Environ. Res. Public Health* **2023**, *20*, 5025. [CrossRef]
29. Plura, J.; Vykydal, D.; Tošenovský, F.; Klaput, P. Graphical Tools for Increasing the Effectiveness of Gage Repeatability and Reproducibility Analysis. *Process.* **2023**, *11*, 1. [CrossRef]

30. Fiore, G.; Pascuzzi, M.C.; Di Profio, E.; Corsello, A.; Agostinelli, M.; La Mendola, A.; Milanta, C.; Campoy, C.; Calcaterra, V.; Zuccotti, G.; et al. Bioactive compounds in childhood obesity and associated metabolic complications: Current evidence, controversies and perspectives. *Pharmacol. Res.* **2023**, *187*, 106599. [CrossRef] [PubMed]

31. Buckley, J.P.; Kim, H.; Wong, E.; Rebholz, C.M. Ultra-processed food consumption and exposure to phthalates and bisphenols in the US National Health and Nutrition Examination Survey, 2013–2014. *Environ. Int.* **2019**, *131*, 105057. [CrossRef] [PubMed]

32. Anderson, B.; Rafferty, A.P.; Lyon-Callo, S.; Fussman, C.; Imes, G. Fast-food consumption and obesity among Michigan adults. *Prev. Chronic Dis.* **2011**, *8*, A71.

33. Vetrani, C.; Di Nisio, A.; Paschou, S.A.; Barrea, L.; Muscogiuri, G.; Graziadio, C.; Savastano, S.; Colao, A.; on behalf of the Obesity Programs of Nutrition, Education, Research and Assessment (OPERA) Group. From Gut Microbiota through Low-Grade Inflammation to Obesity: Key Players and Potential Targets. *Nutrients* **2022**, *14*, 2103. [CrossRef]

34. Xu, Y.-S.; Liu, X.-J.; Liu, X.-X.; Chen, D.; Wang, M.-M.; Jiang, X.; Xiong, Z.-F. The Roles of the Gut Microbiota and Chronic Low-Grade Inflammation in Older Adults with Frailty. *Front. Cell. Infect. Microbiol.* **2021**, *11*, 675414. [CrossRef]

35. Cheng, Z.; Zheng, L.; Almeida, F.A. Epigenetic reprogramming in metabolic disorders: Nutritional factors and beyond. *J. Nutr. Biochem.* **2018**, *54*, 1–10. [CrossRef]

36. González-Muniesa, P.; Mártinez-González, M.A.; Hu, F.B.; Després, J.P.; Matsuzawa, Y.; Loos, R.J.F.; Moreno, L.A.; Bray, G.A.; Martinez, J.A. Obesity. *Nat. Rev. Dis. Prim.* **2017**, *3*, 17034. [CrossRef]

37. Stelmach-Mardas, M.; Rodacki, T.; Dobrowolska-Iwanek, J.; Brzozowska, A.; Walkowiak, J.; Wojtanowska-Krosniak, A.; Zagrodzki, P.; Bechthold, A.; Mardas, M.; Boeing, H. Link between Food Energy Density and Body Weight Changes in Obese Adults. *Nutrients* **2016**, *8*, 229. [CrossRef]

38. Tristan Asensi, M.; Napoletano, A.; Sofi, F.; Dinu, M. Low-Grade Inflammation and Ultra-Processed Foods Consumption: A Review. *Nutrients* **2023**, *15*, 1546. [CrossRef]

39. Wu, E.; Ni, J.; Zhou, W.; You, L.; Tao, L.; Xie, T. Consumption of fruits, vegetables, and legumes are associated with overweight/obesity in the middle- and old-aged Chongqing residents: A case-control study. *Medicine* **2022**, *101*, e29749. [CrossRef] [PubMed]

40. Nour, M.; Lutze, S.A.; Grech, A.; Allman-Farinelli, M. The Relationship between Vegetable Intake and Weight Outcomes: A Systematic Review of Cohort Studies. *Nutrients* **2018**, *10*, 1626. [CrossRef] [PubMed]

41. Kelly, R.K.; Calhoun, J.; Hanus, A.; Payne-Foster, P.; Stout, R.; Sherman, B.W. Increased dietary fiber is associated with weight loss among Full Plate Living program participants. *Front. Nutr.* **2023**, *10*, 1110748. [CrossRef]

42. Jorge, T.; Sousa, S.; do Carmo, I.; Lunet, N.; Padrão, P. Accuracy of Assessing Weight Status in Adults by Structured Observation. *Appl. Sci.* **2023**, *13*, 8185. [CrossRef]

43. Pedrianes-Martin, P.B.; Perez-Valera, M.; Morales-Alamo, D.; Martin-Rincon, M.; Perez-Suarez, I.; Serrano-Sanchez, J.A.; Gonzalez-Henriquez, J.J.; Galvan-Alvarez, V.; Acosta, C.; Curtelin, D.; et al. Resting metabolic rate is increased in hypertensive patients with overweight or obesity: Potential mechanisms. *Scand. J. Med. Sci. Sport.* **2021**, *31*, 1461–1470. [CrossRef]

44. Baradaran, A.; Dehghanbanadaki, H.; Naderpour, S.; Pirkashani, L.M.; Rajabi, A.; Rashti, R.; Riahifar, S.; Moradi, Y. The association between Helicobacter pylori and obesity: A systematic review and meta-analysis of case–control studies. *Clin. Diabetes Endocrinol.* **2021**, *7*, 15. [CrossRef]

45. Crane, M.M.; Jeffery, R.W.; Sherwood, N.E. Exploring Gender Differences in a Randomized Trial of Weight Loss Maintenance. *Am. J. Men's Health* **2017**, *11*, 369–375. [CrossRef]

46. Kim, J.Y. Optimal Diet Strategies for Weight Loss and Weight Loss Maintenance. *J. Obes. Metab. Syndr.* **2021**, *30*, 20–31. [CrossRef]

47. Harvard Health Publishing. Should you Try the Keto Diet? Available online: https://www.health.harvard.edu/staying-healthy/should-you-try-the-keto-diet (accessed on 18 September 2023).

48. Batch, J.T.; Lamsal, S.P.; Adkins, M.; Sultan, S.; Ramirez, M.N. Advantages and Disadvantages of the Ketogenic Diet: A Review Article. *Cureus* **2020**, *12*, e9639. [CrossRef]

49. Clark, J.E. Diet, exercise or diet with exercise: Comparing the effectiveness of treatment options for weight-loss and changes in fitness for adults (18–65 years old) who are overfat, or obese; systematic review and meta-analysis. *J. Diabetes Metab. Disord.* **2015**, *14*, 31, Erratum in *J. Diabetes Metab. Disord.* **2015**, *14*, 73. [CrossRef] [PubMed]

50. Gerić, M.; Matković, K.; Gajski, G.; Rumbak, I.; Štancl, P.; Karlić, R.; Bituh, M. Adherence to Mediterranean Diet in Croatia: Lessons Learned Today for a Brighter Tomorrow. *Nutrients* **2022**, *14*, 3725. [CrossRef] [PubMed]

51. Ayele, G.M.; Atalay, R.T.; Mamo, R.T.; Hussien, S.; Fissha, A.; Michael, M.B.; Nigussie, B.G.; Behailu, A. Is Losing Weight Worth Losing Your Kidney: Keto Diet Resulting in Renal Failure. *Cureus* **2023**, *15*, e36546. [CrossRef] [PubMed]

52. Gomez-Arbelaez, D.; Bellido, D.; Castro, A.I.; Ordonez-Mayan, L.; Carreira, J.; Galban, C.; Martinez-Olmos, A.A.; Crujeiras, A.B.; Sajoux, I.; Casenueva, F.F. Body composition changes after very-low-calorie ketogenic diet in obesity evaluated by 3 standardized methods. *J. Clin. Endocrinol. Metab.* **2017**, *102*, 488–498. [CrossRef]

53. Gomez-Arbelaez, D.; Crujeiras, A.B.; Castro, A.I.; Goday, A.; Mas-Lorenzo, A.; Bellon, A.; Tejera, C.; Bellido, D.; Galban, C.; Sajoux, I.; et al. Acid–base safety during the course of a very low-calorie-ketogenic diet. *Endocrine* **2017**, *58*, 81–90. [CrossRef]

applied sciences

Article

Implementation of a Nutrition-Oriented Clinical Decision Support System (CDSS) for Weight Loss during the COVID-19 Epidemic in a Hospital Outpatient Clinic: A 3-Month Controlled Intervention Study

Paraskevi Detopoulou [1], Panos Papandreou [2], Lida Papadopoulou [3] and Maria Skouroliakou [3,*]

[1] Department of Clinical Nutrition, General Hospital Korgialenio Benakio, 11526 Athens, Greece; viviandeto@gmail.com
[2] Department of Nutrition, IASO Hospital, 15123 Athens, Greece; ppapandreou@cibusmed.com
[3] Department of Dietetics and Nutritional Science, School of Health Science and Education, Harokopio University, 17671 Athens, Greece; dp4521823@hua.gr
* Correspondence: mskour@hua.gr

Citation: Detopoulou, P.; Papandreou, P.; Papadopoulou, L.; Skouroliakou, M. Implementation of a Nutrition-Oriented Clinical Decision Support System (CDSS) for Weight Loss during the COVID-19 Epidemic in a Hospital Outpatient Clinic: A 3-Month Controlled Intervention Study. *Appl. Sci.* **2023**, *13*, 9448. https://doi.org/10.3390/app13169448

Academic Editor: Wojciech Kolanowski

Received: 12 June 2023
Revised: 24 July 2023
Accepted: 25 July 2023
Published: 21 August 2023

Abstract: Clinical Decision Support Systems (CDSSs) facilitate evidence-based clinical decision making for health professionals. Few studies have applied such systems enabling distance monitoring in the COVID-19 epidemic, especially in a hospital setting. The purpose of the present work was to assess the clinical efficacy of CDSS-assisted dietary services at a general hospital for patients intending to lose weight during the COVID-19 pandemic. Thirty-nine patients (28 men, 71.8%) comprised the intervention group and 21 patients (four men, 16%) of the control group. After a 3-month CDSS intervention, reductions in both body weight (mean ± standard deviation (SD): 95.5 ± 21.8 vs. 90.6 ± 19.9 kg, $p < 0.001$) and body mass index (BMI) (median, interquartile range (IQR): 35.2, 28.4–37.5 vs. 33.2, 27.4–35.4 kg/m^2, $p < 0.001$) were observed. Beneficial effects were also recorded for total body fat (44.9 ± 11.3 vs. 41.9 ± 10.5%, $p < 0.001$), glycated hemoglobin (5.26 ± 0.55 vs. 4.97 ± 0.41%, $p = 0.017$) (mean ± SD) and triglycerides (137, 115–152 vs. 130, 108–160 mg/dL, $p = 0.005$) (medians, IQR). Lean tissue was borderline decreased (25.4, 21.7–29.1 vs. 24.6, 21.8–27.9 kg, $p = 0.050$). No changes were documented in the control group. In multivariate linear regression models, serum triglycerides were inversely associated with % absolute weight loss (B = −0.018, standard error (SE) = 0.009, $p = 0.050$) in the CDSS intervention group. In women, a principal component analysis-derived pattern characterized by high BMI/lean tissue was positively related to % absolute weight loss (B = 20.415, SE = 0.717, $p = 0.028$). In conclusion, a short-term CDSS-facilitated intervention beneficially affected weight loss and other cardiovascular risk factors.

Keywords: clinical decision support system; CDSS; weight loss; COVID-19; nutrition intervention; Mediterranean diet

1. Introduction

Clinical Decision Support Systems (CDSSs) constitute a useful tool in reducing human-related errors and facilitating evidence-based clinical decision making for medical doctors, pharmacists, and other health professionals [1,2]. Patients' characteristics are entered in the system, and they are compared to recommendations. Then, patient-specific evaluations are provided to the clinician, who makes the final decision [1]. Similarly, a nutrition-oriented CDSS helps dietitians to calculate patients' nutritional needs more quickly and easily compared to paper or simply computerized methods [3]. Such a system reduces the time required for data calculation and analysis during nutritional assessment [3]. Moreover, it ensures the implementation of a step-by-step nutrition care process (NCP) including nutritional assessment, diagnosis, intervention, monitoring and evaluation [4,5]. All records are standardized and are stored and accessible, allowing clinical dietitians and health professionals to conduct clinical research as well as to assess the improvement in patients with long-term nutritional care [1]. Indeed, the use of structured records and the incorporation of alerts/reminders have been shown to enhance the quality of medical records and provision of evidence-based care adherent to guidelines [6].

During the COVID-19 era in Greece, certified dietitian and nutritionist services were restricted during certain quarantine periods (between March and June of 2020), due to state measures to control the spread of COVID-19. Similar measures were also in force for hospital outpatients while the health system was "under pressure". In parallel, good nutritional status and obesity prevention was of outmost importance for COVID-19 prevention and management [7,8]. Hospital dietitians took care of hospitalized patients, and the special needs of whom were covered via menu modifications [9] and oral nutritional supplements [10]. During this difficult time, the use of electronic medical records [11] and tele-health were recommended to provide and monitor nutritional care [7]. Within this scope, the CDSS developed by our group, named Nutrinet®, aimed to provide distance lifestyle advice while avoiding visits to healthcare providers and minimizing the risk of COVID-19 infection [3]. The clinical efficacy of the CDSS used in the present study was recently demonstrated in patients with breast cancer [4], multiple sclerosis [12] and rheumatoid arthritis [13], as well as pregnant women [14]. Several programs based on tele-health and remote consultations were developed to assist patients with obesity during weight loss, especially during COVID-19 [15–17]. However, to our knowledge, there is no study using CDSS in a hospital setting to enable distance counseling and monitoring of overweight and obese patients.

The purpose of the present work was to assess the application and clinical efficacy of CDSS (Nutrinet®)-assisted dietary services for patients at a general hospital who were intending to lose weight during the COVID-19 era.

2. Materials and Methods

2.1. Participants

The study was carried out at a general hospital from March 2020 to September 2021 and consisted mainly of patients who visited outpatient clinics or were hospitalized and needed nutritional intervention and follow-up. More particularly, participants screened for participating in the study were inpatients or outpatients being referred to the dietitian department. The inclusion criteria for the volunteers were (i) age equal to or over 18 years, (ii) ability to be systematically followed up by the dietitian in the time frame that was set and (iii) familiarization with computer systems. The exclusion criteria were (i) pregnancy or breastfeeding, (ii) presence of cancer, liver disease, kidney disease or celiac disease, (iii) alcohol abuse or specific dietary preferences, such as ketogenic or vegetarian diet, (iv) recent change in glucose and lipid-lowering therapy, (v) bariatric or other type of surgery, (vi) computer illiteracy (for the CDSS intervention group) and (vii) inability to read and understand the Greek or English language. In the intervention group, 28 men participated (71.8%). A control group was identified from the outpatients of the hospital's clinic and received only dietary advice (n = 4 men, 16%). All subjects gave their verbal informed consent for inclusion in the study. The study was performed according to the principles of the Helsinki Declaration (1964). Ethical approval was obtained from the hospital's Institutional Review Board (number f310519).

2.2. Study Design

A 3-month single-centered controlled intervention was performed. In the CDSS intervention group, the participants received a personalized eating plan based on the principles of the Mediterranean diet. The energy and macro-nutrient content of the dietary plan were created using a CDSS, as previously described [4]. An example of the dietary plan is given in Supplementary Table S1. The control group received simple lifestyle advice (dietary advice group). The CDSS was accessible to both patients and dietitians, so that bi-directional feedback was possible.

2.3. Anthropometric and Body Composition Measurements

Height was assessed with a stadiometer (Seca 216) in cm to the nearest 0.1 cm. Measurements were taken without shoes, with a straight back, relaxed shoulders and looking straight ahead at a horizontal Frankfurt imaginary line. Body weight was measured with the SOEHNLE Fitness Scale 7850 in kg to the nearest 0.1 kg. Waist circumference was measured with a GIMA tape measure (Gessate, Italy) in cm to the nearest 0.1 cm. All measurements were performed by the same researcher-dietitian.

The participants underwent a body composition analysis using the bioelectrical impedance method. Body composition analysis was performed with the digital scale-body fat analyzer (SOEHNLE Fitness Scale 7850) (Soehnle GmbH, Murrhardt, Germany) and included body fat (%) and fat free mass (% and kg). Measurements were taken without shoes and socks and with light clothing according to the manufacturer's recommendations. Lean mass index was then calculated as lean mass divided by squared height (kg/m^2).

2.4. Nutrition-Oriented CDSS

In the present study, the Nutrinet® CDSS was used. The system was initially launched in the pre-COVID era. However, the functions included facilitated a distance-based intervention and follow-up. Basic information (patient's name, date of birth, gender), anthropometric, medical, nutritional and pharmaceutical data were deposited in the database during the visit (Figure 1). More particularly, the CDSS assisted in the following procedures: medical and pharmaceutical history, dietary assessment, physical activity assessment, formulation of personalized tailored dietary plans, provision of guidelines, definition of nutritional and physical activity goals, and monitoring nutritional and physical activity goals.

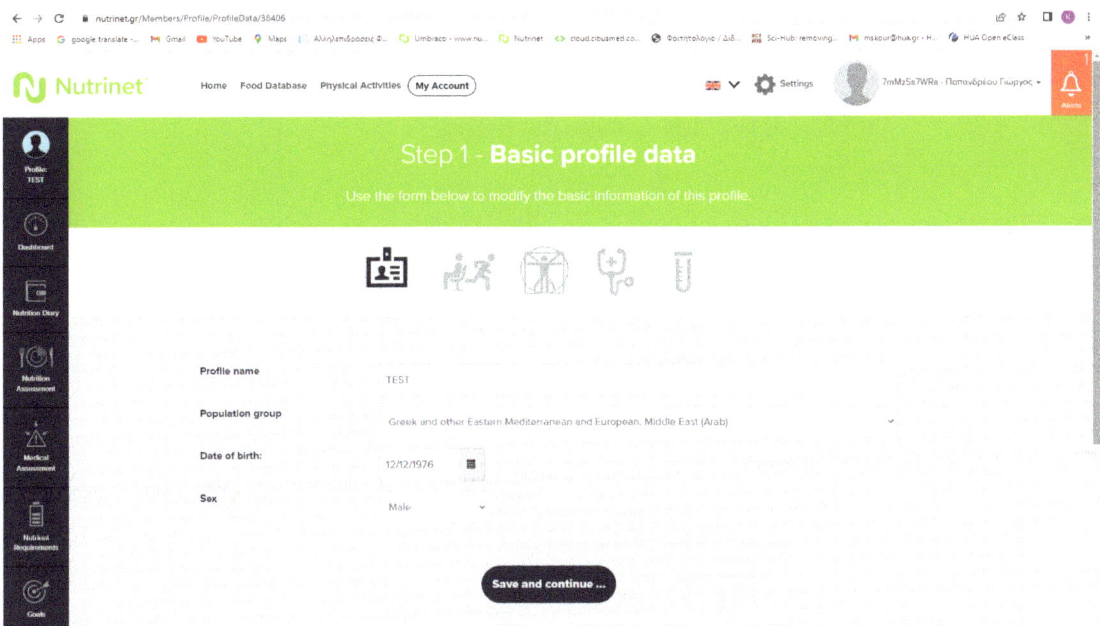

Figure 1. The Nutrinet® CDSS: basic profile data.

2.4.1. Medical and Pharmaceutical History

Regarding the medical history, information was recorded about health state and the presence of diseases or chronic conditions. A history of surgical procedures affecting nutrient absorption was also included. In addition, information was collected on smoking habits, drinking habits, and unintentional weight loss in the past 3–6 months. In the pharmaceutical history, the intake of medicines, herbal products and dietary supplements was recorded.

At the first session, the volunteers were asked to present their most recent laboratory tests performed in the last 4 months. Laboratory tests were repeated under the direction of the dietitian in order to detect possible differences at follow-up. The laboratory data requested and recorded were blood tests of fasting glucose (mg/dL), glycated hemoglobin (HbA1c), total-cholesterol, HDL-cholesterol, and triglycerides. LDL-cholesterol was calculated based on the Fridewald formula, provided that triglycerides were <400 mg/dL [18]. It is noted that with the aid of the CDSS the assessment of blood lipid and glucose values was automatically performed according to international standards. For lipids assessment, the NIH/NHLBI/NCEP (ATP III) 2002 criteria were used [19], and for the assessment of blood glucose levels, the American Diabetes Association (ADA) 2014 criteria were used [20].

2.4.2. Dietary Assessment

The assessment of the participants' dietary intake in terms of energy, macronutrients, micronutrients and eating habits was conducted using a 24 h recall method, by a dietitian. The 24 h recall was carried out on the day of admission to the study. Essentially, the volunteers were asked to describe all foods and beverages they consumed the previous day in detail, and their corresponding quantity. Data from the 24 h recall were immediately entered into the CDSS nutritional program to perform the analysis. The embedded nutrition database used by the CDSS uses data from the USDA database (Releases 27 and 28) [21], the food database created by Trichopoulou [22], and information provided by food manufacturers. The dietary reference intakes are used to assess and design personal dietary plans [23].

The ideal body weight was automatically calculated with the aid of the CDSS with the use of the Hamwi equation [24]. The body mass index was calculated by the division of weight with squared meters (in kg/m^2) for each person who participated in the study. Moreover, an assessment of waist circumference was provided based on the criteria of the International Diabetes Federation [25] (Figure 2).

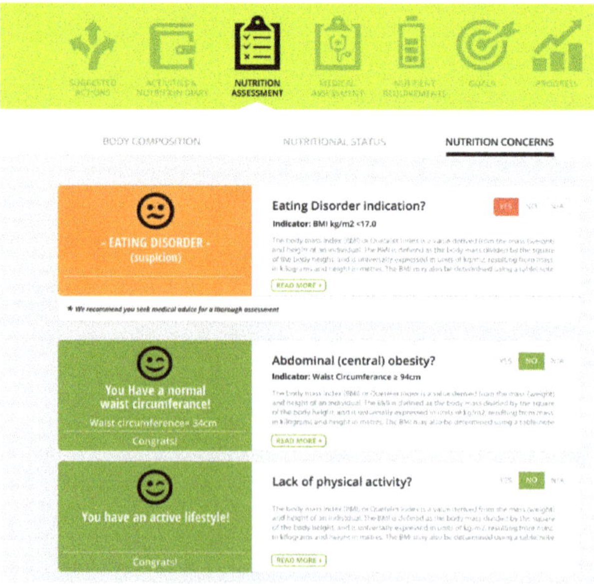

Figure 2. The Nutrinet® CDSS: Nutrition assessment.

2.4.3. Physical Activity Assessment

Four physical activity categories were available to select through the CDSS: (i) mildly active, (ii) moderately active, (iii) active, and (iv) intensely active or athlete. Physical activity intensity was evaluated based on the corresponding metabolic equivalents (MET) [26]. Moreover, an assessment of physical activity status was provided (Figure 2).

2.4.4. Formulation of Personalized Tailored Dietary Plans

In the intervention group, patients were prescribed a CDSS-generated individualized daily dietary plan (food portions in grams, recipes, etc.) based on the Mediterranean diet. Moreover, the presence of gastro-intestinal issues, in which medical nutrition therapy is indicated for long with other therapies (such as constipation, diarrhea or gastroesophageal reflux) was considered in the formulation of CDSS diet plans.

2.4.5. Dietary Monitoring and Evaluation

The ability to track patient progress per session was also provided by the CDSS. The application allowed the dietitian to see the metrics of previous meetings in a graphical way and to evaluate progress. Depending on the case and health problem, the dietitian was able to focus on different points each time. Moreover, login passwords were provided to patients in the intervention group to allow remote access to the CDSS. In this way, patients had the opportunity to also self-track their progress relating to body weight, physical activity, fruits consumption, etc. It is noted that Nutrinet® helped people focus on enhancing self-awareness, being confident and understanding behavioral patterns, through small steps.

2.5. Statistical Analysis

The Kolmogorov–Smirnov test was used to test for data normality. Continuous normal variables were expressed as mean $\pm$ standard deviation (SD), and continuous non normal variables were expressed as median, interquartile range (IQR). Dichotomous variables were expressed as counts and frequencies (n, %). To assess the differences between control and experimental groups, the Student's *t*-test or Mann–Whitney U-test was used for normally distributed and non-normally distributed variables, respectively. The paired samples *t*-test or the Wilcoxon test was performed to test pre- and post-intervention differences within groups, for normally distributed and not-normally distributed variables, respectively. Principal component analysis (PCA) was performed to identify body weight and body composition patterns. To choose the number of components to keep from the PCA analysis, the eigenvalues derived from the correlation matrix of the standardized variables were assessed, and the scree plot was checked for confirmation. Components with eigenvalue higher than one were retained for data analyses. Based on the concept that the component scores are interpreted as correlation coefficients (i.e., higher absolute scores indicate that the variable characterizes most to the component), the bodyweight and body composition patterns were defined in relation to variables that correlated most with the component (absolute factor loadings > 0.45). Moreover, the orthogonal varimax rotation was used to derive optimal patterns, which were non-correlated.

Spearman coefficients were calculated in order to identify the factors that correlated mostly with % weight loss. Then, linear regression analysis was carried out to identify the factors that can "predict" weight loss after taking into account basic confounders. More particularly, the absolute % weight loss was set as dependent variable and several variables were set as independent ones, such as age, body weight patterns and blood exams. Sex-specific analysis was also performed. All analyses were carried out with the SPSS statistical software (Version 21.0, SPSS, Inc., IBM, Chicago, IL, USA). Statistical significance was set at *p*-value < 0.05.

3. Results

In Table 1, the general characteristics of participants are shown as well as the pre- and post-intervention differences in the CDSS intervention group versus dietary advice group. It is noted that the CDSS intervention group consisted mainly of males (~70%). The mean age and body mass index (BMI) of the CDSS intervention group was 48.9 years and 35.2 kg/m^2, respectively. The percentage of overweight subjects was similar between the two groups (30.8% in the CDSS intervention group vs. 32.3% in the dietary advice group, $p = 0.153$), while the intervention group consisted of more obese subjects (17.9% in the CDSS intervention group vs. 9.5% in the dietary advice group). The levels of several biochemical markers of glucose and lipid metabolism were not different between the CDSS-intervention group and the control group. Moreover, it is noted that the pre- and post-effects in the CDSS intervention and dietary advice groups are seen in Table 1. In the CDSS intervention arm, body weight, BMI, total body fat, glycated hemoglobin and triglycerides were significantly decreased at follow-up. Lean tissue was borderline decreased, and lean mass index was decreased, while glucose and cholesterol (total, LDL- and HDL-cholesterol) did not change. No difference was observed in the dietary advice group. It is noted that grouped results for both genders may be misleading, since anthropometric and body composition measurements are different in men and women. For this reason, sex-specific analysis was also conducted. In sex-specific analysis, body weight, BMI and total body fat were reduced in both sexes (Table 2). More particularly, body weight was reduced from 87.9 ± 14.9 kg to 83.3 ± 13.9 kg in men and from 114.6 ± 25.4 kg to 109.2 ± 21.4 kg in women. BMI was reduced from 35.2 (28.4–37.5) kg/m^2 to 33.2 (27.4–35.4) kg/m^2 in men and from 26.4 (22.4–28.5) kg/m^2 to 24.9 (21.7–27.4) kg/m^2 in women. Total body fat was reduced from 46.9 ± 9.8% to 43.7 ± 9.7% in men and from 39.9 ± 13.8% to 37.3 ± 11.5% in women. Lean tissue and lean mass index remained unchanged in both men and women. In addition, beneficial changes were documented in glucose, total cholesterol and triglycerides in male participants (Table 2). No sex-specific analysis was conducted in the dietary advice group, since a low number of men were recruited.

Table 1. Baseline and follow-up measurements of the patients in the CDSS intervention and dietary advice group.

	CDSS Intervention Group (n = 39)			Dietary Advice Group (n = 21)			
	Baseline	Follow-Up	p-Value §	Baseline	Follow-Up	p-Value ‡	p-Value ∫
Age (y)	48.9 ± 13.4	NA	NA	67.0 ± 16.4	NA	NA	<0.001
Sex (men), n (%)	28 (71.8%)	NA	NA	4 (16%)	NA	NA	<0.001
Body weight (kg)	95.5 ± 21.8	90.6 ± 19.9	<0.001	73.4 ± 10.1	72.8 ± 11.3	0.563	<0.001
BMI (kg/m^2)	35.2 (28.4–37.5)	33.2 (27.4–35.4)	<0.001	26.4 (22.4–28.5)	24.9 (21.7–27.4)	0.434	<0.001
Total body fat (%)	44.9 ± 11.3	41.9 ± 10.5	<0.001	35.7 ± 5.7	35.6 ± 5.9	0.469	0.001
Lean tissue (kg)	25.4 (21.7–29.1)	24.6 (21.8–27.9)	0.050	NA	NA	NA	NA
Lean mass index (kg/m^2)	9.31 ± 1.29	8.98 ± 1.31	0.048	NA	NA	NA	NA
Glucose (mg/dL) †	89 (80–97)	87 (85–97)	0.478	91 ± 12.4	91.2 ± 15.6	0.612	0.733
Glycated hemoglobin, HbA1c (%)	5.26 ± 0.55	4.97 ± 0.41	0.017	5.10 ± 0.84	NA	NA	0.573
Total-cholesterol (mg/dL)	186 ± 45	164 ± 37	0.628	193 ± 38	197 ± 40	0.582	0.554
LDL-cholesterol (mg/dL)	111 (91–120)	108 (77–118)	0.919	107 (100–143)	109 (83–135)	0.748	0.638
HDL-cholesterol (mg/dL)	55 (46–67)	50 (40–65)	0.569	49 (56–64)	53 (65–74)	0.125	0.552
Triglycerides (mg/dL)	137 (115–152)	130 (108–160)	0.005	95 (65–155)	97 (70–121)	0.808	0.554

Data are presented as mean ± standard deviation or as median and interquartile range. † Values logarithmized prior to statistical comparison to achieve normality. NA: Not available or not applicable. § Comparison between baseline and follow-up values in the CDSS intervention group. ‡ Comparison between baseline and follow-up values in the dietary advice group. ∫ Comparison of baseline values between CDSS intervention and dietary advice group.

Body weight patterns identified using PCA analysis are shown in Table 3. More particularly, two distinct patterns were identified: a high BMI–high % body fat pattern explaining 60.4% of the total variance and a high BMI–high lean tissue pattern explaining 38.4% of the total variance. The total variance explained was as high as 98.9%. In Table 4, Spearman correlation coefficients are shown between the absolute % weight loss and several variables of body composition (raw data or body weight patterns), as well as several biochemical variables. In detail, lean mass, lean mass index and a high BMI–high

% body fat pattern were positively related to % weight loss in women. Baseline glucose levels were inversely related to % weight loss in women. In men, as well as in the total sample, baseline triglyceride levels were inversely related to % weight loss. Furthermore, multiple linear regression models were created to identify the factors that can independently "predict" weight loss (Table 5). As can be seen in the total sample, serum triglycerides were inversely associated with the magnitude of weight loss, irrespective of age, gender, physical activity at baseline and identified body weight patterns. In women, a pattern with high BMI and high lean tissue was consistently positively related to the % weight loss irrespectively of age, physical activity and triglyceride levels (Table 5).

Table 2. Differences pre- and post-CDSS intervention stratified by sex.

	Men (n = 28)			Women (n = 11)		
	Baseline	Follow-Up	*p*-Value	Baseline	Follow-Up	*p*-Value
Body weight (kg)	87.9 ± 14.9	83.3 ± 13.9	<0.001	114.6 ± 25.4	109.2 ± 21.4	<0.001
BMI (kg/m^2)	31.9 (28.3–36.2)	30.3 (26.7–34.6)	<0.001	36.2 (28.8–40.1)	35.1 (29.4–36.5)	0.021
Total body fat (%)	46.9 ± 9.8	43.7 ± 9.7	<0.001	39.9 ± 13.8	37.3 ± 11.5	<0.001
Lean tissue (kg)	22.4 (21.5–26.1)	23.5 (19.9–25.9)	0.167	35.4 (29.1–41.6)	32.8 (28.0–38.5)	0.109
Lean mass index (kg/m^2)	8.83 (8.03–9.22)	8.80 (7.64–9.16)	0.178	10.83 (9.92–11.67)	9.90 (9.30–11.76)	0.091
Glucose (mg/dL) †	89 (80–93)	87 (85–97)	0.010	75 (92–98)	88 (85–96)	0.220
Glycated hemoglobin, HbA1c (%)	5.25 ± 0.44	5.06 ± 0.46	0.151	5.31 ± 0.83	4.83 ± 0.29	0.077
Total-cholesterol (mg/dL)	182 ± 38	164 ± 39	<0.001	197 ± 61	164.2 ± 38.9	0.069
LDL-cholesterol (mg/dL)	111 (91–119)	109 (35–121)	0.237	109 (87–138)	107 (99–130)	0.109
HDL-cholesterol (mg/dL)	55 (45–67)	50 (40–52)	0.343	59 (45–68)	57 (41–65)	0.705
Triglycerides (mg/dL)	137 (110–144)	130 (108–138)	0.018	135 (116–189)	140 (107–194)	0.109

Data are presented as mean ± standard deviation or as median and interquartile range. † Values logarithmized prior to statistical comparison to achieve normality.

Table 3. Factor loadings using principal component analysis for the identification of body weight patterns in the CDSS intervention group.

	Pattern 1: High BMI–High % Body Fat	Pattern 2: High BMI–High Lean Tissue
BMI at baseline (kg/m^2)	**0.873**	**0.473**
Lean mass at baseline (kg)	0.014	**0.997**
Body fat at baseline (%)	**0.975**	−0.195
% variance explained	60.4%	38.4 %
Total variance explained 98.9%		

The factor loadings (component scores) are interpreted as correlation coefficients (r). Higher absolute values of the loadings indicate that the variable is correlated with the respective component. Numbers in bold indicate absolute loadings greater than 0.45.

Table 4. Spearman correlation coefficients between % weight loss and several variables in the CDSS intervention group for the total sample and separately for men and women.

	% Weight Loss (r)		
	Total (n = 39)	Women (n = 11)	Men (n = 28)
Age (y)	0.092 (*p* = 0.576)	0.050 (*p* = 0.884)	0.130 (*p* = 0.511)
BMI (kg/m^2)	0.292 (*p* = 0.072)	0.400 (*p* = 0.223)	0.222 (*p* = 0.256)
Body fat (%)	0.118 (*p* = 0.473)	0.127(*p* = 0.709)	0.191 (*p* = 0.332)
Lean mass (kg)	0.108 (*p* = 0.514)	0.618 (*p* = 0.043)	0.079 (*p* = 0.690)
Lean mass index (kg/m^2)	0.158 (*p* = 0.337)	0.618 (*p* = 0.043)	0.147 (*p* = 0.456)
High BMI–high % body fat pattern	0.164 (*p* = 0.318)	0.200 (*p* = 0.555)	0.185 (*p* = 0.346)
High BMI–high lean tissue pattern	0.166 (*p* = 0.313)	0.618 (*p* = 0.043)	0.149 (*p* = 0.450)
Glucose (mg/dL)	−0.270 (*p* = 0.128)	−0.714 (*p* = 0.047)	−0.154 (*p* = 0.462)
Glycated hemoglobin, HbA1c (%)	−0.145 (*p* = 0.428)	−0.143 (*p* = 0.736)	−0.139 (*p* = 0.518)
Total-cholesterol (mg/dL)	−0.150 (*p* = 0.377)	0.231 (*p* = 0.521)	−0.218 (*p* = 0.274)
HDL-cholesterol (mg/dL)	−0.026 (*p* = 0.877)	−0.588 (*p* = 0.074)	0.199 (*p* = 0.321)
Triglycerides (mg/dL)	−0.410 (*p* = 0.012)	−0.067 (*p* = 0.855)	−0.601 (*p* = 0.001)

Spearman correlation coefficients (r) are shown. *p*-value is shown in parenthesis. BMI: body mass index; HDL: high-density lipoprotein; y: years.

Table 5. Linear regression analysis with % weight loss after a CDSS intervention as dependent variable in the total sample, men and women.

	Total (n = 39)			Women (n = 11)			Men (n = 28)		
	B	**SE**	*p*	**B**	**SE**	*p*	**B**	**SE**	*p*
Basic Model (R^2 = 0.2%, W: 0%, M: 0.1%)									
Age (y)	0.005	0.041	0.903	0.004	0.086	0.967	0.006	0.047	0.908
Gender (men vs. women *)	0.320	1.208	0.792	NA	NA	NA	NA	NA	NA
Model 1 (R^2 = 1.1%, W: 18.6%, M: 2%)									
Age (y)	0.002	0.042	0.967	−0.009	0.083	0.913	0.007	0.049	0.888
Gender (men vs. women *)	0.228	10.232	0.854	NA	NA	NA	NA	NA	NA
Physical activity at baseline §	−0.393	0.724	0.591	−1.915	10.419	0.214	0.160	0.849	0.852
Model 2 (R^2 T: 13.5%, W: 83.6%, M: 4.6%)									
Age (y)	0.004	0.041	0.928	−0.069	0.049	0.207	0.023	0.053	0.665
Gender (men vs. women *)	2.757	1.804	0.136	NA	NA	NA	NA	NA	NA
Physical activity at baseline §	−0.017	0.785	0.983	−2.568	1.093	0.057	0.524	0.941	0.583
Pattern 1: High BMI–high% fat	0.282	0.628	0.657	−0.469	0.738	0.549	0.522	0.869	0.554
Pattern 2: High BMI–high lean tissue	1.611	0.803	0.050	3.353	0.692	0.003	0.769	1.167	0.517
Model 3 (R^2 T: 20.6%, W: 88.5%, M: 18.6%)									
Age (y)	−0.006	0.040	0.887	−0.040	0.044	0.424	0.003	0.054	0.949
Gender (men vs. women *)	2.579	1.734	0.147	NA	NA	NA	NA	NA	NA
Physical activity at baseline §	−0.290	0.762	0.706	−30.416	0.979	0.025	0.248	0.915	0.789
Pattern 1: High BMI–high % fat	0.128	0.610	0.835	−10.245	0.703	0.151	0.411	0.840	0.630
Pattern 2: High BMI–high lean tissue	10.307	0.818	0.121	20.415	0.717	0.028	0.727	10.123	0.524
Triglycerides (mg/dL)	−0.018	0.009	0.050	−0.005	0.009	0.618	−0.022	0.011	0.069

T: Total sample, M: Men, W: Women, NA: Not applicable; y = years, * Men = 1, Women = 0, § 1 = mildly active, 2 = moderately active, 3 = active, 4 = intensely active or athlete.

4. Discussion

The present study documented that a 3-month nutrition-oriented CDSS intervention during the COVID-19 period reduced body weight, body fat, glucose and lipid biomarkers, with borderline reductions in lean body mass. In multivariate linear regression models, serum triglycerides (total sample) were inversely associated to % weight loss, while a pattern characterized by high BMI/lean tissue (women) was positively related to % weight loss.

The importance of our findings lies in the fact that a hospital-based weight loss intervention during the COVID-19 epidemic was possible with the use of new technologies and evidence-based procedures, such as those ensured by the use of CDSS. Indeed, several programs based on tele-health and remote consultations were developed to assist patients with obesity during weight loss, especially during COVID-19 [15–17]. However, the present study firstly reports the use of a CDSS in a hospital setting for distance counseling in obesity. In parallel, electronic medical records have been useful so far to screen individuals for obesity [27] and CDSS-created alerts have been used to further refer patients to professionals for obesity management [28]. Several CDSS programs have been also created for childhood obesity prevention and treatment [29–31]. In parallel, through the use of CDSS, cost and time are reduced, and thus, more time can be dedicated to each patient [32]. However, as recently reviewed, many contemporary CDSS are still focused on alerts and reminders [33], and only ~24% of them include patient education features, dietary advice and other lifestyle recommendations [33]. It has been suggested that future CDSS should target screening, diagnosis and management of multiple chronic diseases and that user-friendly interfaces on top of alerts should be developed [33]. Hopefully, the Nutrinet® CDSS used in the present study fulfills the aforementioned aspirations. Indeed, it ensures the implementation of all the steps of the NCP including nutritional assessment, diagnosis, intervention, monitoring and evaluation [4,5]. It facilitates dietary assessment, physical activity assessment, formulation of personalized tailored dietary plans, provision of guidelines, definition of nutritional and physical activity goals and monitoring of nutritional and physical activity goals [4,5]. It thus aims at assisting nutritional care provision at all stages and is not only centered on alerts. Moreover, the Nutrinet® CDSS can be used by both the dietitian and the patient (after appropriate training), since it "allows" patients to visit their personal profile and record/track their personal progress, e.g., weight goals, physical activity goals, etc. [4,12]. The tracking feature of the present CDSS also enables a "continuous treatment model" approach, which is crucial to combat a chronic condition, such as obesity [34]. In parallel, the Nutrinet® CDSS endorses patient stimulation, a powerful mechanism for addressing health inequalities. The use of CDSS tools by patients themselves has shown promising results in a hospital-based intervention [35]. Similarly, an electronic nutritional assessment tool improved dietary habits, as well as motivation and self-efficiency, when used by patients [36]. A bi-directional communication of the patient and health provider can be achieved through the Nutrinet® CDSS, which is particularly important in the COVID-19 era [4].

In a wider perspective, new technology-assisted dietary assessment has been connected to higher acceptability, self-adherence and self-monitoring compared to paper records in persons with obesity [16,37–39] or type 2 diabetes [40,41]. In line with our results suggesting a decrease in body weight and body fat, website-assisted interventions combining nutrition intervention and dietary self-monitoring have shown higher rates of weight loss and body fat loss compared to standard treatment [37,38], although not all studies agree [42]. A borderline reduction of lean mass was observed in the present study. Of note, in sex-specific analysis, lean mass remained unchanged. This issue is important, since the loss of lean mass can lower resting metabolic rate, cause fatigue and increase injury risk [43,44]. Moreover, it can increase the risk of sarcopenia, which worsens quality of life and prognosis, especially in disease states [45]. It is possible that the Mediterranean diet pattern followed, along with enhancement of physical activity through the CDSS, minimized the risk of lean mass loss [46].

Reductions in HbA1c and triglycerides were also documented in the present work, underlying the importance of Mediterranean diet adherence in glycemia management [47], cardiovascular disease prevention [48], and the contribution of technology to the same direction [49]. Indeed, a diet rich in antioxidants, fiber and mono-unsaturated and unsaturated fatty acids and low in saturated fatty acids, as in the Mediterranean dietary pattern [50], exerts anti-inflammatory actions by targeting several mediators, such as the Platelet Activating Factor (PAF) [51,52] and can improve glycemia [53,54]. Several CDSS have been used for glucose management in a hospital setting and for cardiovascular disease prevention and management, as recently reviewed elsewhere [33,55]. However, the Nutrinet® CDSS was administered by a professional dietitian, while most CDSS were administered by medical doctors and nurses [55], even if they were directly related to food consumption, such as "MyFood" [35].

In addition, the identification of weight loss predictors may assist in ascertaining the probable outcomes of lifestyle interventions [56]. Paying attention to baseline subjects' characteristics affecting weight loss is thus important, and possibly population- and situation-specific. A recent review identified that several patient-related variables affect weight loss, such as male gender, older age, presence of cardiometabolic diseases and low fat intake [56]. Although sex-specific predictors of weight loss have been reported [57], in most studies, sex is only included as a covariate in models and sex-specific models are not applied [58], with some exceptions [59,60]. In the present study, sex-specific analysis was performed in both pre- and post-intervention results as well as in the models for prediction of weight loss. However, age and gender were not identified as predictors of % weight loss. A pattern characterized by increased BMI and lean mass was positively related to the extent of weight loss in women. In another study, pre-treatment fat mass was related to weight loss in males [59]. It is thus possible that high BMI is connected to high lean mass and increases energy expenditure, facilitating weight loss [61]. Serum triglycerides were inversely related to % weight loss in the total sample and men as a possible trend. This finding is in line with data suggesting that triglycerides changes have a positive relationship with weight change (r = 0.82) [62]. In addition, sex differences have been reported in the relationship of weight and lipoproteins [63], with triglyceride kinetics being affected mostly in men than women [64,65], which is in line with the trend observed in the present work.

In addition, the present study was conducted in the COVID-19 pandemic, which has been per se associated with weight changes. According to a recent meta-analysis, self-reported body weight increased in 11.1–72.4% of individuals and decreased in 7.2–51.4% of individuals [66]. Moreover, predictors of body weight may be different during COVID-19 compared to other periods, since emotional stress, telework and confinement took place [67]. For example, a study in Greek volunteers showed that sleep impairment was present during the lockdown, and it was related to changes in dietary and physical activity habits [68]. In addition, health issues affecting food intake and activity patterns following acute or long COVID-19 infection are not to be excluded.

A major limitation of the present study was the fact that several differences between the control and experimental group existed, including age and sex of the participants. This was partially due to the fact that during COVID-19 a low number of patients were treated in the out-patient clinic. It is also possible that the younger individuals in the intervention group were more used to technology and computer environment. Moreover, in the control group, several measurements, such as lean mass, were not available. The sample size and the duration of the study were relatively small. However, changes were documented in weight, body fat and biochemical measurements. In addition, biochemical exams were undertaken in several laboratories. However, all laboratories were licensed by state, which assures the accuracy of measurements. Regarding LDL-cholesterol, it was not measured but estimated, which may be related to differentiated cardiovascular risk at the individual level [69]. Regarding "predictors" of weight loss, several variables were not recorded, such as previous weight loss interventions, socio-economic history, marital status and motivation level.

5. Conclusions

In conclusion, a short-term CDSS-facilitated intervention beneficially affected weight loss and other cardiovascular risk factors, such as glucose and lipid indices. Further research is required in larger samples and with longer intervention periods and follow-ups to identify if the beneficial results persist over time. Moreover, several issues should be addressed in the future, such as the interoperability of health-related systems and the use of artificial intelligence or complex networks to optimize and tailor-cut nutritional advice.

Supplementary Materials: The following supporting information can be downloaded at: https://www.mdpi.com/article/10.3390/app13169448/s1, Supplementary Table S1: Diet plan for weight loss.

Author Contributions: Conceptualization, M.S. and P.P.; methodology, M.S. and P.P.; software, M.S. and P.P.; formal analysis, P.D. and P.P.; investigation, L.P.; data curation, P.D. and L.P.; writing—original draft preparation, P.D.; writing—review and editing, P.D., P.P. and M.S.; supervision, M.S.; project administration, M.S. and L.P. All authors have read and agreed to the published version of the manuscript.

Funding: This research received no external funding.

Institutional Review Board Statement: The study was conducted in accordance with the Declaration of Helsinki and approved by the Institutional Review Board of Iaso Hospital, number f310519.

Informed Consent Statement: Oral informed consent was obtained from all subjects involved in the study.

Data Availability Statement: Data are unavailable due to privacy restrictions.

Conflicts of Interest: The authors declare no conflict of interest.

References

1. Sutton, R.T.; Pincock, D.; Baumgart, D.C.; Sadowski, D.C.; Fedorak, R.N.; Kroeker, K.I. An Overview of Clinical Decision Support Systems: Benefits, Risks, and Strategies for Success. *NPJ Digit. Med.* **2020**, *3*, 17. [CrossRef] [PubMed]
2. Papandreou, P.; Nousiou, K.; Papandreou, G.; Steier, J.; Skouroliakou, M.; Karageorgopoulou, S. The Use of a Novel Clinical Decision Support System for Reducing Medication Errors and Expediting Care in the Provision of Chemotherapy. *Health Technol.* **2022**, *12*, 515–521. [CrossRef]
3. Nimee, F.; Gioxari, A.; Steier, J.; Skouroliakou, M. Bridging the Gap: Community Pharmacists' Burgeoning Role as Point-Of-Care Providers During the COVID-19 Pandemic Through the Integration of Emerging Technologies. *J. Nutr. Health Food Sci.* **2021**, *9*, 1–9. [CrossRef]
4. Papandreou, P.; Gioxari, A.; Nimee, F.; Skouroliakou, M. Application of Clinical Decision Support System to Assist Breast Cancer Patients with Lifestyle Modifications during the COVID-19 Pandemic: A Randomised Controlled Trial. *Nutrients* **2021**, *13*, 2115. [CrossRef]
5. Swan, W.I.; Vivanti, A.; Hakel-Smith, N.A.; Hotson, B.; Orrevall, Y.; Trostler, N.; Beck Howarter, K.; Papoutsakis, C. Nutrition Care Process and Model Update: Toward Realizing People-Centered Care and Outcomes Management. *J. Acad. Nutr. Diet.* **2017**, *117*, 2003–2014. [CrossRef]
6. Henry, S.B.; Douglas, K.; Galzagorry, G.; Lahey, A.; Holzemer, W.L. A Template-Based Approach to Support Utilization of Clinical Practice Guidelines Within an Electronic Health Record. *J. Am. Med. Inform. Assoc.* **1998**, *5*, 237–244. [CrossRef]
7. Detopoulou, P.; Tsouma, C.; Papamikos, V. COVID-19 and Nutrition: Summary of Official Recommendations. *Top. Clin. Nutr.* **2022**, *37*, 187–202. [CrossRef]
8. Detopoulou, P.; Demopoulos, C.A.; Antonopoulou, S. Micronutrients, Phytochemicals and Mediterranean Diet: A Potential Protective Role against COVID-19 through Modulation of PAF Actions and Metabolism. *Nutrients* **2021**, *13*, 462. [CrossRef]
9. Detopoulou, P.; Al-Khelefawi, Z.H.; Kalonarchi, G.; Papamikos, V. Formulation of the Menu of a General Hospital After Its Conversion to a "COVID Hospital": A Nutrient Analysis of 28-Day Menus. *Front. Nutr.* **2022**, *9*, 833628. [CrossRef]
10. Detopoulou, P.; Panoutsopoulos, G.I.; Kalonarchi, G.; Alexatou, O.; Petropoulou, G.; Papamikos, V. Development of a Tool for Determining the Equivalence of Nutritional Supplements to Diabetic Food Exchanges. *Nutrients* **2022**, *14*, 3267. [CrossRef]
11. Wells Mulherin, D.; Walker, R.; Holcombe, B.; Guenter, P. ASPEN Report on Nutrition Support Practice Processes with COVID-19: The First Response. *Nutr. Clin. Pract.* **2020**, *35*, 783–791. [CrossRef]
12. Papandreou, P.; Gioxari, A.; Daskalou, E.; Vasilopoulou, A.; Skouroliakou, M. Personalized Nutritional Intervention to Improve Mediterranean Diet Adherence in Female Patients with Multiple Sclerosis: A Randomized Controlled Study. *Dietetics* **2022**, *1*, 25–38. [CrossRef]
13. Papandreou, P.; Gioxari, A.; Daskalou, E.; Grammatikopoulou, M.G.; Skouroliakou, M.; Bogdanos, D.P. Mediterranean Diet and Physical Activity Nudges versus Usual Care in Women with Rheumatoid Arthritis: Results from the MADEIRA Randomized Controlled Trial. *Nutrients* **2023**, *15*, 676. [CrossRef] [PubMed]
14. Papandreou, P.; Amerikanou, C.; Vezou, C.; Gioxari, A.; Kaliora, A.C.; Skouroliakou, M. Improving Adherence to the Mediterranean Diet in Early Pregnancy Using a Clinical Decision Support System; A Randomised Controlled Clinical Trial. *Nutrients* **2023**, *15*, 432. [CrossRef] [PubMed]
15. Tchang, B.G.; Morrison, C.; Kim, J.T.; Ahmed, F.; Chan, K.M.; Alonso, L.C.; Aronne, L.J.; Shukla, A.P. Weight Loss Outcomes with Telemedicine During COVID-19. *Front. Endocrinol.* **2022**, *13*, 793290. [CrossRef]
16. Ufholz, K.; Werner, J. The Efficacy of Mobile Applications for Weight Loss. *Curr. Cardiovasc. Risk Rep.* **2023**, *17*, 83–90. [CrossRef]
17. Bailly, S.; Fabre, O.; Legrand, R.; Pantagis, L.; Mendelson, M.; Terrail, R.; Tamisier, R.; Astrup, A.; Clément, K.; Pépin, J.-L. The Impact of the COVID-19 Lockdown on Weight Loss and Body Composition in Subjects with Overweight and Obesity Participating in a Nationwide Weight-Loss Program: Impact of a Remote Consultation Follow-Up—The CO-RNPC Study. *Nutrients* **2021**, *13*, 2152. [CrossRef] [PubMed]
18. Friedewald, W.T.; Levy, R.I.; Fredrickson, D.S. Estimation of the Concentration of Low-Density Lipoprotein Cholesterol in Plasma, without Use of the Preparative Ultracentrifuge. *Clin. Chem.* **1972**, *18*, 499–502. [CrossRef]
19. Pasternak, R.C. 2001 National Cholesterol Education Program (NCEP) Guidelines on the Detection, Evaluation and Treatment of Elevated Cholesterol in Adults: Adult Treatment Panel III (ATP III). *ACC Curr. J. Rev.* **2002**, *11*, 37–45. [CrossRef]
20. American Diabetes Association Standards of Medical Care in Diabetes—2014. *Diabetes Care* **2014**, *37* (Suppl. S1), S14–S80. [CrossRef]
21. U.S. Department of Agriculture (USDA), Agricultural Research Service. FoodData Central. 2021. Available online: https://fdc.nal.usda.gov/ (accessed on 8 July 2023).
22. Trichopoulou, A.; Georga, K. *Composition Tables of Food and Greek Dishes*; Parisianos SA.: Athens, Greece, 2004.
23. National Academy of Sciences. Nutrition—Dietary Reference Intakes. Available online: https://nap.nationalacademies.org/collection/57/dietary-reference-intakes (accessed on 8 July 2023).
24. Hamwi, G.J. Therapy: Changing Dietary Concepts. In *Diabetes Mellitus: Diagnosis and Treatment*; Danowski, T.S., Ed.; American Diabetes Association: New York, NY, USA, 1964; Volume 1, pp. 73–78.
25. Alberti, K.G.; Zimmet, P.; Shaw, J. Metabolic syndrome—A new world-wide definition. A Consensus Statement from the International Diabetes Federation. *Diabet. Med.* **2006**, *23*, 469–480. [CrossRef] [PubMed]

26. Ainsworth, B.E.; Haskell, W.L.; Herrmann, S.D.; Meckes, N.; Bassett, D.R.; Tudor-Locke, C.; Greer, J.L.; Vezina, J.; Whitt-Glover, M.C.; Leon, A.S. 2011 Compendium of Physical Activities: A Second Update of Codes and MET Values. *Med. Sci. Sports Exerc.* **2011**, *43*, 1575–1581. [CrossRef] [PubMed]
27. Salinas, J.J.; Sheen, J.; Shokar, N.; Wright, J.; Vazquez, G.; Alozie, O. An Electronic Medical Records Study of Population Obesity Prevalence in El Paso, Texas. *BMC Med. Inf. Decis. Mak.* **2022**, *22*, 46. [CrossRef]
28. Alexeeva, O.; Keswani, R.N.; Pandolfino, J.E.; Liebovitz, D.; Gregory, D.; Yadlapati, R. Electronic Clinical Decision Support Tools for Obesity and Gastroesophageal Reflux Disease: The Provider's Perspective. *Am. J. Gastroenterol.* **2018**, *113*, 916. [CrossRef] [PubMed]
29. Naureckas, S.M.; Zweigoron, R.; Haverkamp, K.S.; Kaleba, E.O.; Pohl, S.J.; Ariza, A.J. Developing an Electronic Clinical Decision Support System to Promote Guideline Adherence for Healthy Weight Management and Cardiovascular Risk Reduction in Children: A Progress Update. *Behav. Med. Pract. Policy Res.* **2011**, *1*, 103–107. [CrossRef]
30. Rattay, K.T.; Ramakrishnan, M.; Atkinson, A.; Gilson, M.; Drayton, V. Use of an Electronic Medical Record System to Support Primary Care Recommendations to Prevent, Identify, and Manage Childhood Obesity. *Pediatrics* **2009**, *123*, S100–S107. [CrossRef]
31. Skiba, D.J.; Gance-Cleveland, B.; Gilbert, K.; Gilbert, L.; Dandreaux, D. Comparing the Effectiveness of CDSS on Provider's Behaviors to Implement Obesity Prevention Guidelines. In Proceedings of the NI 2012: 11th International Congress on Nursing Informatics, Montreal, QC, Canada, 23–27 June 2012; Volume 2012, p. 376.
32. Rossi, M.; Campbell, K.L.; Ferguson, M. Implementation of the Nutrition Care Process and International Dietetics and Nutrition Terminology in a Single-Center Hemodialysis Unit: Comparing Paper vs. Electronic Records. *J. Acad. Nutr. Diet.* **2014**, *114*, 124–130. [CrossRef]
33. Chen, W.; Howard, K.; Gorham, G.; O'Bryan, C.M.; Coffey, P.; Balasubramanya, B.; Abeyaratne, A.; Cass, A. Design, Effectiveness, and Economic Outcomes of Contemporary Chronic Disease Clinical Decision Support Systems: A Systematic Review and Meta-Analysis. *J. Am. Med. Inform. Assoc.* **2022**, *29*, 1757–1772. [CrossRef]
34. Bray, G.A.; Ryan, D.H. Evidence-based Weight Loss Interventions: Individualized Treatment Options to Maximize Patient Outcomes. *Diabetes Obes. Metab.* **2021**, *23*, 50–62. [CrossRef]
35. Paulsen, M.M.; Paur, I.; Gjestland, J.; Henriksen, C.; Varsi, C.; Tangvik, R.J.; Andersen, L.F. Effects of Using the MyFood Decision Support System on Hospitalized Patients' Nutritional Status and Treatment: A Randomized Controlled Trial. *Clin. Nutr.* **2020**, *39*, 3607–3617. [CrossRef]
36. Bonilla, C.; Brauer, P.; Royall, D.; Keller, H.; Hanning, R.M.; DiCenso, A. Use of Electronic Dietary Assessment Tools in Primary Care: An Interdisciplinary Perspective. *BMC Med. Inf. Decis. Mak.* **2015**, *15*, 14. [CrossRef]
37. Burke, L.E.; Conroy, M.B.; Sereika, S.M.; Elci, O.U.; Styn, M.A.; Acharya, S.D.; Sevick, M.A.; Ewing, L.J.; Glanz, K. The Effect of Electronic Self-Monitoring on Weight Loss and Dietary Intake: A Randomized Behavioral Weight Loss Trial. *Obesity* **2011**, *19*, 338–344. [CrossRef] [PubMed]
38. Carter, M.C.; Burley, V.J.; Nykjaer, C.; Cade, J.E. Adherence to a Smartphone Application for Weight Loss Compared to Website and Paper Diary: Pilot Randomized Controlled Trial. *J. Med. Internet Res.* **2013**, *15*, e32. [CrossRef] [PubMed]
39. Raaijmakers, L.C.H.; Pouwels, S.; Berghuis, K.A.; Nienhuijs, S.W. Technology-Based Interventions in the Treatment of Overweight and Obesity: A Systematic Review. *Appetite* **2015**, *95*, 138–151. [CrossRef] [PubMed]
40. Hansel, B.; Giral, P.; Gambotti, L.; Lafourcade, A.; Peres, G.; Filipecki, C.; Kadouch, D.; Hartemann, A.; Oppert, J.-M.; Bruckert, E.; et al. A Fully Automated Web-Based Program Improves Lifestyle Habits and HbA1c in Patients with Type 2 Diabetes and Abdominal Obesity: Randomized Trial of Patient E-Coaching Nutritional Support (The ANODE Study). *J. Med. Internet Res.* **2017**, *19*, e360. [CrossRef]
41. American Diabetes Association Professional Practice Committee. 5. Facilitating Behavior Change and Well-Being to Improve Health Outcomes: Standards of Medical Care in Diabetes—2022. *Diabetes Care* **2022**, *45* (Suppl. S1), S60–S82. [CrossRef]
42. Yon, B.A.; Johnson, R.K.; Harvey-Berino, J.; Gold, B.C.; Howard, A.B. Personal Digital Assistants Are Comparable to Traditional Diaries for Dietary Self-Monitoring During a Weight Loss Program. *J. Behav. Med.* **2007**, *30*, 165–175. [CrossRef]
43. Tsai, A.G.; Wadden, T.A. Systematic Review: An Evaluation of Major Commercial Weight Loss Programs in the United States. *Ann. Intern. Med.* **2005**, *142*, 56. [CrossRef]
44. Ravussin, E.; Lillioja, S.; Knowler, W.C.; Christin, L.; Freymond, D.; Abbott, W.G.H.; Boyce, V.; Howard, B.V.; Bogardus, C. Reduced Rate of Energy Expenditure as a Risk Factor for Body-Weight Gain. *N. Engl. J. Med.* **1988**, *318*, 467–472. [CrossRef]
45. Detopoulou, P.; Voulgaridou, G.; Papadopoulou, S. Cancer, Phase Angle and Sarcopenia: The Role of Diet in Connection with Lung Cancer Prognosis. *Lung* **2022**, *200*, 347–379. [CrossRef]
46. Papadopoulou, S.K.; Detopoulou, P.; Voulgaridou, G.; Tsoumana, D.; Spanoudaki, M.; Sadikou, F.; Papadopoulou, V.G.; Zidrou, C.; Chatziprodromidou, I.P.; Giaginis, C.; et al. Mediterranean Diet and Sarcopenia Features in Apparently Healthy Adults over 65 Years: A Systematic Review. *Nutrients* **2023**, *15*, 1104. [CrossRef] [PubMed]
47. AlAufi, N.S.; Chan, Y.M.; Waly, M.I.; Chin, Y.S.; Mohd Yusof, B.-N.; Ahmad, N. Application of Mediterranean Diet in Cardiovascular Diseases and Type 2 Diabetes Mellitus: Motivations and Challenges. *Nutrients* **2022**, *14*, 2777. [CrossRef] [PubMed]
48. Rees, K.; Takeda, A.; Martin, N.; Ellis, L.; Wijesekara, D.; Vepa, A.; Das, A.; Hartley, L.; Stranges, S. Mediterranean-Style Diet for the Primary and Secondary Prevention of Cardiovascular Disease. *Cochrane Database Syst. Rev.* **2019**, *2019*, CD009825. [CrossRef]
49. Wu, X.; Guo, X.; Zhang, Z. The Efficacy of Mobile Phone Apps for Lifestyle Modification in Diabetes: Systematic Review and Meta-Analysis. *JMIR Mhealth Uhealth* **2019**, *7*, e12297. [CrossRef] [PubMed]

50. Detopoulou, P.; Aggeli, M.; Andrioti, E.; Detopoulou, M. Macronutrient Content and Food Exchanges for 48 Greek Mediterranean Dishes: Macronutrient Content and Exchanges for 48 Greek Dishes. *Nutr. Diet.* **2017**, *74*, 200–209. [CrossRef] [PubMed]
51. Detopoulou, P.; Fragopoulou, E.; Nomikos, T.; Yannakoulia, M.; Stamatakis, G.; Panagiotakos, D.B.; Antonopoulou, S. The Relation of Diet with PAF and Its Metabolic Enzymes in Healthy Volunteers. *Eur. J. Nutr.* **2015**, *54*, 25–34. [CrossRef]
52. Fragopoulou, E.; Detopoulou, P.; Alepoudea, E.; Nomikos, T.; Kalogeropoulos, N.; Antonopoulou, S. Associations between Red Blood Cells Fatty Acids, Desaturases Indices and Metabolism of Platelet Activating Factor in Healthy Volunteers. *Prostaglandins Leukot. Essent. Fat. Acids* **2021**, *164*, 102234. [CrossRef]
53. Psaltopoulou, T.; Panagiotakos, D.B.; Pitsavos, C.; Chrysochoou, C.; Detopoulou, P.; Skoumas, J.; Stefanadis, C. Dietary Antioxidant Capacity Is Inversely Associated with Diabetes Biomarkers: The ATTICA Study. *Nutr. Metab. Cardiovasc. Dis.* **2011**, *21*, 561–567. [CrossRef]
54. Nomikos, T.; Detopoulou, P.; Fragopoulou, E.; Pliakis, E.; Antonopoulou, S. Boiled Wild Artichoke Reduces Postprandial Glycemic and Insulinemic Responses in Normal Subjects but Has No Effect on Metabolic Syndrome Patients. *Nutr. Res.* **2007**, *27*, 741–749. [CrossRef]
55. Mebrahtu, T.F.; Skyrme, S.; Randell, R.; Keenan, A.-M.; Bloor, K.; Yang, H.; Andre, D.; Ledward, A.; King, H.; Thompson, C. Effects of Computerised Clinical Decision Support Systems (CDSS) on Nursing and Allied Health Professional Performance and Patient Outcomes: A Systematic Review of Experimental and Observational Studies. *BMJ Open* **2021**, *11*, e053886. [CrossRef]
56. Chopra, S.; Malhotra, A.; Ranjan, P.; Vikram, N.K.; Sarkar, S.; Siddhu, A.; Kumari, A.; Kaloiya, G.S.; Kumar, A. Predictors of Successful Weight Loss Outcomes amongst Individuals with Obesity Undergoing Lifestyle Interventions: A Systematic Review. *Obes. Rev.* **2021**, *22*, e13148. [CrossRef] [PubMed]
57. Stubbs, J.; Whybrow, S.; Teixeira, P.; Blundell, J.; Lawton, C.; Westenhoefer, J.; Engel, D.; Shepherd, R.; Mcconnon, Á.; Gilbert, P.; et al. Problems in Identifying Predictors and Correlates of Weight Loss and Maintenance: Implications for Weight Control Therapies Based on Behaviour Change: Predicting Weight Outcomes. *Obes. Rev.* **2011**, *12*, 688–708. [CrossRef] [PubMed]
58. Teixeira, P.J.; Going, S.B.; Sardinha, L.B.; Lohman, T.G. A Review of Psychosocial Pre-Treatment Predictors of Weight Control. *Obes. Rev.* **2005**, *6*, 43–65. [CrossRef] [PubMed]
59. Morgan, P.J.; Hollis, J.L.; Young, M.D.; Collins, C.E.; Teixeira, P.J. Workday Sitting Time and Marital Status: Novel Pretreatment Predictors of Weight Loss in Overweight and Obese Men. *Am. J. Men's Health* **2018**, *12*, 1431–1438. [CrossRef] [PubMed]
60. Mroz, J.E.; Pullen, C.H.; Hageman, P.A. Health and Appearance Reasons for Weight Loss as Predictors of Long-Term Weight Change. *Health Psychol. Open* **2018**, *5*, 1–8. [CrossRef]
61. Speakman, J.R.; Selman, C. Physical Activity and Resting Metabolic Rate. *Proc. Nutr. Soc.* **2003**, *62*, 621–634. [CrossRef]
62. Poobalan, A.; Aucott, L.; Smith, W.C.S.; Avenell, A.; Jung, R.; Broom, J.; Grant, A.M. Effects of Weight Loss in Overweight/Obese Individuals and Long-Term Lipid Outcomes—A Systematic Review. *Obes. Rev.* **2004**, *5*, 43–50. [CrossRef]
63. Mittendorfer, B.; Patterson, B.W.; Klein, S. Effect of Sex and Obesity on Basal VLDL-Triacylglycerol Kinetics. *Am. J. Clin. Nutr.* **2003**, *77*, 573–579. [CrossRef]
64. Ginsberg, H.N.; Le, N.A.; Gibson, J.C. Regulation of the Production and Catabolism of Plasma Low Density Lipoproteins in Hypertriglyceridemic Subjects. Effect of Weight Loss. *J. Clin. Investig.* **1985**, *75*, 614–623. [CrossRef]
65. Jourdan, M.; Margen, S.; Bradfield, R.B. The Turnover Rate of Serum Glycerides in the Lipoproteins of Fasting Obese Women during Weight Loss. *Am. J. Clin. Nutr.* **1974**, *27*, 850–858. [CrossRef]
66. Bakaloudi, D.R.; Barazzoni, R.; Bischoff, S.C.; Breda, J.; Wickramasinghe, K.; Chourdakis, M. Impact of the First COVID-19 Lockdown on Body Weight: A Combined Systematic Review and a Meta-Analysis. *Clin. Nutr.* **2022**, *41*, 3046–3054. [CrossRef] [PubMed]
67. Kuk, J.L.; Christensen, R.A.G.; Kamran Samani, E.; Wharton, S. Predictors of Weight Loss and Weight Gain in Weight Management Patients during the COVID-19 Pandemic. *J. Obes.* **2021**, *2021*, 4881430. [CrossRef] [PubMed]
68. Papazisis, Z.; Nikolaidis, P.T.; Trakada, G. Sleep, Physical Activity, and Diet of Adults during the Second Lockdown of the COVID-19 Pandemic in Greece. *Int. J. Environ. Res. Public Health* **2021**, *18*, 7292. [CrossRef] [PubMed]
69. Nomikos, T.; Georgoulis, M.; Chrysohoou, C.; Damigou, E.; Barkas, F.; Skoumas, I.; Liberopoulos, E.; Pitsavos, C.; Tsioufis, C.; Sfikakis, P.P.; et al. Comparative Performance of Equations to Estimate Low-density Lipoprotein Cholesterol Levels and Cardiovascular Disease Incidence: The ATTICA Study (2002–2022). *Lipids* **2023**, *58*, 159–170. [CrossRef] [PubMed]

applied sciences

Article

Hydroxytyrosol-Rich Olive Extract for Plasma Cholesterol Control

Arrigo F. G. Cicero [1,2,*], Federica Fogacci [1,2], Antonio Di Micoli [1], Maddalena Veronesi [1], Elisa Grandi [1] and Claudio Borghi [1]

[1] Hypertension and Cardiovascular Risk Research Group, Medical and Surgical Sciences Department, University of Bologna, 40100 Bologna, Italy
[2] Italian Nutraceutical Society (SINut), 40100 Bologna, Italy
* Correspondence: arrigo.cicero@unibo.it

Abstract: Emerging research and epidemiological studies established the health benefits of the Mediterranean diet, whose hallmark is the high consumption of olives and olive oil as the primary source of dietary fatty acids and major sources of antioxidants. The aim of this study was to evaluate the effect of daily dietary supplementation with highly standardized polyphenols—mainly hydroxytyrosol—which are derived from olive oil production by-products of an Italian olive variety (Coratina Olive) on the plasma cholesterol of a sample of hypercholesterolemic individuals. This single-arm, non-controlled, non-randomized, prospective pilot clinical study involved a sample of 30 volunteers with polygenic hypercholesterolemia. The study design included a 2-week run-in and a 4-week intervention period. Patients were evaluated for their clinical status and by the execution of a physical examination and laboratory analyses before and after the treatment. The intervention effect was assessed using Levene's test followed by the independent Student's *t* test after the log-transformation of the non-normally distributed continuous variables. Dietary supplementation with highly standardized polyphenols that are derived from Coratina Olive (namely SelectSIEVE® OptiChol) was associated with a significant improvement in systolic blood pressure, pulse pressure, total cholesterol, high-density lipoprotein cholesterol (HDL-C), low-density lipoprotein cholesterol, non-HDL-C, fasting plasma glucose, and uric acid compared to baseline values. Furthermore, SelectSIEVE® OptiChol was well tolerated by volunteers. We acknowledge that the study has some limitations, namely the small patient sample, the short follow-up, and the lack of randomization and control procedures. However, these results are consistent with previous literature that referred to extracts from different olive varieties. Definitely, our observations lay further foundations for the use of polyphenolic-rich olive extract from Coratina Olive in the prevention and treatment of first-stage metabolic syndrome.

Keywords: olive; Coratina Olive; olive extract; polyphenols; hydroxytyrosol; cholesterol

Citation: Cicero, A.F.G.; Fogacci, F.; Di Micoli, A.; Veronesi, M.; Grandi, E.; Borghi, C. Hydroxytyrosol-Rich Olive Extract for Plasma Cholesterol Control. *Appl. Sci.* **2022**, *12*, 10086. https://doi.org/10.3390/app121910086

Academic Editor: Alessandro Genovese

Received: 12 August 2022
Accepted: 5 October 2022
Published: 7 October 2022

Publisher's Note: MDPI stays neutral with regard to jurisdictional claims in published maps and institutional affiliations.

1. Introduction

The Mediterranean diet is typically high in fat and there is evidence about its role in the prevention of cardiovascular disease [1,2], in which the proportions of unsaturated and saturated fatty acids have been shown to play an important role [3,4]. One of the hallmarks of the Mediterranean diet is the high consumption of olives and olive oil as the primary source of dietary fatty acids and major sources of antioxidants [5]. The high nutritional value of these products is due to their richness in monounsaturated fatty acids (MUFA), fiber, vitamin E, and a number of phytochemicals [6]. In effect, olive oil provides an exceptional lipid matrix that is rich in molecules with different bioactive chemical entities [7]. The main phytochemicals that have been identified and quantified in olives and olive oil are phenolic and non-phenolic compounds [8]. Olive phenolic compounds belong to six different classes, including phenolic alcohols (hydroxytyrosol and tyrosol), flavones (luteolin, luteolin-7-*O*-glucoside, apigenin, and apigenin-7-*O*-glucoside), flavonols (rutin), anthocyanins (cyanidin-3-*O*-glucoside), phenolic acids (5-*O*-caffeoylquinic acid) and a hydroxycinnamic acid derivative (verbascoside), while triterpenic acids (notably maslinic and oleanolic acids) are the main subclass of non-phenolic compounds that have been identified in olives [6].

In 2011, the European Food Safety Authority (EFSA) published a health claim that was related to polyphenols in olive oil and their possible protection of blood lipids against oxidative stress, stating that 5 mg of hydroxytyrosol and its derivatives (e.g., oleuropein complex and tyrosol) should be consumed daily in the context of a balanced diet for the sufficient avoidance of oxidative damage [9,10]. In effect, according to the most recent observations, olive hydroxytyrosol acts as a free-scavenger and metal-chelator [11], and it reduces the levels of low-density lipoprotein cholesterol (LDL-C) oxidation, platelet aggregation, and chronic inflammation, thus counteracting atherosclerosis-related cardiovascular (CV) disease (ASCVD) through the prevention of endothelial dysfunction and macrophages activation [12–14].

In light of this evidence, the aim of this study was to evaluate the effect of daily dietary supplementation with highly standardized polyphenols—mainly hydroxytyrosol—which were derived from the olive oil production by-products of an Italian olive variety (Coratina Olive) on the plasma cholesterol of a sample of hypercholesterolemic individuals.

2. Materials and Methods

2.1. Study Design and Participants

This was designed as a single-arm, non-controlled, non-randomized, prospective pilot clinical study, and it involved a sample of 30 Italian free-living volunteers who were consecutively recruited from the Lipid Clinic of the S. Orsola-Malpighi University Hospital (Bologna, Italy) among patients referring for polygenic hypercholesterolemia.

The participants were required to be aged 20–70 years, with LDL-C > 115 mg/dL and <190 mg/dL, an estimated 10-year ASCVD risk <5% based on the SCORE (Systematic COronary Risk Evaluation) risk charts, and not requiring lipid-lowering treatment according to the relevant International guidelines [15]. The exclusion criteria included having a previous history of ASCVD, diabetes mellitus, uncontrolled hypertension, obesity (defined as body mass index (BMI) > 30 Kg/m^2), TG > 400 mg/dL, a positive test for human immunodeficiency virus (HIV) or viral hepatitis (HBC, HCV, and HEV), uncontrolled thyroid diseases, history of malignancies, using either medications and dietary supplements that altered BP levels or plasma lipids (e.g., anti-proprotein convertase subtilisin/kexin type 9 (PCSK9) inhibitors, statins, ezetimibe, bile acid resins, omega-3 fatty acids, and fibrates), having alcoholism, being pregnant, and breastfeeding.

The enrolled subjects adhered to a low-fat low-sodium Mediterranean diet for two weeks before and for the entire duration of the study. The intervention period lasted 4 weeks. Before and after the dietary supplementation with highly standardized polyphenols, the patients were evaluated for their clinical status and by the execution of laboratory analyses. The study timeline is reported in detail below (Figure 1).

The study fully complied with the Ethical Principles for Medical Research Involving Human Subjects of the Declaration of Helsinki and with the International Council for Harmonization of Technical Requirements for Registration of Pharmaceuticals for Human Use (ICH) Harmonized Tripartite Guideline for Good Clinical Practice (GCP). The study's protocol was approved by the Local Ethical Committee, and all of the patients signed a written informed consent to participate.

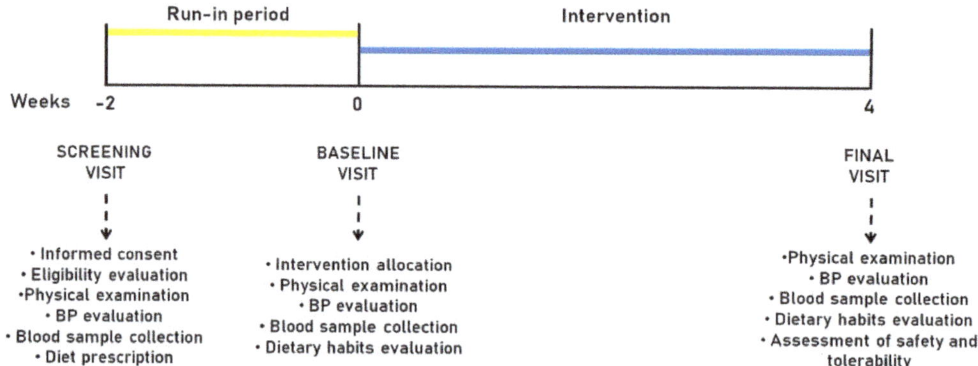

Figure 1. Timeline of the study. BP = Blood pressure.

2.2. Treatment

After a 2-week period of diet standardization, the enrolled subjects were instructed to take a capsule/day of SelectSIEVE® OptiChol containing 100 mg olive-derived extract that was standardized in hydroxytyrosol (Table 1) and acacia gum and maltodextrin as inactive carriers.

Table 1. Analytical specifications of the dietary supplement—namely SelectSIEVE® OptiChol—used in the clinical study.

Active Components Per Capsule (100 mg)
4–9% Hydroxytyrosol 6–15% Other olive polyphenols

The study product was manufactured and packaged by Roelmi HPC (Milan, Italy) in accordance with Quality Management System ISO 9001:2008 and the European Good Manufacturing Practices (GMP), thereby satisfying the requirements in the "Code of Federal Regulation" title 21,volume 2, part 111.

After the run-in, each patient was provided with boxes containing 30 capsules. For the entire duration of the study, the patients were instructed to take a capsule of SelectSIEVE® OptiChol once daily before their breakfast early in the morning. At the end of the study, all of the unused capsules were retrieved for inventory, and the participants' compliance was assessed by counting the number of returned capsules. A pill-count (dispensed pills—remaining pills)/(pills to be consumed between the visits) value of 0.85 to ≤1.15 was recorded as appropriate compliance. Underdose (<0.85) and overdose (>1.15) were labeled as non-compliance.

2.3. Assessments

2.3.1. Clinical Data and Anthropometric Measurements

The information that was gathered in the patients' history included the presence of ASCVD, other systemic diseases, and medications. The validated semi-quantitative questionnaires including a Food Frequency Questionnaire (FFQ) were used to assess the demographic variables of smoking and dietary habits and leisure time and physical activities [16].

The quantification and analysis of the energy intake and daily diet composition was performed using the MètaDieta® software (INRAN/IEO 2008 revision/ADI), and the data were handled in compliance with the company procedure IOA87.

Waist circumference (WC) was measured in the minimum perimeter at the end of a normal expiration and with arms being relaxed at the sides. Height and weight were respectively measured to the nearest 0.1 cm and 0.1 Kg with the patients standing erect with their eyes directed straight and wearing light clothes and having bare feet. BMI was calculated as body weight [Kg], which was divided by the height squared [m^2] (Kg/m^2). Finally, the index of central obesity (ICO) was calculated from the WC-to-height ratio.

2.3.2. Laboratory Parameters Measurements

The biochemical analyses were carried out on venous blood that was withdrawn after an overnight fasting period (~12 h). The plasma was obtained by the addition of disodium ethylenediaminetetraacetate (Na_2EDTA) and blood centrifugation at 3000 RPM for 15 min.

Immediately after the centrifugation, trained personnel performed laboratory analyses according to standardized methods [17]. The following parameters were directly assessed: total cholesterol (TC), high-density lipoprotein cholesterol (HDL-C), triglycerides (TG), serum uric acid (SUA), creatinine, fasting plasma glucose (FPG), alanine transaminase (ALT), aspartate transaminase (AST), and gamma-glutamyl transferase (gGT).

Non-HDL cholesterol (Non-HDL-C) resulted from the difference between the TC and the HDL-C. The LDL-C was obtained by the use of the Friedewald formula [LDL-C = TC − HDL-C − TG/5]. The glomerular filtration rate (eGFR) was estimated by the Chronic Kidney Disease Epidemiology Collaboration (CKD-epi) equation [18].

2.3.3. Blood Pressure Measurements

The arterial blood pressure (BP) was assessed in accordance with the recommendations of the International Guidelines for the management of arterial hypertension [19]. The resting systolic (SBP) and diastolic BP (DBP) were measured by the use of a validated oscillometric device, while they were in a sitting position and wearing a cuff of the appropriate size which was applied on the right upper arm. To improve the detection accuracy, three BP readings were sequentially obtained at 1 min intervals [20]. The first measurement was performed after 10 to 15 min of rest, and this was discarded. The average between the second and the third readings was recorded as the study variable. Pulse pressure was calculated as the difference between the SBP and the DBP.

2.3.4. Assessment of Safety and Tolerability

The level of safety and tolerability were evaluated through continuous monitoring during the study in order to detect any adverse event (AE) by taking the vital sign measurements, and employing the laboratory findings, clinical safety procedures, and physical examinations [21]. All of the reports of AEs were collected from the time of them giving their informed consent until the end of the study. A 10-point visual analog scale (VAS) was used to measure the patients' acceptability of SelectSIEVE® OptiChol.

2.4. Statistical Analysis

Statistical analysis was performed with intention to treat by means of the Statistical Package for Social Science (SPSS) 25.0, version for Windows.

The Kolmogorov–Smirnov test was used to test the normality distribution of the recorded data. The non normally distributed variables were log-transformed before further statistical testing was conducted. The intervention effect was assessed using Levene's test which was followed by the independent Student's *t* test. All of the data were expressed as means and standard deviations (SDs). A 2-tailed $p < 0.05$ was considered as statistically significant for all of the tests.

3. Results

3.1. Efficacy Analysis

A total of 36 volunteers was screened, and 30 subjects (Men: 16; Women: 14) were enrolled and successfully completed the study according to its design (Figure 2).

No patient was found to be noncompliant with the study protocol (Figure 2). No statistically significant changes were recorded in dietary habits during the study, with no changes in the total energy and the macronutrient intake occurring (Table 2).

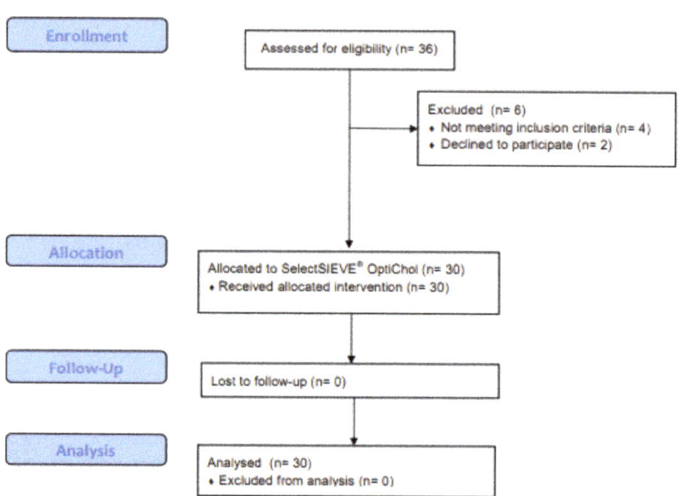

Figure 2. CONSORT (Consolidated Standards of Reporting Trials) flow diagram.

Table 2. Diet composition (g/day) at enrollment and at the end of the intervention. Values are reported as mean $\pm$ SD.

Parameters	Baseline	Week 4	*p*-Value
Total energy (Kcal/day)	1627 $\pm$ 121	1630 $\pm$ 105	n.s.
Carbohydrates (% of total energy)	55.2 $\pm$ 2.7	54.8 $\pm$ 3.1	n.s.
Proteins (% of total energy)	17.8 $\pm$ 2.1	18.1 $\pm$ 1.9	n.s.
Animal protein (% of total energy)	11.1 $\pm$ 0.7	10.9 $\pm$ 0.7	n.s.
Vegetal protein (% of total energy)	6.7 $\pm$ 0.4	7.2 $\pm$ 0.2	n.s.
Total fats (% of total energy)	26.9 $\pm$ 2.4	27.1 $\pm$ 1.8	n.s.
Saturated fatty acids (% of total energy)	8.9 $\pm$ 0.5	9.1 $\pm$ 0.2	n.s.
MUFA (% of total energy)	13.3 $\pm$ 1.4	12.9 $\pm$ 1.5	n.s.
PUFA (% of total energy)	4.7 $\pm$ 0.8	5.1 $\pm$ 0.9	n.s.
Total dietary fibers (g/day)	15.7 $\pm$ 1.8	15.9 $\pm$ 1.3	n.s.
Cholesterol (mg/day)	193.1 $\pm$ 12.7	192.8 $\pm$ 11.9	n.s.

MUFA = Monounsaturated fatty acids; N = Number of individuals; n.s. = not significant; PUFA = Polyunsaturated fatty acids.

At the end of the study, the dietary supplementation with SelectSIEVE® OptiChol was associated with a significant improvement in SBP, PP, FPG, TC, HDL-C, LDL-C, non-HDL-C, and SUA when they were compared to the baseline values (Table 3).

3.2. Safety Analysis

All of the participants completed the study according to its design (dropout rate = 0%). No treatment-emergent adverse events were reported, nor did any laboratory abnormality occurred. The volunteers' acceptability of SelectSIEVE® OptiChol was good.

Table 3. Anthropometric, hemodynamic, and blood chemistry parameters from the baseline to the end of the clinical trial.

Parameters	SelectSIEVE® OptiChol (N. 30)			
	Pre-Run-in	Baseline	Week 4	
	Mean ± SD	Mean ± SD	Mean ± SD	*p*-Value versus Baseline
Age (years)	53 ± 5			
WC (cm)	89.8 ± 5.3	88.9 ± 5.1	87.7 ± 5.5	n.s.
ICO	0.56 ± 0.08	0.54 ± 0.07	0.53 ± 0.08	n.s.
BMI (Kg/m^2)	24.8 ± 2.2	24.6 ± 2.2	24.3 ± 2.3	n.s.
SBP (mmHg)	134 ± 5	133 ± 5	130 ± 2	<0.05
DBP (mmHg)	87 ± 2	86 ± 3	86 ± 2	n.s.
PP (mmHg)	47 ± 2	47 ± 2	44 ± 2	<0.05
HR (bpm)	74 ± 4	74 ± 4	75 ± 5	n.s.
FPG (mg/dL)	88 ± 3	90 ± 3	85 ± 2	<0.05
TC (mg/dL)	248 ± 13	238 ± 12	225 ± 7	<0.05
HDL-C (mg/dL)	44 ± 3	44 ± 3	48 ± 2	<0.05
LDL-C (mg/dL)	161 ± 8	155 ± 8	145 ± 5	<0.05
Non HDL-C (mg/dL)	204 ± 11	198 ± 11	177 ± 8	<0.05
TG (mg/dL)	216 ± 19	197 ± 16	186 ± 18	n.s.
AST (mg/dL)	23 ± 3	25 ± 4	24 ± 3	n.s.
ALT (mg/dL)	22 ± 3	22 ± 3	23 ± 4	n.s.
gGT (mg/dL)	32 ± 2	33 ± 2	30 ± 5	n.s.
SUA (mg/dL)	8.5 ± 1.8	8.6 ± 1.5	7.7 ± 1.1	<0.05
Creatinine (mg/dL)	0.8 ± 0.1	0.8 ± 0.1	0.8 ± 0.2	n.s.
eGFR (ml/min)	88 ± 4	89 ± 4	87 ± 5	n.s.

ALT = Alanine aminotransferase; AST = Aspartate aminotransferase; BMI = Body mass index; DBP = Diastolic blood pressure; eGFR = Estimated glomerular filtration rate; FPG = Fasting plasma glucose; gGT = Gamma-glutamyl transferase; HDL-C = High-density lipoprotein cholesterol; HR = Heart rate; ICO = Index of central obesity; LDL-C = Low-density lipoprotein cholesterol; N = Number of individuals; n.s. = Not significant; PP = Pulse pressure; SD = Standard deviation; SUA = Serum uric acid; SBP = Systolic blood pressure; TG = Triglycerides; WC = Waist circumference.

4. Discussion

In vitro and in vivo studies have previously found that hydroxytyrosol from *Olea europaea* extract exerts a number of potential clinical benefits with putative anti-atherosclerotic and anti-ischemic properties [22]. The mechanisms underlying the vascular protective effect of hydroxytyrosol consist in the prevention of LDL-C oxidation, in the reversion of angiogenesis through the inhibition of the activity of matrix metalloproteinase-2 (MMP-2) and 9 (MMP-9), in the reduction of inflammatory damage and eicosanoid formation, and in the expression of the vascular cell adhesion molecule1 (VCAM-1) and the intercellular adhesion molecule 1 (ICAM-1) [14,23]. Moreover, a significant reduction in E-selectin, P-selectin, ICAM-1, and VCAM-1 secretions were found in the human aortic endothelial cells that were treated with physiological concentrations of hydroxytyrosol and co-incubated with tumor necrosis factor alpha (TNF-α) [24]. Finally, hydroxytyrosol exerts antithrombotic properties by decreasing the platelet aggregation and the expression of the cell adhesion molecules, and by reducing the synthesis of thromboxane B2 and leukotriene B4 and their capacity to reduce cyclic adenosine (cAMP) and guanosine (cGMP) monophosphate platelet phosphodiesterase [8].

In humans, a dietary supplementation with polyphenolic-rich olive extract exerts antioxidant properties, thereby resulting in a number of cardioprotective effects whose extents are heterogeneous across the published studies and are largely dependent on the differences in the phenolic tested doses and the treatment duration [25]. Starting from these observations, a recently released systematic review and meta-analysis of the controlled clinical trials has aimed to assess whether the effects on the single components of metabolic syndrome (MetS)—including obesity, glucose intolerance, dyslipidemia, and high BP—were related to hydroxytyrosol or oleic acid contents or their combination in olive oil [26]. By summarizing the available evidence, this pooled analysis did not show any significant effects of the hydroxytyrosol consumption on the MetS components, while the polyphenolic-rich olive oil was as good as a standard care for MetS management. These findings are definitely conclusive in determining that the supplementation with a phenolic-rich olive extract (also known as "virgin olive oil") is an effective tool for the prevention and treatment of MetS, unlike the isolated polyphenolic fractions. Of course, there is no doubt that Mediterranean dietary patterns that are rich in olive oil exert per se different protective action against CV aging. As a matter of fact, in the Seguimiento University of Navarra (SUN) project, the hazard ratio for CVD for olive oil consumption $\geq$30 g/day (versus <10 g/day) was 0.57 (95% Confidence Interval (CI): 0.34, 0.96) in a follow-up that was conducted over 10.8 years [27]. Moreover, in the European Prospective Investigation into Cancer and Nutrition (EPIC, Spain) that had a follow-up of 22.8 years, the hazard ratios for stroke according to the amount of olive oil consumption were 0.84 (0.70, 1.02), 0.80 (0.66, 0.96), 0.89 (0.74, 1.07) for 0 to <10, 10 to <20, 20 to <30, and $\geq$30 g/day of olive oil, respectively [27]. However, it is necessary to specify that not all of the commercially available olive oils have the same polyphenols content. Furthermore, olive oil intake is hardly predictable because it is a seasoning, not a food, and its antioxidant effect is variably impaired by the exposition of it at cooking temperature. It should also be taken into account that olive oil is a source of lipids, so the recommended daily amount of olive oil in weight-management programs should be less than that which is recommended for the maintenance of CV health in normal-weight individuals [28]. Last but not least, not everyone likes the taste of olive oil and, of consequence, not everyone is able to adhere to a Mediterranean dietary pattern on the long-term. For all of these reasons, an olive-derived extract that is standardized in hydroxytyrosol to be consumed on top of Mediterranean diet could be of particular interest, taking into consideration that olive oil's supposed beneficial effects on the vascular and metabolic parameters depend only on its polyphenolic fraction and not MUFA.

According to our observations, 100 mg olive-derived extracts that are standardized in hydroxytyrosol are able to yield a broad-spectrum of activity with significant improvement in SBP, PP, FPG, TC, HDL-C, LDL-C, non-HDL-C, and SUA after one month of its dietary supplementation. Previously, evidence from clinical trials had been mostly limited to highly selected demographic groups such as elderly people who were either free-living or living in protective residences [29,30]. A randomized, double-blind, placebo-controlled, clinical study had already showed that one-year consumption of a polyphenol extract from *Olea europaea* improved the lipid profile of postmenopausal women (pre-post treatment changes: TC = −26.2 mg/dL in the active group, p = 0.01 versus placebo; LDL-C = −34.7 mg/dL in the active group, p =0.02 versus placebo; TG = −4.15 mg/dL in the active group, p = 0.01 versus placebo) [29]. Another randomized, double-blind, clinical study involving institutionalized individuals that were aged 65–96 years had already revealed that a nutritional intervention testing a 6-week daily dietary supplementation with virgin oil improved the antioxidant status in the elderly by reducing the serum lipid levels, the serum total antioxidant capacity (TAC), and the superoxide dismutase (SOD) and glutathione peroxidase (GH-PX) activity, and by increasing the catalase (CAT) in the erythrocytes [30]. Most recently, the lipid-lowering effect of dietary supplementation with phenolic-rich olive extract has been investigated as it is associated with red yeast rice. In this context, a nutraceutical compound containing red yeast rice and olive fruit extract that is highly concentrated in hydroxytyrosol (5 mg hydroxytyrosol equivalent) has been found to be effective in reducing the serum lipid levels and also well tolerated by hypercholesterolemic patients with a history of statin-associated muscle symptoms (SAMS) [31]. Another randomized, double-blind, placebo-controlled, clinical study has showed that a food supplement combining red yeast rice and olive fruit extract (for a daily intake of 10.82 mg of monacolins and 9.32 mg of hydroxytyrosol) was able to exert beneficial effects in patients with MetS even in the short term by lowering the LDL-C by 24% (+1% in the placebo group) and also improving the TC

(-17% versus $+2$% in the control group), apolipoprotein B (-15% versus $+6$%, $p < 0.001$), TG (-9% versus $+16$%, $p = 0.02$), oxidized LDL (-20% versus $+5$% in the control group. $p < 0.001$), SBP (mean difference versus placebo = -0.3 mmHg, $p = 0.001$), and DBP (mean difference versus placebo = -0.4 mmHg, $p = 0.05$) after only 6 weeks of dietary supplementation taking place [32]. In effect, the preclinical evidence suggests that olive polyphenols are able to partially inhibit the activity of 3-hydroxy-3-methylglutaryl-Coenzyme A (HMG-CoA) reductase and acetyl-Coenzyme A cholesterol acyltransferase (ACAT), thus resulting in a decreased cholesterol biosynthesis [33]. Moreover, olive polyphenols might have effect on the bile flow and secondarily promote lipid fecal excretion by increasing the biliary cholesterol and the bile acid concentrations [34]. Previous evidence has shown that diet is one of the most important contributors to the balance of both the gut microbiota and bile acid homeostasis [35]. In particular, population studies that demonstrate a higher consumption of fruits and vegetables with a high content in polyphenols are associated with the enhancement of the growth of the probiotic bacteria that actively interact with the bile acid metabolising activity [36]. In addition, to date, a number of polyphenols have been reported to exert bile acid sequestering activity [37,38]. As intervention studies have showed that the pharmaceutical sequestering agents decrease the circulating LDL-C, reduce obesity, improve insulin sensitivity, and induce thermogenesis [39], it is likely that polyphenols too—as bile acid sequestering agents of a natural origin—have potential for the treatment of the metabolic disease.

In addition to the LDL-C lowering effect, SelectSIEVE® OptiChol increases the plasma levels of HDL-C. It is well known that the increase of TC and particularly of LDL-C is positively associated with the risk of ASCVD, while higher values of HDL-C are inversely correlated with the risk of ASCVD [40,41]. A large meta-analysis of four prospective studies (namely the Lipid Research Clinics Prevalence Mortality Follow-Up Study, the Multiple Risk Factor Intervention Trial (MRFIT), the Framingham Heart Study, and the Lipid Research Clinics Coronary Primary Prevention Trial (LRC-CPPT)) showed that every 1 mg/dL increase in the HDL-C was associated with a significant coronary heart disease (CHD) risk reduction by3% in women and 2% in men [42]. For this reason, the favorable change in the HDL-C after the supplementation with SelectSIEVE® OptiChol is clinically relevant and deserves consideration. In effect, a systematic review on this topic has recently provided evidence that polyphenol-rich olive oil favors the enhancement of the HDL-C through the promotion of the HDL cholesterol efflux capacity, which is HDL-C main antiatherogenic function [43]. Based on the published evidence, olive polyphenols increase the HDL size and improve the HDL oxidative status and composition (i.e., promote a greater HDL stability, which is reflected as a TG-poor core) [44]. In this regard, a randomized crossover clinical trial involving 200 healthy male volunteers has recently showed that a 15-week dietary supplementation with olive oil with high phenolic content is able to exert further benefits on the HDL-C serum levels and the oxidative damage compared to refined olive oil, and it has definitely provided additional evidence to recommend the use of virgin olive oil as a source of bioactive compounds that are able to promote the optimization of some CV risk factors, including the increase in the HDL-C [45].

As expected [26], following the dietary supplementation with SelectSIEVE® OptiChol, also FPG and SUA mildly but significantly improved. This is particularly interesting and lays further conceptual foundations for the use of this extract in the prevention and treatment of first-stage MetS, considering that high SUA levels predict the incidence of MetS in populations [46]. In the past, a number of studies focused on the importance of the insulin resistance for hyperuricemia, bringing as link a fructose intake excess [47–49]. To date, several polyphenols have been tested as SUA-lowering agents and have been recognized for their ability to reduce the SUA levels through the inhibition of xanthine oxidase and renal urate transporters [50–52].

Finally, the observed effect on BP following the dietary supplementation with SelectSIEVE® OptiChol also needs to be discussed. Previously, dietary supplementation with a phenolic-rich olive leaf extract had been already showed to significantly reduce in patients with pre-hypertension either 24-h and daytime SBP and DBP, with an extent being potentially associated to a 9–14% risk reduction in developing CHD and a 20–22.5% risk reduction in stroke and heart attack, based on published literature [53,54]. However, according to our current observations, the effects on SBP and PP can be also detectable in non-hypertensive patients and after their dietary supplementation with olive fruit extract that usually contains fewer bioactive compounds than olive leaf extract does [55]. Certainly, it must also be noted that the positive effects that were observed in our study were obtained after the dietary supplementation with an extract that was derived from olives and that it is, of consequence, ecofriendly.

Of course, we acknowledge that our study has some limitations, namely, the small patient sample and the short study duration. Moreover, the absence of randomization and control procedures cannot preclude the positive role of potential confounding variables. However, despite the preliminary nature of the study findings, our observations are relevant and consistent with the previous literature, and certainly deserve to be further investigated in placebo-controlled clinical trials.

From a toxicological point of view, the tested extract has not to be considered as a novel food. As a matter of fact, it has been derived from a well-known food source with a very-well characterized method of extraction and standardization, and it contains polyphenols with non-dramatic concentrations. [56]. Moreover, the single components that are included in the extract are the same as those that are found in olives and olive oil, whose safety has been largely shown in the everyday use byentire population and by a number of preclinical trials [57]. Of course, longer term clinical trials on larger population samples wouldbe able to confirm the high tolerability and safety of the extract.

5. Conclusions

In conclusion, according to the findings of this non-randomized, prospective pilot clinical study, the tested dietary supplement (namely SelectSIEVE® OptiChol) containing 100 mg olive-derived extracts that were standardized in hydroxytyrosol is well tolerated and able to improve a number of metabolic parameters (i.e., BP, FPG, SUA, and plasma lipids) in the individuals with polygenic hypercholesterolemia. Even though we acknowledge the limitations of the present study, of course these results are particularly interesting and consistent with the previous literaturethat, however, referred to extracts from different olive varieties. Our observations lay further foundations for the use of polyphenolic-rich olive extract from Coratina Olive in the prevention and treatment of first-stage MetS.

Author Contributions: Conceptualization, A.F.G.C.; methodology, A.F.G.C. and F.F.; software, A.F.G.C.; formal analysis, A.F.G.C.; investigation, A.F.G.C., F.F., A.D.M., M.V. and E.G.; data curation, F.F., A.D.M., M.V. and E.G.; writing—original draft preparation, A.F.G.C. and F.F.; writing—review and editing, A.D.M., M.V., E.G. and C.B.; supervision, C.B.; project administration, A.F.G.C.; funding acquisition, A.F.G.C. All authors have read and agreed to the published version of the manuscript.

Funding: This research received no external funding.

Institutional Review Board Statement: The study was conducted in accordance with the Declaration of Helsinki, and it was approved by the Local Institutional Review Board.

Informed Consent Statement: Informed consent was obtained from all subjects involved in the study.

Data Availability Statement: Data supporting study's findings are available from the Corresponding Author with the permission of the University of Bologna.

Conflicts of Interest: The authors declare no conflict of interest.

References

1. Ros, E.; Martínez-González, M.A.; Estruch, R.; Salas-Salvadó, J.; Fitó, M.; Martínez, J.A.; Corella, D. Mediterranean diet and cardiovascular health: Teachings of the PREDIMED study. *Adv. Nutr.* **2014**, *5*, 330S–336S. [CrossRef] [PubMed]
2. Sofi, F.; Abbate, R.; Gensini, G.F.; Casini, A. Accruing evidence on benefits of adherence to the Mediterranean diet on health: An updated systematic review and meta-analysis. *Am. J. Clin. Nutr.* **2010**, *92*, 1189–1196. [CrossRef] [PubMed]

3. Beulen, Y.; Martínez-González, M.A.; van de Rest, O.; Salas-Salvadó, J.; Sorlí, J.V.; Gómez-Gracia, E.; Fiol, M.; Estruch, R.; Santos-Lozano, J.M.; Schröder, H.; et al. Quality of Dietary Fat Intake and Body Weight and Obesity in a Mediterranean Population: Secondary Analyses within the PREDIMED Trial. *Nutrients* **2018**, *10*, 2011. [CrossRef] [PubMed]

4. Guasch-Ferré, M.; Babio, N.; Martínez-González, M.A.; Corella, D.; Ros, E.; Martín-Peláez, S.; Estruch, R.; Arós, F.; Gómez-Gracia, E.; Fiol, M.; et al. Dietary fat intake and risk of cardiovascular disease and all-cause mortality in a population at high risk of cardiovascular disease. *Am. J. Clin. Nutr.* **2015**, *102*, 1563–1573. [PubMed]

5. Nan, J.N.; Ververis, K.; Bollu, S.; Rodd, A.L.; Swarup, O.; Karagiannis, T.C. Biological effects of the olive polyphenol, hydroxytyrosol: An extra view from genome-wide transcriptome analysis. *Hell. J. Nucl. Med.* **2014**, *17*, 62–69. [PubMed]

6. Uylaşer, V.; Yildiz, G. The historical development and nutritional importance of olive and olive oil constituted an important part of the Mediterranean diet. *Crit. Rev. Food Sci. Nutr.* **2014**, *54*, 1092–1101. [CrossRef] [PubMed]

7. Claro-Cala, C.M.; Jiménez-Altayó, F.; Zagmutt, S.; Rodriguez-Rodriguez, R. Molecular Mechanisms Underlying the Effects of Olive Oil Triterpenic Acids in Obesity and Related Diseases. *Nutrients* **2022**, *14*, 1606. [CrossRef] [PubMed]

8. Rocha, J.; Borges, N.; Pinho, O. Table olives and health: A review. *J. Nutr. Sci.* **2020**, *9*, e57. [CrossRef] [PubMed]

9. Rizwan, S.; Benincasa, C.; Mehmood, K.; Anjum, S.; Mehmood, Z.; Alizai, G.H.; Azam, M.; Perri, E.; Sajjad, A. Fatty Acids and Phenolic Profiles of Extravirgin Olive Oils from Selected Italian Cultivars Introduced in Southwestern Province of Pakistan. *J.Oleo. Sci.* **2019**, *68*, 33–43. [CrossRef] [PubMed]

10. EFSA Panel on Dietetic Products, Nutrition and Allergies (NDA). Scientific opinion on the substantiation of health claims related to polyphenols in olive and protection of LDL particles from oxidative damage (Id 1333, 1638, 1639, 1696, 2865), maintenance of normal blood HDL cholesterol concentrations (Id 1639), maintenance of normal blood pressure (Id 3781), "anti-inflammatory properties" (id 1882), "contributes to the upper respiratory tract health" (Id 3467) pursuant to article 13(1) of regulation (ec) no 1924/2006. *EFSA J.* **2011**, *9*, 2033.

11. Visioli, F.; Poli, A.; Gall, C. Antioxidant and other biological activities of phenols from olives and olive oil. *Med. Res. Rev.* **2002**, *22*, 65–75. [CrossRef] [PubMed]

12. KarkovićMarković, A.; Torić, J.; Barbarić, M.; JakobušićBrala, C. Hydroxytyrosol, Tyrosol and Derivatives and Their Potential Effects on Human Health. *Molecules* **2019**, *24*, 2001. [CrossRef] [PubMed]

13. Tejada, S.; Pinya, S.; Del Mar Bibiloni, M.; Tur, J.A.; Pons, A.; Sureda, A. Cardioprotective Effects of the Polyphenol Hydroxytyrosol from Olive Oil. *Curr. Drug Targets* **2017**, *18*, 1477–1486. [CrossRef] [PubMed]

14. Vilaplana-Pérez, C.; Auñón, D.; García-Flores, L.A.; Gil-Izquierdo, A. Hydroxytyrosol and potential uses in cardiovascular diseases, cancer, and AIDS. *Front. Nutr.* **2014**, *1*, 18. [CrossRef] [PubMed]

15. Authors/Task Force Members; ESC Committee for Practice Guidelines (CPG); ESC National Cardiac Societies. 2019 ESC/EAS guidelines for the management of dyslipidaemias: Lipid modification to reduce cardiovascular risk. *Atherosclerosis* **2019**, *290*, 140–205. [CrossRef]

16. Cicero, A.F.G.; Fogacci, F.; Bove, M.; Giovannini, M.; Borghi, C. Impact of a short-term synbiotic supplementation on metabolic syndrome and systemic inflammation in elderly patients: A randomized placebo-controlled clinical trial. *Eur. J. Nutr.* **2021**, *60*, 655–663. [CrossRef]

17. Cicero, A.F.G.; Fogacci, F.; Rosticci, M.; Parini, A.; Giovannini, M.; Veronesi, M.; D'Addato, S.; Borghi, C. Effect of a short-term dietary supplementation with phytosterols, red yeast rice or both on lipid pattern in moderately hypercholesterolemic subjects: A three-arm, double-blind, randomized clinical trial. *Nutr. Metab.* **2017**, *14*, 61. [CrossRef]

18. Levey, A.S.; Stevens, L.A.; Schmid, C.H.; Zhang, Y.L.; Castro, A.F., III; Feldman, H.I.; Kusek, J.W.; Eggers, P.; Van Lente, F.; CKD-EPI (Chronic Kidney Disease Epidemiology Collaboration); et al. A new equation to estimate glomerular filtration rate. *Ann. Intern. Med.* **2009**, *150*, 604–612. [CrossRef]

19. Williams, B.; Mancia, G.; Spiering, W.; AgabitiRosei, E.; Azizi, M.; Burnier, M.; Clement, D.; Coca, A.; De Simone, G.; Dominiczak, A.; et al. Practice Guidelines for the management of arterial hypertension of the European Society of Hypertension and the European Society of Cardiology: ESH/ESC Task Force for the Management of Arterial Hypertension. *J. Hypertens.* **2018**, *36*, 2284–2309. [CrossRef]

20. Cicero, A.F.G.; Fogacci, F.; Veronesi, M.; Grandi, E.; Dinelli, G.; Hrelia, S.; Borghi, C. Short-Term Hemodynamic Effects of Modern Wheat Products Substitution in Diet with Ancient Wheat Products: A Cross-Over, Randomized Clinical Trial. *Nutrients* **2018**, *10*, 1666. [CrossRef]

21. Cicero, A.F.G.; Fogacci, F.; Veronesi, M.; Strocchi, E.; Grandi, E.; Rizzoli, E.; Poli, A.; Marangoni, F.; Borghi, C. A randomized Placebo-Controlled Clinical Trial to Evaluate the Medium-Term Effects of Oat Fibers on Human Health: The Beta-Glucan Effects on Lipid Profile, Glycemia and inTestinal Health (BELT) Study. *Nutrients* **2020**, *12*, 686. [CrossRef] [PubMed]

22. Efentakis, P.; Iliodromitis, E.K.; Mikros, E.; Papachristodoulou, A.; Dagres, N.; Skaltsounis, A.L.; Andreadou, I. Effects of the olive tree leaf constituents on myocardial oxidative damage and atherosclerosis. *Planta Med.* **2015**, *81*, 648–654. [CrossRef] [PubMed]

23. Granados-Principal, S.; Quiles, J.L.; Ramirez-Tortosa, C.L.; Sanchez-Rovira, P.; Ramirez-Tortosa, M.C. Hydroxytyrosol: From laboratory investigations to future clinical trials. *Nutr. Rev.* **2010**, *68*, 191–206. [CrossRef] [PubMed]

24. Catalán, Ú.; López de Las Hazas, M.C.; Rubió, L.; Fernández-Castillejo, S.; Pedret, A.; de la Torre, R.; Motilva, M.J.; Solà, R. Protective effect of hydroxytyrosol and its predominant plasmatic human metabolites against endothelial dysfunction in human aortic endothelial cells. *Mol. Nutr. Food Res.* **2015**, *59*, 2523–2536. [CrossRef]

25. Romani, A.; Ieri, F.; Urciuoli, S.; Noce, A.; Marrone, G.; Nediani, C.; Bernini, R. Health Effects of Phenolic Compounds Found in Extra-Virgin Olive Oil, By-Products, and Leaf of *Olea europaea* L. *Nutrients* **2019**, *11*, 1776. [CrossRef]

26. Pastor, R.; Bouzas, C.; Tur, J.A. Beneficial effects of dietary supplementation with olive oil, oleic acid, or hydroxytyrosol in metabolic syndrome: Systematic review and meta-analysis. *Free Radic. Biol. Med.* **2021**, *172*, 372–385. [CrossRef]

27. Hooper, L.; Abdelhamid, A.S.; Jimoh, O.F.; Bunn, D.; Skeaff, C.M. Effects of total fat intake on body fatness in adults. *Cochr. Database Syst. Rev.* **2020**, *6*, CD013636. [CrossRef]

28. Donat-Vargas, C.; Sandoval-Insausti, H.; Peñalvo, J.L.; Moreno Iribas, M.C.; Amiano, P.; Bes-Rastrollo, M.; Molina-Montes, E.; Moreno-Franco, B.; Agudo, A.; Mayo, C.L.; et al. Olive oil consumption is associated with a lower risk of cardiovascular disease and stroke. *Clin. Nutr.* **2022**, *41*, 122–130. [CrossRef]

29. Filip, R.; Possemiers, S.; Heyerick, A.; Pinheiro, I.; Raszewski, G.; Davicco, M.J.; Coxam, V. Twelve-month consumption of a polyphenol extract from olive (Olea europaea) in a double blind, randomized trial increases serum total osteocalcin levels and improves serum lipid profiles in postmenopausal women with osteopenia. *J. Nutr. Health Aging* **2015**, *19*, 77–86. [CrossRef]

30. Oliveras-López, M.J.; Molina, J.J.; Mir, M.V.; Rey, E.F.; Martín, F.; de la Serrana, H.L. Extra virgin olive oil (EVOO) consumption and antioxidant status in healthy institutionalized elderly humans. *Arch. Gerontol. Geriatr.* **2013**, *57*, 234–242. [CrossRef]

31. TshongoMuhindo, C.; Ahn, S.A.; Rousseau, M.F.; Dierckxsens, Y.; Hermans, M.P. Efficacy and safety of a combination of red yeast rice and olive extract in hypercholesterolemic patients with and without statin-associated myalgia. *Complement. Ther. Med.* **2017**, *35*, 140–144. [CrossRef] [PubMed]

32. Verhoeven, V.; Van der Auwera, A.; Van Gaal, L.; Remmen, R.; Apers, S.; Stalpaert, M.; Wens, J.; Hermans, N. Can red yeast rice and olive extract improve lipid profile and cardiovascular risk in metabolic syndrome? A double blind, placebo controlled randomized trial. *BMC Complement. Altern. Med.* **2015**, *15*, 52. [CrossRef]

33. Lee, J.S.; Choi, M.S.; Jeon, S.M.; Jeong, T.S.; Park, Y.B.; Lee, M.K.; Bok, S.H. Lipid-lowering and antioxidative activities of 3,4-di(OH)-cinnamate and 3,4-di(OH)-hydrocinnamate in cholesterol-fed rats. *Clin. Chim. Acta* **2001**, *314*, 221–229. [CrossRef]

34. Krzeminski, R.; Gorinstein, S.; Leontowicz, H.; Leontowicz, M.; Gralak, M.; Czerwinski, J.; Lojek, A.; Cíz, M.; Martin-Belloso, O.; Gligelmo-Miguel, N.; et al. Effect of different olive oils on bile excretion in rats fed cholesterol-containing and cholesterol-free diets. *J. Agric. Food Chem.* **2003**, *51*, 5774–5779. [CrossRef] [PubMed]

35. Pushpass, R.G.; Alzoufairi, S.; Jackson, K.G.; Lovegrove, J.A. Circulating bile acids as a link between the gut microbiota and cardiovascular health: Impact of prebiotics, probiotics and polyphenol-rich foods. *Nutr. Res. Rev.* **2021**, 1–20. [CrossRef] [PubMed]

36. Koutsos, A.; Tuohy, K.M.; Lovegrove, J.A. Apples and cardiovascular health–is the gut microbiota a core consideration? *Nutrients* **2015**, *7*, 3959–3998. [CrossRef] [PubMed]

37. Hylemon, P.B.; Zhou, H.; Pandak, W.M.; Ren, S.; Gil, G.; Dent, P. Bile acids as regulatory molecules. *J. Lipid Res.* **2009**, *50*, 1509–1520. [CrossRef] [PubMed]

38. Li, T.; Chiang, J.Y. Bile acids as metabolic regulators. *Curr. Opin. Gastroenterol.* **2015**, *31*, 159–165. [CrossRef] [PubMed]

39. Watanabe, M.; Morimoto, K.; Houten, S.M.; Kaneko-Iwasaki, N.; Sugizaki, T.; Horai, Y.; Mataki, C.; Sato, H.; Murahashi, K.; Arita, E.; et al. Bile acid binding resin improves metabolic control through the induction of energy expenditure. *PLoS ONE* **2012**, *7*, e38286. [CrossRef] [PubMed]

40. Allard-Ratick, M.P.; Kindya, B.R.; Khambhati, J.; Engels, M.C.; Sandesara, P.B.; Rosenson, R.S.; Sperling, L.S. HDL: Fact, fiction, or function? HDL cholesterol and cardiovascular risk. *Eur. J. Prev. Cardiol.* **2021**, *28*, 166–173. [CrossRef] [PubMed]

41. Fogacci, F.; Borghi, C.; Cicero, A.F.G. New evidences on the association between high-density lipoprotein cholesterol and cardiovascular risk: A never ending research story. *Eur. J. Prev. Cardiol.* **2022**, *29*, 842–843. [CrossRef] [PubMed]

42. Gordon, D.J.; Probstfield, J.L.; Garrison, R.J.; Neaton, J.D.; Castelli, W.P.; Knoke, J.D.; Jacobs, D.R., Jr.; Bangdiwala, S.; Tyroler, H.A. High-density lipoprotein cholesterol and cardiovascular disease. Four prospective American studies. *Circulation* **1989**, *79*, 8–15. [CrossRef]

43. Rondanelli, M.; Giacosa, A.; Morazzoni, P.; Guido, D.; Grassi, M.; Morandi, G.; Bologna, C.; Riva, A.; Allegrini, P.; Perna, S. MediterrAsian Diet Products That Could Raise HDL-Cholesterol: A Systematic Review. *Biomed. Res. Int.* **2016**, *2016*, 2025687. [CrossRef] [PubMed]

44. Hernáez, Á.; Fernández-Castillejo, S.; Farràs, M.; Catalán, Ú.; Subirana, I.; Montes, R.; Solà, R.; Muñoz-Aguayo, D.; Gelabert-Gorgues, A.; Díaz-Gil, Ó.; et al. Olive oil polyphenols enhance high-density lipoprotein function in humans: A randomized controlled trial. *Arterioscler. Thromb. Vasc. Biol.* **2014**, *34*, 2115–2119. [CrossRef]

45. Covas, M.I.; Nyyssönen, K.; Poulsen, H.E.; Kaikkonen, J.; Zunft, H.J.; Kiesewetter, H.; Gaddi, A.; de la Torre, R.; Mursu, J.; EUROLIVE Study Group; et al. The effect of polyphenols in olive oil on heart disease risk factors: A randomized trial. *Ann. Intern. Med.* **2006**, *145*, 333–341. [CrossRef]

46. Cicero, A.F.G.; Fogacci, F.; Giovannini, M.; Grandi, E.; Rosticci, M.; D'Addato, S.; Borghi, C. Serum uric acid predicts incident metabolic syndrome in the elderly in an analysis of the Brisighella Heart Study. *Sci. Rep.* **2018**, *8*, 11529. [CrossRef]

47. Facchini, F.; Chen, Y.D.; Hollenbeck, C.B.; Reaven, G.M. Relationship between resistance to insulin-mediated glucose uptake, urinary uric acid clearance, and plasma uric acid concentration. *J. Am. Med. Assoc.* **1991**, *266*, 3008–3011. [CrossRef]

48. Vuorinen-Markkola, H.; Yki-Järvinen, H. Hyperuricemia and insulin resistance. *J. Clin. Endocrinol. Metab.* **1994**, *78*, 25–29. [PubMed]

49. Zhu, Y.; Hu, Y.; Huang, T.; Zhang, Y.; Li, Z.; Luo, C.; Luo, Y.; Yuan, H.; Hisatome, I.; Yamamoto, T.; et al. High uric acid directly inhibits insulin signalling and induces insulin resistance. *Biochem. Biophys. Res. Commun.* **2014**, *447*, 707–714. [CrossRef]

50. Wu, D.; Chen, R.; Zhang, W.; Lai, X.; Sun, L.; Li, Q.; Zhang, Z.; Cao, J.; Wen, S.; Lai, Z.; et al. Tea and its components reduce the production of uric acid by inhibiting xanthine oxidase. *Food Nutr. Res.* **2022**, *66*, 8239. [CrossRef] [PubMed]

51. Cicero, A.F.G.; Caliceti, C.; Fogacci, F.; Giovannini, M.; Calabria, D.; Colletti, A.; Veronesi, M.; Roda, A.; Borghi, C. Effect of apple polyphenols on vascular oxidative stress and endothelium function: A translational study. *Mol. Nutr. Food Res.* **2017**, *61*, 1700373. [CrossRef] [PubMed]

52. Olechno, E.; Puścion-Jakubik, A.; Zujko, M.E. Chokeberry (A. melanocarpa (Michx.) Elliott)-A Natural Product for Metabolic Disorders? *Nutrients* **2022**, *14*, 2688. [CrossRef] [PubMed]

53. Lockyer, S.; Rowland, I.; Spencer, J.P.E.; Yaqoob, P.; Stonehouse, W. Impact of phenolic-rich olive leaf extract on blood pressure, plasma lipids and inflammatory markers: A randomised controlled trial. *Eur. J. Nutr.* **2017**, *56*, 1421–1432. [CrossRef]

54. Cook, N.R.; Cohen, J.; Hebert, P.R.; Taylor, J.O.; Hennekens, C.H. Implications of small reductions in diastolic blood pressure for primary prevention. *Arch. Intern. Med.* **1995**, *155*, 701–709. [CrossRef]

55. Nocella, C.; Cammisotto, V.; Fianchini, L.; D'Amico, A.; Novo, M.; Castellani, V.; Stefanini, L.; Violi, F.; Carnevale, R. Extra Virgin Olive Oil and Cardiovascular Diseases: Benefits for Human Health. *Endocr. Metab. Immune Disord. Drug Targets* **2018**, *18*, 4–13. [CrossRef] [PubMed]

56. Konstantinidou, V.; Garcia-Santamarina, S. Moving forward the Effects of Gene-Diet Interactions on Human Health. *Nutrients* **2022**, *14*, 3782. [CrossRef]

57. Del Saz-Lara, A.; López de Las Hazas, M.C.; Visioli, F.; Dávalos, A. Nutri-epigenetic Effects of Phenolic Compounds from Extra Virgin Olive Oil: A Systematic Review. *Adv. Nutr.* **2022**, *13*, 2039–2060. [CrossRef] [PubMed]

Communication

Black Garlic and Pomegranate Standardized Extracts for Blood Pressure Improvement: A Non-Randomized Diet-Controlled Study

Federica Fogacci [1,2], Antonio Di Micoli [1], Elisa Grandi [1], Giulia Fiorini [1], Claudio Borghi [1,†] and Arrigo F. G. Cicero [1,2,*,†]

[1] Hypertension and Cardiovascular Risk Research Group, Medical and Surgical Sciences Department, University of Bologna, 40100 Bologna, Italy
[2] Italian Nutraceutical Society (SINut), 40100 Bologna, Italy
* Correspondence: arrigo.cicero@unibo.it
† These authors contributed equally to this work.

Abstract: Recently released position papers by the European Society of Hypertension (ESH) and the Italian Society of Hypertension (SIIA) provide therapeutic recommendations for the use of nutraceuticals in the management of high blood pressure (BP) and hypertension, opening up new perspectives in the field. This not-randomized diet-controlled clinical study aimed to evaluate if daily dietary supplementation with black garlic and pomegranate (namely SelectSIEVE® SlowBeat) could advantageously affect BP in individuals with high-normal BP or stage I hypertension. Enrolled subjects were adhering to a Mediterranean DASH (Dietary Approaches to Stop Hypertension) diet for two weeks before deciding whether to continue following Mediterranean DASH diet alone or in association with SelectSIEVE® SlowBeat. At the end of the study, dietary supplementation with SelectSIEVE® SlowBeat was associated with significant improvement in systolic blood pressure (SBP) and diastolic blood pressure (DBP) compared to baseline (Pre-treatment: SBP = 134.3 ± 4.2 and DBP = 88.2 ± 3.4; 4-Week Follow-up: SBP = 130.1 ± 2.8 and DBP= 83.7 ± 2.6). SBP improved also in comparison with control. In conclusion, the study shows that dietary supplementation with extracts from black garlic and pomegranate safely exert significant improvements in BP in healthy individuals adhering to a Mediterranean DASH diet.

Keywords: black garlic; pomegranate; angiotensin-converting enzyme; DASH diet; blood pressure; hypertension

Citation: Fogacci, F.; Di Micoli, A.; Grandi, E.; Fiorini, G.; Borghi, C.; Cicero, A.F.G. Black Garlic and Pomegranate Standardized Extracts for Blood Pressure Improvement: A Non-Randomized Diet-Controlled Study. *Appl. Sci.* **2022**, *12*, 9673. https://doi.org/10.3390/app12199673

Academic Editors: Theodoros Varzakas and Maria Antoniadou

Received: 31 August 2022
Accepted: 23 September 2022
Published: 26 September 2022

Publisher's Note: MDPI stays neutral with regard to jurisdictional claims in published maps and institutional affiliations.

1. Introduction

High blood pressure (BP) and hypertension substantially contribute to the global burden of atherosclerotic cardiovascular disease (ASCVD) [1], and despite affordable and effective antihypertensive treatments being nowadays available, the rates of hypertension treatment and control remain largely perfectible [2]. Current international guidelines emphasize the importance of following the dietary approach to stop hypertension (DASH) and a low-salt Mediterranean diet to improve BP control [3]. DASH and a low-salt Mediterranean diet are not only characterized by sodium restriction (whose effect is largely mediated by ethnicity, age, and presence of specific comorbidities, such as chronic kidney disease or diabetes), but also by a high intake of vegetables (natural sources of nitric oxide and polyphenols), whole grains and low-fat dairy products, and a low intake of red meat, sugar, and trans-hydrogenated fats [4]. Some position papers recently released by the European Society of Hypertension (ESH) and the Italian Society of Hypertension (SIIA) open up new perspectives in the field, providing therapeutic recommendations also for the use of nutraceuticals in the management of high BP and hypertension [5,6]. However, in addition to nutraceuticals that have been traditionally associated with significant improvement in BP control [7], some other foods and natural compounds have recently shown to improve BP levels. Among them are black garlic and pomegranate [8,9].

Allium sativum L. (garlic) anti-hypertensive properties are mainly related to the production of endothelium-active molecules—namely nitric oxide and hydrogen sulfide (H_2S)—leading to vasodilation [10]. Moreover, garlic is a source of gamma-glutamylcysteine that inhibits the activity of angiotensin-converting enzyme (ACE), reduces the formation of ACE-II, and protects the activity of bradykinin [11].

Punica granatum L. (pomegranate) juice is rich in soluble polyphenols comprising anthocyanins and tannins (e.g., ellagitannins—mainly punicalagin—ellagic acid, gallic acid, and catechins) with antioxidant and anti-inflammatory properties, and recognized biological activities, such as ACE inhibition [12,13].

Starting from these assumptions, we aimed to evaluate if daily dietary supplementation with SelectSIEVE® SlowBeat (SelectSIEVE® SlowBeat was manufactured and kindly provided by Roelmi HPC (Milan, Italy)) could advantageously affect BP in individuals with high-normal BP or stage I hypertension as an add-on lifestyle intervention to DASH diet.

2. Methods

2.1. Study Design and Participants

This not-randomized diet-controlled clinical study involved a sample of Italian free-living individuals recruited between April and June 2021 from the Internal Medicine Division's Hypertension Clinic of the S. Orsola Malpighi University Hospital, Bologna, Italy.

Participants were required to be aged 30–70 years, with high-normal BP (systolic blood pressure (SBP)= 130–139 mmHg and/or diastolic blood pressure (DBP)= 85–89 mmHg) or stage I hypertension (SBP= 140–159 mmHg and/or DBP= 90–99 mmHg) [3], and an estimated 10-year CV risk <5% based on SCORE (Systematic COronary Risk Evaluation) risk charts [14]. Exclusion criteria included diabetes, previous history of ASCVD, obesity (body mass index (BMI) > 30 Kg/m^2), uncontrolled thyroid diseases, history of malignancies, use of medication or dietary supplement affecting BP, alcoholism, pregnancy, and breastfeeding.

Enrolled individuals adhered to a Mediterranean Dietary Approaches to Stop Hypertension (DASH) diet for two weeks before allocation and were clinically evaluated at baseline (week 0) and at weeks 2 and 4. The study timeline is described in detail in Figure 1.

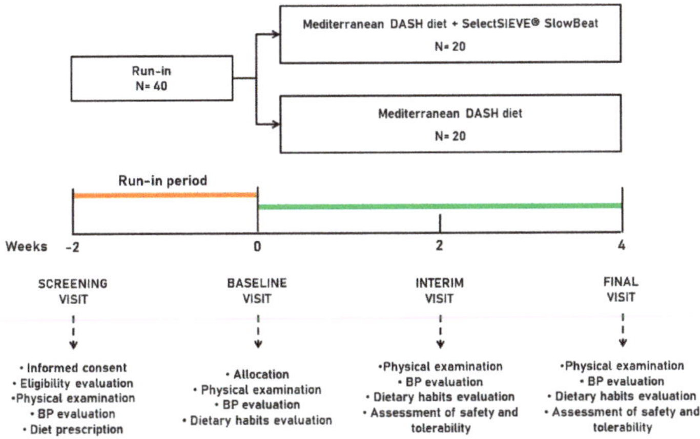

Figure 1. Study timeline. BP = Blood pressure; DASH = Dietary Approaches to Stop Hypertension; N = Number of individuals.

The study fully complied with the ethical principles of the Declaration of Helsinki and later amendments, and with The International Council for Harmonization of Technical Requirements for Registration of Pharmaceuticals for Human Use (ICH) Harmonized Tripartite Guideline for Good Clinical Practice (GCP), and its protocol was approved by the Ethical Committee of the University of Bologna. All patients signed a written informed consent to participate. The registration of the study on ClinicalTrials.gov is not mandatory because of its non-randomized design.

2.2. Dietary Supplementation

After a 2-week period of standardization, enrolled subjects decided whether to continue following the Mediterranean DASH diet alone or in association with SelectSIEVE® SlowBeat, containing black garlic (bulb extract) and pomegranate (fruit extract) as active ingredients, and maltodextrin (from corn) and arabic gum as inactive carriers (Table 1).

Table 1. Quantitative composition of the dietary supplement—namely SelectSIEVE®SlowBeat—used in the clinical study.

Active Ingredients	Quantity Per Capsule
Black garlic (*Allium Sativum* L.)	140 mg
Pomegranate (*Punica Granatum* L.)	60 mg

Patients allocated to SelectSIEVE® SlowBeat were provided with 2 boxes each containing 30 capsules, and were instructed to take 2 capsules of the dietary supplement once daily in the morning, for the entire duration of the study. At the end of the clinical study, all unused capsules were retrieved for inventory and participants' compliance was assessed by counting the number of returned capsules.

The study's product was manufactured and packaged by Roelmi HPC (Milan, Italy) following the Quality Management System ISO 9001:2008 and the European Good Manufacturing Practices (GMP), satisfying requirements in the "Code Of Federal Regulation" title 21,volume 2, part 111.

2.3. Assessments

2.3.1. Clinical Information

Data regarding demographic issues, medical history, allergies, and current medications were collected from volunteers. Validated semi-quantitative questionnaires—including Food Frequency Questionnaire (FFQ)—were used to assess demographic variables, recreational physical activity, and dietary and smoking habits [15].

Height and weight were respectively measured to the nearest 0.1 cm and 0.1 kg, with subjects standing erect with eyes directed straight wearing light clothes and bare feet. Waist circumference (WC) was measured in a horizontal plane at the end of a normal expiration, at the midpoint between the inferior margin of the last rib and the superior iliac crest. Body mass index (BMI) was calculated as body weight in kilograms, divided by height squared in meters (kg/m^2).

2.3.2. Blood Pressure Measurements

Resting SBP and DBP were measured with a validated oscillometric device, in accordance with the recommendations of the "European Guidelines for the management of arterial hypertension" [3]. Three BP readings were obtained. The first reading was discarded and the average between the second and the third reading was collected as a study variable. Pulse pressure (PP) was calculated as the difference between SBP and DBP.

2.3.3. Assessment of Safety and Tolerability

Safety and tolerability were evaluated through a continuous monitoring during the study [16]. A 10-point visual analog scale (VAS) was used to measure patients' acceptability of SelectSIEVE® SlowBeat.

2.4. Statistical Analysis

Kolmogorov–Smirnov test was used to test the normality distribution of the studied variables. Parameters collected at baseline were compared using Levene's test and Student's

T test, and by the χ^2 test and Fisher's exact test. Between-group differences were evaluated by repeated-measures analysis of variance (ANOVA) and Tukey's post hoc test. All data were reported as means and related standard deviations (SD). A two-tailed p level of <0.05 was considered statistically significant for all tests.

Data were analyzed using intention to treat by means of the Statistical Package for Social Science (SPSS) 26.0, version for Windows.

3. Results

3.1. Analysis of Efficacy

A total of 69 volunteers were assessed for eligibility, and 40 individuals (Men: 23; Women: 17) were enrolled and successfully completed the study according to its design (Figure 2).

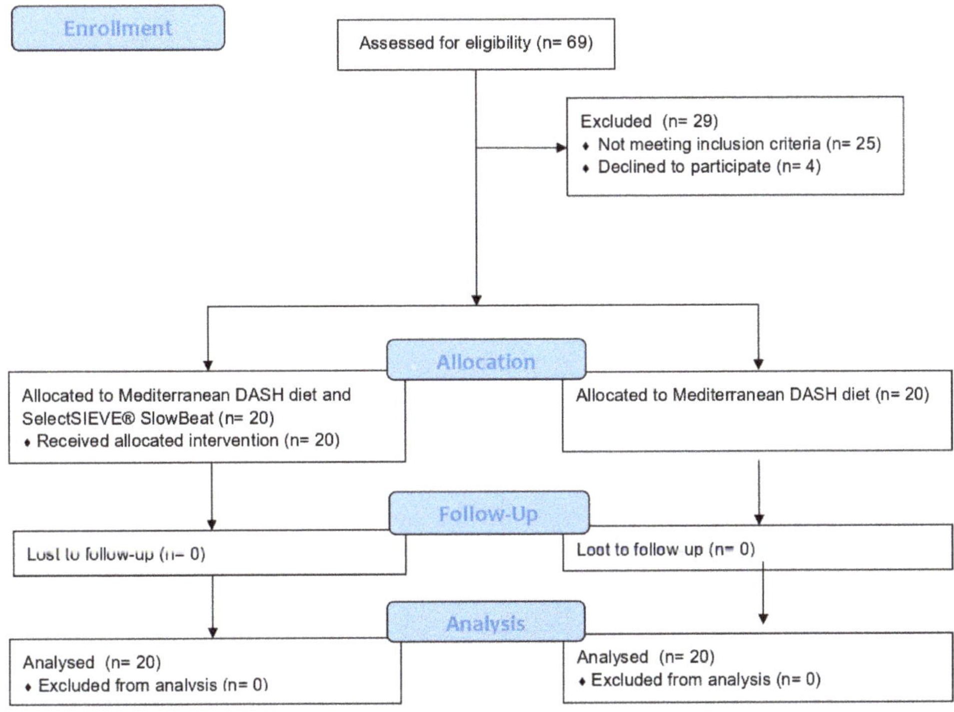

Figure 2. CONSORT (Consolidated Standards of Reporting Trials) flow diagram.

At baseline, the groups were cross-matched in regards to age and the main anthropometric and hemodynamic parameters (Table 2).

Table 2. Anthropometric and hemodynamic parameters from the baseline to the end of the study, reported as means and related standard deviations.

Parameters	Mediterranean DASH Diet + SelectSIEVE® SlowBeat (N. 20)			Mediterranean DASH Diet (N. 20)		
	Baseline	Week 2	Week 4	Baseline	Week 2	Week 4
Age (years)	54.4 ± 3.3			54.9 ± 3.8		
WC (cm)	84.5 ± 6.9	84.2 ± 6.8	84.3 ± 6.6	84.8 ± 6.1	84.4 ± 6.8	84.5 ± 6.7
BMI (kg/m^2)	25.2 ± 1.6	25.1 ± 1.7	25.3 ± 1.8	25.3 ± 1.9	25.2 ± 1.8	25.4 ± 1.6
SBP (mmHg)	134.3 ± 4.2	131.4 ± 3.9	130.1 ± 2.8 *§	135.4 ± 3.9	132.9 ± 5.6	133.5 ± 4.3
DBP (mmHg)	88.2 ± 3.4	84.4 ± 4.1	83.7 ± 2.6 *	87.4 ± 4	85.1 ± 4.8	84.4 ± 4.3
PP (mmHg)	46.4 ± 2.2	47.1 ± 1.9	46.4 ± 2.2	48.0 ± 2.9	48.2 ± 3.7	47.3 ± 2.4
HR (bpm)	68.3 ± 4.1	69.4 ± 4.6	67.2 ± 5.6	66.3 ± 4.7	68.1 ± 3.4	65.2 ± 4.9

* $p < 0.05$ versus baseline; § $p < 0.05$ versus control. BMI = Body mass index; DASH = Dietary Approaches to Stop Hypertension; DBP = Diastolic blood pressure; HR = Heart rate; PP = Pulse pressure; SBP = Systolic blood pressure; WC = Waist circumference.

At the end of the study, dietary supplementation with SelectSIEVE® SlowBeat resulted in significant improvements in SBP and DBP compared to baseline. SBP improved also in comparison with control (Table 3).

Table 3. Diet composition at enrollment and at the end of the study. Values are expressed as mean ± SD.

Parameters	Baseline	Week 4
Total energy (Kcal/day)	1724 ± 142	1718 ± 157
Carbohydrates(% of total energy)	56.7 ± 3.4	55.9 ± 4.6
Proteins (% of total energy)	18.4 ± 3.2	17.7 ± 3.8
Animal protein (% of total energy)	12.7 ± 1.1	12.4 ± 1.3
Vegetal protein (% of total energy)	5.6 ± 0.6	5.3 ± 1.2
Total fats (% of total energy)	24.9 ± 5.4	26.1 ± 1.8
Saturated fatty acids (% of total energy)	7.1 ± 0.2	7.6 ± 0.1
Monounsaturated fatty acids (% of total energy)	12.9 ± 0.8	13.2 ± 1.4
Polyunsaturated fatty acids (% of total energy)	4.8 ± 0.4	5.3 ± 1.7
Total dietary fibers (g/day)	26.4 ± 2.6	27.7 ± 4.1
Cholesterol (mg/day)	190.9 ± 14.2	189.8 ± 15.4

No statistically significant changes were recorded in dietary habits during the study with any changes in total energy and macronutrient intake (Table 3).

3.2. Analysis of Safety

All participants completed the clinical study according to its design (dropout rate = 0%). No treatment-emergent adverse events were reported during the study. Patients' acceptability of SelectSIEVE® SlowBeat was good and comparable to that of standard care.

4. Discussion

During the last decades, CV epidemiology has focused the attention on normal-high BP and stage I hypertension as an emergent public health issue. In effect, according to the most recent observations, the prevalence of these conditions is about 30% in the general adult population, and they have been clearly associated with an increased risk of developing hypertension (8–20%) [17], hypertension-related target organ damage [18,19], and ASCVD [20], whereas reduced BP associates a decreased risk of CV complications [1].

The most recently released International Guidelines suggest to firstly manage normal-high BP values and stage I hypertension by increasing weekly aerobic physical activity and reducing the overall energy intake with the ultimate aim to maintain the BMI between 20 and 25 kg/m^2 (or keeping the waist size to less than 94 cm for men and 80 cm for women) [5]. Other recommendations from ESH include the urgency to reduce the daily intake of alcohol, salt (<5 gr/day), and processed meat, by increasing the intake of vegetables, olive oil, and low-fat dairy products, and refraining from passive and active cigarette smoking [5].

Even though the aforementioned lifestyle measures are undoubtedly effective, they are not always sufficient to manage BP levels, even in low-risk individuals. For this reason, during the last years, pre-clinical and clinical investigators have been searching for natural compounds with a detectable BP lowering effect in humans [21]. Therefore, a constantly growing body of evidence supports the use of a number of nutraceuticals in reducing BP, with some specificities relating to different mechanisms of action and pleiotropic effects [21].

Previous studies on black garlic indicated a food with higher levels of reducing sugars, organic acids, and bioactive compounds (particularly S-allyl-cysteine and coumaric acid) than fresh garlic [22]. In effect, the consumption of aged black garlic has been recently associated to clinically detectable improvements in a number of CV risk factors, as long as the extent of these effects has been shown to be dependent on the garlic aging process and the amount and type of accumulated chemical compounds [23]. Recently, a randomized double blind crossover clinical study involving a sample of 67 hypercholesterolemic individuals showed that the consumption of 250 mg aged black garlic extract with 1.25 mg S-allyl-L-cysteine was able to significantly decrease DBP after 6 weeks of daily dietary supplementation [8]. As suggested by the authors, the mechanisms of action by which aged black garlic could modulate BP involve antioxidant activity by organosulfur compounds, the regulation of transcription factors involved in hypertension, and ACE regulation [24]. In this context, animal studies had already shown that black garlic extracts and their bioactive compounds have a greater inhibitory effect on endothelial ACE activity than normal garlic extracts (88.8% versus 52.7%) [25].

In the latest years, pomegranate has also gained widespread popularity as a functional food and nutraceutical source, and its consumption has been studied in relation to a variety of chronic diseases including CV diseases [26]. A systematic review and meta-analysis of randomized clinical studies suggested consistent benefits of pomegranate juice consumption on both SBP and DBP, with an effect on SBP that was independent from the duration of dietary supplementation [9]. Moreover, pomegranate juice was shown to significantly reduce intensity, occurrence, and duration of angina pectoris in patients with unstable angina [27].

In our study, combined dietary supplementation with aged black garlic and pomegranate extracts was associated with significant improvements in SBP and DBP. These findings are particularly interesting, since the treatment-related decreases in BP levels were experienced by individuals who were already following a Mediterranean DASH diet, that represents the gold-standard treatment for patients with high-normal BP and the first-stage treatment for patients with stage I hypertension and low-to-intermediate ASCVD risk, as collegially recommended by the European Society of Cardiology (ESC) and ESH [3]. Moreover, the BP-lowering effect that dietary supplementation with aged black garlic and pomegranate extracts can bring about is clinically relevant in addition to being statistically significant, since previous evidence showed that prolonged reduction in SBP also by 2 mmHg result in a 7% reduced incidence of death secondary to stroke and in a 10% reduced incidence of death secondary to other vascular etiology [28].

Certainly, some limitations need to be acknowledged, such as the small sample size, the short follow-up and—definitely—the pilot and non-randomized nature of the study design. Moreover, the lack of a placebo group did not allow judgement of nocebo and placebo effects. On the other side, the enrolled individuals maintained a similar dietary pattern during the study, so that we have no reason to think that the observed hemodynamic effects were due to changes in food consumption and nutrient intake rather than SelectSIEVE®

SlowBeat. Additionally, this study design simulates the clinical practice setting, where these kinds of patients are usually managed in a less intensive way than in a standard randomized clinical trial. However, more research is needed that focuses on the underlying reasons and mechanisms of the effects observed during the study. Furthermore, it remains to be assessed whether combined dietary supplementation with aged black garlic and pomegranate extracts could also improve vascular aging, by reducing arterial stiffness and increasing endothelial reactivity.

The relevance of our findings lies in the fact that an increased consumption of food bioactive compounds configures as a well-tolerated and effective strategy to improve BP control. Moreover, this result has to be interpreted in light of previous evidence showing that effective BP control can only be achieved through a comprehensive strategy of prevention at the individual and population levels, with the use of non-pharmacological interventions throughout the life course [29].

5. Conclusions

Finally, the study shows that dietary supplementation with extracts from black garlic and pomegranate safely exert significant improvements in BP in healthy individuals adhering to a Mediterranean DASH diet.

Author Contributions: Conceptualization, A.F.G.C.; methodology, A.F.G.C. and F.F.; software, A.F.G.C.; formal analysis, A.F.G.C.; investigation, A.F.G.C., F.F., A.D.M., E.G. and G.F.; data curation, F.F., A.D.M., E.G. and G.F.; writing—original draft preparation, A.F.G.C. and F.F.; writing—review and editing, A.D.M., E.G., G.F. and C.B.; supervision, C.B.; project administration, A.F.G.C.; funding acquisition, A.F.G.C. All authors have read and agreed to the published version of the manuscript.

Funding: This research received no external funding.

Institutional Review Board Statement: The study (BP_open2021) was approved by the Local Institutional Review Board (Code: LLD-RP2018) and was conducted following the Declaration of Helsinki and later amendments.

Informed Consent Statement: Informed consent was obtained from all volunteers involved in the study.

Data Availability Statement: Data supporting the study's findings are available upon request from the corresponding author with the permission of the University of Bologna.

Conflicts of Interest: The authors declare no conflict of interest.

References

1. Del Pinto, R.; Grassi, G.; Muiesan, M.L.; Borghi, C.; Carugo, S.; Cicero, A.F.G.; Di Meo, L.; Iaccarino, G.; Minuz, P.; Mulatero, P.; et al. World Hypertension Day 2021 in Italy: Results of a Nationwide Survey. *High Blood Press. Cardiovasc. Prev.* **2022**, *29*, 353–359. [CrossRef]
2. Borghi, C.; Fogacci, F.; Agnoletti, D.; Cicero, A.F.G. Hypertension and Dyslipidemia Combined Therapeutic Approaches. *High Blood Press. Cardiovasc. Prev.* **2022**, *29*, 221–230. [CrossRef] [PubMed]
3. Williams, B.; Mancia, G.; Spiering, W.; Agabiti Rosei, E.; Azizi, M.; Burnier, M.; Clement, D.; Coca, A.; De Simone, G.; Dominiczak, A.; et al. List of authors/Task Force members:. 2018 Practice Guidelines for the management of arterial hypertension of the European Society of Hypertension and the European Society of Cardiology: ESH/ESC Task Force for the Management of Arterial Hypertension. *J. Hypertens.* **2018**, *36*, 2284–2309. [CrossRef] [PubMed]
4. Cicero, A.F.G.; Veronesi, M.; Fogacci, F. Dietary Intervention to Improve Blood Pressure Control: Beyond Salt Restriction. *High Blood Press. Cardiovasc. Prev.* **2021**, *28*, 547–553. [CrossRef] [PubMed]
5. Borghi, C.; Tsioufis, K.; Agabiti Rosei, E.; Burnier, M.; Cicero, A.F.G.; Clement, D.; Coca, A.; Desideri, G.; Grassi, G.; Lovic, D.; et al. Nutraceuticals and blood pressure control: A European Society of Hypertension position document. *J. Hypertens.* **2020**, *38*, 799–812. [CrossRef]
6. Cicero, A.F.G.; Grassi, D.; Tocci, G.; Galletti, F.; Borghi, C.; Ferri, C. Nutrients and Nutraceuticals for the Management of High Normal Blood Pressure: An Evidence-Based Consensus Document. *High Blood Press. Cardiovasc. Prev.* **2019**, *26*, 9–25. [CrossRef]
7. Cicero, A.F.G.; Fogacci, F.; Colletti, A. Food and plant bioactives for reducing cardiometabolic disease risk: An evidence based approach. *Food Funct.* **2017**, *8*, 2076–2088. [CrossRef]

8.   Valls, R.M.; Companys, J.; Calderón-Pérez, L.; Salamanca, P.; Pla-Pagà, L.; Sandoval-Ramírez, B.A.; Bueno, A.; Puzo, J.; Crescenti, A.; Bas, J.M.D.; et al. Effects of an Optimized Aged Garlic Extract on Cardiovascular Disease Risk Factors in Moderate Hypercholesterolemic Subjects: A Randomized, Crossover, Double-Blind, Sustained and Controlled Study. *Nutrients* **2022**, *14*, 405. [CrossRef]

9.   Sahebkar, A.; Ferri, C.; Giorgini, P.; Bo, S.; Nachtigal, P.; Grassi, D. Effects of pomegranate juice on blood pressure: A systematic review and meta-analysis of randomized controlled trials. *Pharmacol. Res.* **2017**, *115*, 149–161. [CrossRef]

10.   Zadhoush, R.; Alavi-Naeini, A.; Feizi, A.; Naghshineh, E.; Ghazvini, M.R. The effect of garlic (*Allium sativum*) supplementation on the lipid parameters and blood pressure levels in women with polycystic ovary syndrome: A randomized controlled trial. *Phytother. Res.* **2021**, *35*, 6335–6342. [CrossRef]

11.   Gao, X.; Xue, Z.; Ma, Q.; Guo, Q.; Xing, L.; Santhanam, R.K.; Zhang, M.; Chen, H. Antioxidant and antihypertensive effects of garlic protein and its hydrolysates and the related mechanism. *J. Food Biochem.* **2020**, *44*, e13126. [CrossRef] [PubMed]

12.   Sohrab, G.; Roshan, H.; Ebrahimof, S.; Nikpayam, O.; Sotoudeh, G.; Siasi, F. Effects of pomegranate juice consumption on blood pressure and lipid profile in patients with type 2 diabetes: A single-blind randomized clinical trial. *Clin. Nutr. ESPEN* **2019**, *29*, 30–35. [CrossRef] [PubMed]

13.   Stowe, C.B. The effects of pomegranate juice consumption on blood pressure and cardiovascular health. *Complement. Ther. Clin. Pract.* **2011**, *17*, 113–115. [CrossRef] [PubMed]

14.   Authors/Task Force Members; ESC Committee for Practice Guidelines (CPG); ESC National Cardiac Societies. 2019 ESC/EAS guidelines for the management of dyslipidaemias: Lipid modification to reduce cardiovascular risk. *Atherosclerosis* **2019**, *290*, 140–205. [CrossRef]

15.   Fogacci, F.; Rizzoli, E.; Giovannini, M.; Bove, M.; D'Addato, S.; Borghi, C.; Cicero, A.F.G. Effect of Dietary Supplementation with Eufortyn®Colesterolo Plus on Serum Lipids, Endothelial Reactivity, Indexes of Non-Alcoholic Fatty Liver Disease and Systemic Inflammation in Healthy Subjects with Polygenic Hypercholesterolemia: The ANEMONE Study. *Nutrients* **2022**, *14*, 2099. [CrossRef]

16.   Cicero, A.F.G.; Fogacci, F.; Veronesi, M.; Strocchi, E.; Grandi, E.; Rizzoli, E.; Poli, A.; Marangoni, F.; Borghi, C. A randomized Placebo-Controlled Clinical Trial to Evaluate the Medium-Term Effects of Oat Fibers on Human Health: The Beta-Glucan Effects on Lipid Profile, Glycemia and inTestinal Health (BELT) Study. *Nutrients* **2020**, *12*, 686. [CrossRef]

17.   Vasan, R.S.; Larson, M.G.; Leip, E.P.; Evans, J.C.; O'Donnell, C.J.; Kannel, W.B.; Levy, D. Impact of high-normal blood pressure and the risk of cardiovascular disease. *N. Engl. J. Med.* **2001**, *345*, 1291–1297. [CrossRef]

18.   Cuspidi, C.; Sala, C.; Tadic, M.; Gherbesi, E.; Facchetti, R.; Grassi, G.; Mancia, G. High-normal blood pressure and abnormal left ventricular geometric patterns: A meta-analysis. *J. Hypertens.* **2019**, *37*, 1312–1319. [CrossRef]

19.   Cuspidi, C.; Sala, C.; Tadic, M.; Gherbesi, E.; Grassi, G.; Mancia, G. Pre-hypertension and subclinical carotid damage: A meta-analysis. *J. Hum. Hypertens.* **2019**, *33*, 34–40. [CrossRef]

20.   Egan, B.M.; Stevens-Fabry, S. Prehypertension-prevalence, health risks, and management strategies. *Nat. Rev. Cardiol.* **2015**, *12*, 289–300. [CrossRef]

21.   Strilchuk, L.; Cincione, R.I.; Fogacci, F.; Cicero, A.F.G. Dietary interventions in blood pressure lowering: Current evidence in 2020. *Kardiol. Pol.* **2020**, *78*, 659–666. [CrossRef] [PubMed]

22.   Imaizumi, V.M.; Laurindo, L.F.; Manzan, B.; Guiguer, E.L.; Oshiiwa, M.; Otoboni, A.M.M.B.; Araujo, A.C.; Tofano, R.J.; Barbalho, S.M. Garlic: A systematic review of the effects on cardiovascular diseases. *Crit. Rev. Food Sci. Nutr.* **2022**, *Online ahead of print.* [CrossRef] [PubMed]

23.   Kimura, S.; Tung, Y.C.; Pan, M.H.; Su, N.W.; Lai, Y.J.; Cheng, K.C. Black garlic: A critical review of its production, bioactivity, and application. *J. Food Drug Anal.* **2017**, *25*, 62–70. [CrossRef] [PubMed]

24.   Ried, K.; Fakler, P. Potential of garlic (*Allium sativum*) in lowering high blood pressure: Mechanisms of action and clinical relevance. *Integr. Blood Press. Control* **2014**, *7*, 71–82. [CrossRef]

25.   Ahmed, T.; Wang, C.K. Black Garlic and Its Bioactive Compounds on Human Health Diseases: A Review. *Molecules* **2021**, *26*, 5028. [CrossRef]

26.   Laurindo, L.F.; Barbalho, S.M.; Marquess, A.R.; Grecco, A.I.S.; Goulart, R.A.; Tofano, R.J.; Bishayee, A. Pomegranate (*Punica granatum* L.) and Metabolic Syndrome Risk Factors and Outcomes: A Systematic Review of Clinical Studies. *Nutrients* **2022**, *14*, 1665. [CrossRef]

27.   Razani, Z.; Dastani, M.; Kazerani, H.R. Cardioprotective Effects of Pomegranate (*Punica granatum*) Juice in Patients with Ischemic Heart Disease. *Phytother. Res.* **2017**, *31*, 1731–1738. [CrossRef]

28.   Lewington, S.; Clarke, R.; Qizilbash, N.; Peto, R.; Collins, R.; Prospective Studies Collaboration. Age-specific relevance of usual blood pressure to vascular mortality: A meta-analysis of individual data for one million adults in 61 prospective studies. *Lancet* **2002**, *360*, 1903–1913. [CrossRef]

29.   Zhou, B.; Perel, P.; Mensah, G.A.; Ezzati, M. Global epidemiology, health burden and effective interventions for elevated blood pressure and hypertension. *Nat. Rev. Cardiol.* **2021**, *18*, 785–802. [CrossRef]

Article

Dental Health, Caries Perception and Sense of Discrimination among Migrants and Refugees in Europe: Results from the Mig-HealthCare Project

Pania Karnaki [1,*], Konstantinos Katsas [1], Dimitrios V. Diamantis [1], Elena Riza [2], Maya Simona Rosen [3], Maria Antoniadou [4], Alejandro Gil-Salmerón [5], Igor Grabovac [6] and Athena Linou [1]

[1] Institute of Preventive Medicine, Environmental and Occupational Health, Prolepsis, 15121 Marousi, Greece
[2] Department of Hygiene, Epidemiology and Medical Statistics, Medical School, National Kapodistrian University of Athens, 11527 Athens, Greece
[3] Department of History of Science, Faculty of Arts and Sciences, Harvard College, Cambridge, MA 02138, USA
[4] Dental School, National and Kapodistrian University of Athens, 11527 Athens, Greece
[5] International Foundation for Integrated Care (IFIC), Wolfson College, Linton Rd., Oxford OX2 6UD, UK
[6] Department of Social and Preventive Medicine, Centre for Public Health, Medical University of Vienna, Kinderspitalgasse 15, 1090 Vienna, Austria
* Correspondence: p.karnaki@prolepsis.gr; Tel.: +30-6973094622

Featured Application: Strategic planning of diet and dental caries reduction programs in asylum settings for migrants and refugees.

Citation: Karnaki, P.; Katsas, K.; Diamantis, D.V.; Riza, E.; Rosen, M.S.; Antoniadou, M.; Gil-Salmerón, A.; Grabovac, I.; Linou, A. Dental Health, Caries Perception and Sense of Discrimination among Migrants and Refugees in Europe: Results from the Mig-HealthCare Project. *Appl. Sci.* **2022**, *12*, 9294. https://doi.org/10.3390/app12189294

Academic Editors: Gabriele Cervino and Andrea Scribante

Received: 19 July 2022
Accepted: 14 September 2022
Published: 16 September 2022

Publisher's Note: MDPI stays neutral with regard to jurisdictional claims in published maps and institutional affiliations.

Abstract: Dental and oral health are considered among the main health issues for migrants and refugees, as access to dental health care services is often expensive and difficult. The study investigates dental and oral health determinants among migrants and refugees in 10 European countries (Austria, Bulgaria, Cyprus, France, Germany, Greece, Italy, Malta, Spain, and Sweden), examining how mental health, legal status, discrimination issues and dental services' use frequency affect dental health. Methods: A cross sectional study using a purpose-made questionnaire was carried out to assess health status and access, with a dedicated section to measure self-perceived dental health, prevalence of caries, last visit to dentist and anticipated access to dental health services. Multivariable logistic regression models were performed to investigate the impact of quality of life, discrimination, immigration status, and other demographic factors on dental health. Results: About half of the sample suffered from poor dental condition and 22% had never visited a dentist. Migrants with higher educational levels had higher odds of having good dental health (OR = 1.08; 95%CI (1.03, 1.12)) and brushing their teeth daily (OR = 1.1; 95%CI (1.04, 1.17)). Higher general and mental health scores were indicative of better dental condition (general health: OR = 1.02; 95%CI (1.01, 1.03); mental health: OR = 1.01; 95%CI (1.004, 1.02)) and higher probability of daily teeth brushing (general health: OR = 1.02; 95%CI (1.01, 1.03); mental health: OR = 1.02; 95%CI (1.01, 1.03)). The possession of any kind of legal immigration permission and not having any children showed similar results. Age and discrimination were correlated with decreased likelihood for good dental conditions. Gender was correlated with daily teeth brushing, as female migrants had higher odds of brushing their teeth daily. Conclusions: Many migrants report poor dental health. Nonetheless, migrants with higher education levels, legal immigration status, better general and mental health, no children, lower sense of discrimination, younger age, and regular dental visits were positively correlated to good dental health (perceived as no dental caries).

Keywords: dental caries; diet; food habits; dental health policy; minority groups; vulnerable populations; migrants; refugees; self-rated oral health; Mid-HealthCare project

1. Introduction

Peaking in 2015 and known since as the European refugee crisis, millions of refugees have fled from persecution and war-torn countries to the European continent [1–3]. Refugee flows have been continuous and although the COVID-19 pandemic corresponded to a decrease in arrivals initially, in the beginning of August 2021, arrival numbers once again surged. By the end of 2021, a total of about 10,000 migrants had arrived in Europe in a period of nine months [3]. The war in the Ukraine in the beginning of February 2022, also created an unexpected increase in migrant numbers. One month later, the European Union (EU) faced its most significant refugee crisis since World War II, with more than 10 million people fleeing their homes, 6.5 million displaced within Ukraine and 3.9 million escaping to neighboring countries [4].

Many refugees who have arrived in Europe since 2015 have struggled obtaining asylum or other forms of legal permission [5,6]. Consequently, immigration status insecurity often leads to discrimination [7] and affects access to healthcare services [7–10]. Most of the time, migrants lack pre-departure orientation, the so-called cultural orientation, to help them with a smooth transition [11]. As a result, migrants often face food and housing issues [12,13]. Oral health is a complex process influenced by multiple and interrelated factors. A multitude of factors related to country of origin, urban/rural residence, socio-economic and cultural factors, educational level, racism, sexism, discrimination issues and economic situation in the host country affect oral health outcomes [14–16]. In addition, dietary patterns have shown to negatively affect oral health. For example, studies in elderly populations have shown that diet and oral health coaching can empower the prevention and management of oral diseases and can improve the level of oral health [17,18]. Because of the high cost of dental services, migrants are often in high need of oral health care services and are at a disadvantage when it comes to accessing these services [11,12]. A literature search on the oral health status of migrants related to quality of life (OHRQoL) revealed the dual result of both better and worse dental and overall oral health in migrants as compared to the host populations [19]. Other research has shown definitive negative results that oral health status of recently arrived migrants is inadequate and can pose challenges to the national healthcare system, especially for those without asylum or a permit to stay [11,12,14]. Given that migrants often face limited financial resources and knowledge of the country's healthcare system, in conjunction with food and housing insecurity, limited access to dental care services is a common experience [11]. Furthermore, limited access to quality food and dental services engenders extreme vulnerabilities to dental caries and other oral diseases among migrants [11,12,15].

Furthermore, the emotional and socioeconomic burden of forced migration can substantially impact the quality of life of migrants [19–21], and consequently deteriorate their general [22] and dental health status across all age groups [16,19–26]. Access to dentists or dental health providers is crucial for maintaining good dental health. However, because dental health services are often not included under universal health coverage, unique financial barriers affect migrants who often cannot afford the cost of dental services [27,28]. Thus, health care provision for migrants including oral health is becoming a growing concern for policy makers and researchers, as part of the Universal Health Coverage target 3.8 in Goal 3 of the Sustainable Development Goals [29]. In addition, migrants who have not reached their final destination or are stranded in precarious circumstances (camps, temporary settlements) face difficulties in maintaining their oral hygiene and often visit a dentist only when problems are acute or when they are in pain [30]. Finally, there is a lack of language and culturally sensitive information to share with migrants on oral hygiene measures, new products and relevant diet and oral health habits that exist in the host country [31].

Mig-Healthcare was a 3-year project, co-funded by the European Commission, that aimed to define the elements of best practice to help the health of migrants and refugees at the community level and to develop tools that can assist in this process [30]. On the Mig-Healthcare website (www.mighealthcare.eu, accessed on 9 June 2022) [32], there are tools and applications, including a step-by-step logical plan, that can support health professionals in delivering quality health care to migrants and refugees. The Mig-HealthCare project identified the main health issues of concern to migrants and refugees, which included dental care, and created outlines of the steps required to maintain good oral hygiene. Included in this information are ways to distinguish different dental issues. This resource has patient-empowering potential as it can lead to quicker, more efficient treatment, as well as preventative potential to reduce dietary-induced dental damage. In general, the Mig-HealthCare project aimed to reduce health inequalities and improve the health care services for migrants and refugees through research and the development of tools to facilitate the implementation of community-based care models for basic diseases, such as dental caries [9].

The present study aims to explore dental and oral health among migrants/refugees in 10 European countries (Austria, Bulgaria, Cyprus, France, Germany, Greece, Italy, Malta, Spain, and Sweden), while also examining how various sociodemographic factors, as well as mental health, immigration, discrimination, and dental service indicators, affect dental health. Implications for policymakers and health professionals are discussed.

2. Materials and Methods

2.1. Design and Sample

For this cross-sectional study, 1407 participants were recruited (N_0 = 1407) using a snowball sampling method, whereby the recruited participants are asked to identify and recruit additional participants. The eligibility criteria for recruitment were as follows: must be at least 18 years of age, have resided in the country of interview for 6 months to 5 years, be able to provide consent and attest to understanding the project goals. Participants were asked to complete a 60-item printed purpose-made questionnaire that assessed demographic and health characteristics. Participation was completely voluntary. The questionnaire was translated by official translators into migrant languages. It was then pilot tested in a sample of 10 migrants per language before the initial launch of the questionnaire to assess its originality and reliability in understanding the questions posed in order to give an answer. The focus of the pilot study was to identify potential unclear or confusing wordings that could lead to possible misunderstandings, as well as to measure the time for the questionnaire's administration. The questionnaire was translated into Arabic, Farsi, Dari, Pashto, Somali (consensus between partners based on the most frequent migrant languages spoken in each country), as well as the languages of the partner countries. In each interview setting, a collaborator from the partner country was present along with an interpreter when necessary. All study interviewers were recruited by each partner and received training on the questionnaire prior to the initiation of data collection. In cases where the study participants were able and chose to communicate in the host country's language, the services of an interpreter were not required. In each setting, a special area ensuring privacy was allocated for questionnaire completion.

Participants were first recruited in the 10 Mig-Healthcare project partner countries from reception centers, primary healthcare units and welfare offices from April 2018 to September 2019. The participating countries included Austria, Bulgaria, the Republic of Cyprus, France, Germany, Greece, Italy, Malta, Spain, and Sweden. However, three countries (France, Germany, and Malta) were excluded from the final sample due to insufficient sample size, resulting in a total of 1294 migrants (N_1 = 1294). Ethical review was provided by the National and Kapodistrian University of Athens, Medical School (No. 1718034664), with additional ethical approvals obtained as needed by partner organizations (University of Valencia, French School of Public Health, University of Uppsala). Data collection took place from April 2018 to September 2019, with no identifiable personal data collected. Each participant was assigned an anonymous identity that was available to the main researchers. Further information about the methodology and other published results from the Mig-HealthCare project are also reported in previous publications [7–9].

2.2. Measures

2.2.1. Questionnaire Development and Description

The study questionnaire comprised 60 questions. These questions were presented in 12 sections, including demographics, household, education and employment, access and interaction with healthcare services, screening, dental care, immunization status and perceptions about health Discrimination.

Sociodemographic Measurements

To better understand the study sample, the following sociodemographic characteristics were collected: age, gender, country of origin, country of interview, education level, marital status, immigration status, parental status and fluency in the language of the country of interview.

Discrimination in Medical Settings (DMS)

Discrimination experienced in medical settings was assessed through the Discrimination Scale in Medical Settings, which is based on the article by Peek et al. (2011) (DMS scale) [33]. DMS assessment was based on answers to the following questions: "when getting healthcare of any kind, have you ever had any of the following things happen to you? 1) you are treated with less courtesy than other people, (2) you are treated with less respect than other people, (3) you receive poorer service than others, (4) a doctor or nurse acts as if they think you are not smart, (5) a doctor or nurse acts as if they are afraid of you, (6) a doctor or nurse acts as if they are better than you, and (7) you feel like a doctor or nurse is not listening to what you were saying." Five answer choices were given with each assigned a numeric score (1 = never, 2 = rarely, 3 = sometimes, 4 = most of the time, and 5 = always). The mean of all seven questions was found to assess perceived discrimination, with scores closer to 5 speaking for higher perceived discrimination. The reliability of the DMS scale was demonstrated by Pearson correlations between the DMS score and its seven component items, which were positive and larger than 0.3, and the diagonal Cronbach's results, which scored more than 0.9.

Mental and General Health

Mental and general health were assessed using the mental and general health scale of the Short Form 36 (SF-36) [34], which has been translated into many languages, including the consortium country languages. The SF-36 is a tool to measure health based on 36 questions and 8 domains. Regarding mental health from SF-36, the following five questions were used with scoring from 0 (low) to 100 (high): "Have you been a very nervous person? Have you felt so down in the dumps that nothing could cheer you up? Have you felt calm and peaceful? Have you felt downhearted and blue? Have you been a happy person?". Answer choices were assigned a numeric score of 1 (all of the time) to 6 (none of the time), which were recoded to values of 0, 20, 40, 60, 80 and 100, respectively. Regarding general health from SF-36, the following five questions were used with scoring from 0 (low) to 100 (high): "In general, would you say your general health is excellent/very good/good/fair/poor, I seem to get sick a little easier than other people, I am as healthy as anybody I know, I expect my health to get worse, My health is excellent". Answer choices were assigned a numeric score of 1 (definitely true/poor), 2 (mostly true/fair), 3 (do not know/good), 4 (mostly false/very good), and 5 (definitely false/excellent), which were recoded to values of 0, 25, 50, 75 and 100, respectively. After inversing the score of some questions, the recoded values were summed, and the mean was taken as the score. The reliability of the general and mental health scale was proven by Pearson correlations between the general health score and its five component items and mental health score and its five component items, which were positive and larger than 0.3 and total Cronbach's was also acceptable in both scales (>0.7).

Dental Healthcare Access and Dental Health Perception

Dental healthcare services access was measured through 4 questions. We asked respondents to choose one of the two following statements: "I know where to go when I need a dentist" and "I don't know where to go when I need a dentist." Dental screening frequency was measured with the question of "last visit to a dentist," and respondents were given the following response options: "Never," ">5 years ago," "2–5 years ago," "1–2 years ago," "<12 months ago," and "Don't remember." Perception of dental health was measured by self-assessment of dental health with the following response options: "Poor," "Fair," "Good," "Very Good," and "Excellent." For subsequent analyses, dental health status was coded into the following two categories: "Good dental condition" and "Not good dental condition." "Good dental condition" incorporated migrants who reported "Excellent," "Very Good," and "Good" perceptions of their dental health, whereas "Not Good dental condition" incorporated respondents with "Poor" and "Fair" dental health perceptions. The following three subsequent statements were presented, followed by a yes or no response: "Brush my teeth every day," "I know where to go when I need a dentist," "I have caries" to assess dental health condition and dental healthcare access.

2.2.2. Statistical Analysis

We performed a descriptive analysis for all questionnaire variables. We then ran multivariable logistic regression models to investigate the impact of sociodemographic parameters, sense of discrimination and presence of any kind of permission on perceived dental health in the migrant population of several EU countries in the framework of the Mig-HealthCare European project. In more detail, we ran two multivariable logistic regression models, one with the dependent variable of the perception of dental health ("Good dental condition" and "Not good dental condition") and the second with frequency of brushing their teeth ("daily" and "not every day"). We inserted mental health, general health, sense of discrimination, presence of any kind of permission and sociodemographic parameters in each model as independent variables to investigate their impact in our dependent variables, each adjusted for the others. We added general and mental health in separate models due to collinearity.

For the data analysis, the statistical package for social sciences (IBM SPSS, Chicago) version 20.0 was used and a p value of ≤ 0.05 was regarded as statistically significant. Statistical tests, such as descriptive, chi-square test, Spearman's correlation, and binomial logistic regression, were used.

3. Results

The demographic characteristics of the sample are presented in Table 1.

Table 1. Characteristics of migrants and refugees, except those from Germany, France and Malta (N = 1294).

Country of Interview	N (%)
Austria	126 (9.74%)
Bulgaria	226 (17.47%)
Cyprus	110 (8.5%)
Greece	255 (19.71%)
Italy	271 (20.94%)
Spain	202 (15.61%)
Sweden	104 (8.04%)
Country of origin	
Afghanistan	187 (14.55%)
Iraq	122 (9.49%)
Nigeria	115 (8.95%)
Syria	281 (21.87%)
Other	580 (45.14%)

Table 1. *Cont.*

Country of Interview	N (%)
Gender (males)	816 (63.26%)
Have at least one child	535 (50.71%)
Marital status	
Single	598 (46.36%)
Engaged/married	581 (45.04%)
Separated	55 (4.26%)
Widowed	56 (4.34%)
Possession of any kind of permission (asylum or other kind)	768 (66.4%)
Age (years) (mean $\pm$ SD)	32 $\pm$ 11
Educational level (years) (mean $\pm$ SD)	8.9 $\pm$ 5.1

Countries of origin included mainly Afghanistan, Iraq, Syria and Nigeria. Countries clustered under 'other' included Iran, Venezuela, Somalia, Gambia and other North African countries.

Dental health and frequency of dental health screening are presented in Table 2. The following statement can be concluded from the results: "A total of 44.4% of the respondents reported their dental health as poor or fair", while one out of four had never visited a dentist (22.13%).

Table 2. Dental health and screening frequency of migrants and refugees, except those from Germany, France and Malta (*N* = 1294).

Dental/Teeth Condition	N (%)
Poor	212 (16.72%)
Fair	351 (27.68%)
Good	368 (29.02%)
Very good	209 (16.48%)
Excellent	128 (10.09%)
Last visit to a dentist	
Never	279 (22.13%)
>5 years ago	104 (8.25%)
2–5 years ago	148 (11.74%)
1–2 years ago	226 (17.92%)
<12 months ago	352 (27.91%)
Do not remember	152 (12.05%)
Brush my teeth every day (yes)	1135 (90.15%)
I know where to go when I need a dentist (yes)	894 (70.45%)
With caries (bad teeth)	159 (12.29%)

Migrants with good dental condition reported significantly higher educational levels, better general and mental health, lower age, lower sense of discrimination and better access to dentists (Table 3). The possession of any kind of permission to stay and not having any children were also significantly associated with better dental health ($p < 0.05$).

Migrants who brushed their teeth every day reported significantly higher educational levels, better general and mental health and better access to dentists (Table 4). In addition, more migrants from Afghanistan and less from Syria brushed their teeth daily ($p < 0.05$).

Table 3. Descriptive characteristics and other variables by dental/teeth condition.

	Not in Good Dental Condition ($N = 563$)	In Good Dental Condition ($N = 705$)
Years of education (mean $\pm$ Standard deviation) **	8.3 ± 4.9	9.4 ± 5.1
General health score (units) **	56.7 ± 24.2	69.7 ± 20.6
Mental health score (units) **	55.7 ± 22	65.3 ± 20
Age (years) (median (interquartile range)) **	32 (25–41)	28 (23–37)
DMS scale (units) **	1.3 (1–2.6)	1 (1–2)
Males (%)	60.4	65.1
Marital status (%) *		
Single	42.0	50.1
Engaged/married	48.8	41.8
Separated	4.6	3.8
Widowed	4.6	4.3
Have at least one child (%) **	59.9	42.7
Possession of any kind of permission (asylum or other kind) (%) *	62.4	69.6
I know where to go when I need a dentist (%) *	67.1	73.1
Country of origin (%) *		
Afghanistan	16.1	13.1
Iraq	7.5	10.1
Nigeria	7.1	10.7
Syria	20.5	22.7
Other	48.8	43.3

$* p < 0.05; ** p < 0.001.$

Table 4. Descriptive characteristics and other variables by brushing their teeth.

	Do Not Brush Their Teeth Every Day ($N = 124$)	Brush Their Teeth Every Day ($N = 1135$)
Years of education (mean $\pm$ standard deviation) **	6.9 ± 4.8	9.1 ± 5
General health score (units) (mean $\pm$ standard deviation) *	57.2 ± 21.8	64.7 ± 23.2
Mental health score (units) *	54 ± 20.2	61.5 ± 21.5
Age (years) (median (interquartile range))	30 (24–39)	29 (23–39)
DMS scale (units)	1 (1–2.4)	1 (1–2)
Males (%)	67.7	62.8
Marital status (%)		
Single	38.2	47.7
Engaged/married	52.7	43.6
Separated	5.7	4.2
Widowed	3.3	4.5
Have at least one child (%)	56.2	49.7
Possession of any kind of permission (asylum or other kind) (%)	72.1	66.0
I know where to go when I need a dentist (%) *	58.2	71.6
Country of origin (%) **		

Table 4. *Cont.*

	Do Not Brush Their Teeth Every Day (*N* = 124)	Brush Their Teeth Every Day (*N* = 1135)
Afghanistan	8.2	14.8
Iraq	9.8	9.0
Nigeria	8.2	9.2
Syria	40.2	20.1
Other	33.6	46.8

* $p < 0.05$; ** $p < 0.001$.

Logistic regression analysis was used to investigate the effect of demographic characteristics, quality of life, sense of discrimination and presence of any kind of permission on dental health (Table 5). Two different logistic regression models, one with general health and one with mental health due to collinearity, were used (Pearson's r = 0.55; $p < 0.001$). Higher educational level indicated significantly higher odds of having good dental conditions in both models (Model 1: OR = 1.072; 95%CI (1.028, 1.118); Model 2: OR = 1.075; 95%CI (1.032, 1.120)). Similar positive correlations to good dental condition were observed for general and mental health, having any kind of permission to stay, and not having any children. Migrants with higher age and DMS score had lower odds of having good dental conditions (for age: Model 1: OR = 0.948; 95%CI (0.928, 0.969); Model 2: OR = 0.947; 95%CI (0.927, 0.968), for DMS score: Model 1: OR = 0.786; 95%CI (0.605, 1.021); Model 2: OR = 0.760; 95%CI (0.586, 0.986)). Widowed migrants had approximately 5 times higher odds of having good dental conditions compared with single migrants (Model 1: OR = 5.338; 95%CI (2.016, 14.136); Model 2: OR = 4.387; 95%CI (1.689, 11.391)). In our sample, migrants who never had visited the dentist reported higher odds of having good dental conditions compared with migrants who had visited the dentist <1 year ago (Model 1: OR = 0.400; 95%CI (0.211, 0.758); Model 2: OR = 0.436; 95%CI (0.231, 0.821)) or had visited the dentist >1 year ago (Model 1: OR = 0.294; 95%CI (0.153, 0.565); Model 2: OR = 0.322; 95%CI (0.169, 0.613)).

Table 5. Logistic regression models that investigated the impact of quality of life, sense of discrimination, presence of any kind of permission and demographics on dental health status (dependent variable).

	Odds Ratio	(95% Confidence Interval)
Model 1		
Education (years)	1.072 ***	(1.028, 1.118)
General health (score)	1.020 ***	(1.010, 1.029)
Age (years)	0.948 ***	(0.928, 0.969)
Discrimination Scale (score)	0.786 *	(0.605, 1.021)
Gender (females)	1.134	(0.732, 1.758)
Marital status [+]		
Engaged/married/living with partner	1.531	(0.892, 2.628)
Separated/divorced	1.531	(0.570, 4.108)
Widowed	5.338 ***	(2.016, 14.136)
No possession of any kind of permission	0.631 **	(0.408, 0.978)
Have no children	1.959 **	(1.221, 3.143)
Last visit to the dentist [++]		
<1 year	0.400 **	(0.211, 0.758)
>1 year	0.294 ***	(0.153, 0.565)
Model 2		
Education (years)	1.075 ***	(1.032, 1.120)
Mental health (score)	1.014 **	(1.004, 1.023)
Age (years)	0.947 ***	(0.927, 0.968)
Discrimination Scale (score)	0.760 **	(0.586, 0.986)

Table 5. *Cont.*

	Odds Ratio	(95% Confidence Interval)
Gender (females)	1.162	(0.752, 1.796)
Marital status[+]		
Engaged/married/living with partner	1.311	(0.763, 2.253)
Separated/divorced	1.521	(0.578, 4.004)
Widowed	4.387 **	(1.689, 11.391)
No possession of any kind of permission	0.648 **	(0.419, 1.001)
Have no children	2.043 **	(1.271, 3.283)
Last visit to the dentist [++]		
<1 year	0.436 **	(0.231, 0.821)
>1 year	0.322 ***	(0.169, 0.613)

*** $p < 0.01$; ** $p < 0.05$; * $p < 0.1$. [+] compared with single status. [++] compared with never.

Logistic regression was used to investigate the effect of demographic characteristics, quality of life, sense of discrimination, and legal immigration status on the odds of migrants brushing their teeth daily (Table 6). Migrants with higher educational levels and better general and mental health reported significantly higher odds of brushing teeth daily (Model 1, 2). Female migrants had approximately 2.37 times higher odds of brushing their teeth daily, compared with their male counterparts (Model 1: OR = 2.319; 95%CI (1.236, 4.349); Model 2: OR = 2.429; 95%CI (1.265, 4.663)). Migrants from Afghanistan had also significantly higher odds of brushing their teeth daily compared with all migrants, except those from Iraq. Sense of discrimination was not significantly associated with the odds of migrants brushing their teeth daily ($p > 0.1$), and it was omitted from the final models, due to the poorer fit of the models with the DMS score.

Table 6. Logistic regression models that investigated the impact of quality of life, presence of any kind of permission and demographics on migrants brushing their teeth daily (dependent variable).

	Odds Ratio	(95% Confidence Interval)
Model 1		
Education (years)	1.083 **	(1.019, 1.145)
General health (score)	1.017 **	(1.005, 1.030)
Age (years)	1.023	(0.994, 1.054)
Gender (females)	2.319 **	(1.236, 4.349)
No possession of any kind of permission	1.187	(0.637, 2.212)
Country of origin [+]		
Iraq	0.697	(0.120, 4.067)
Nigeria	0.240 **	(0.067, 0.860)
Syria	0.160 **	(0.050, 0.516)
Other	0.406	(0.132, 1.247)
Have at least one child (no)	1.278	(0.710, 2.297)
Model 2		
Education (years)	1.100 **	(1.036, 1.168)
Mental health (score)	1.020 **	(1.007, 1.034)
Age (years)	1.029 *	(0.997, 1.061)
Gender (females)	2.429 **	(1.265, 4.663)
Possession of any kind of permission	1.137	(0.600, 2.155)
Country of origin [+]		
Iraq	0.470	(0.078, 2.829)
Nigeria	0.208 **	(0.055, 0.786)
Syria	0.120 ***	(0.035, 0.418)
Other	0.299 **	(0.094, 0.951)
		(0.816, 2.694)
Have at least one child (no)	1.482	

*** $p < 0.01$; ** $p < 0.05$; * $p < 0.1$. [+] compared with migrants from Afghanistan.

4. Discussion

This study analyzes data from a questionnaire on self-reported health from migrants and refugees in seven countries in the European Union to assess the effect of sociodemographic factors, as well as other determinants, that may predict poor self-reported dental health of newly arrived migrants, including discrimination and mental and general health. Of the migrants that comprised the sample, the majority identified as male, designated their country of origin as within the Middle East North Africa (MENA) region, and reported their permission status as legal in their country of interview (i.e., granted asylum). Generally, one out of four migrants had never visited a dentist, while almost half of them reported having poor dental health (a finding corroborated in the literature) [35,36]. Previous studies identified barriers to oral healthcare (affordability, awareness, and accommodation) and focused on cultural sensitivity in diets to form recommendations for improving dental health and access [37,38]. Oral health status is influenced by a complex interrelation of factors, as stressed by Pabbla et al. (2021) [38] and this paper tried to explore different sociodemographic factors, as well as other determinants, that may affect oral healthcare uptake in a specific population with specific vulnerabilities, as reported by the migrants themselves in the context of a large European project.

Our study shows that better self-perceived dental health is mainly associated with younger age, higher educational level, legal immigration permission, better general and mental health, childlessness, lower discrimination sense, and never having visited the dentist. More specifically, young age was proven to be a determinant of poor dental health status (in our study). An increased acceptance of dental health practices of the host country among younger migrant populations compared to their older counterparts has been reported [15]. Adolescents generally demonstrate higher uptake of regular brushing as a preventative measure against dental care visits [39]. In terms of gender, migrant women visited dental health providers more frequently compared to their male counterparts, a finding supported by other studies [38]. It seems that gender plays an important role in the perception of general and oral health, dental visits and daily tooth brushing frequency, as well as in the choice of toothbrush and toothpaste. Female migrants, in comparison to males, take significantly better care of their oral health [40]. Our study also showed that female migrants were more likely to brush their teeth daily compared to male migrants. Furthermore, this study reveals that the level of education is another determinant of dental health and may be explained by the previously identified economic and cultural barriers of language and affordability that affect oral healthcare access. Migrants in our study with significantly higher educational levels brushed their teeth every day and were more informed on oral health hygiene issues, as mentioned elsewhere [41]. Permission status in the host country also affects immigrants' perception of dental health status in our study. Those who were legally permitted to stay in the country had a better overview of the dental services access and were more satisfied with their oral health status than those who had no permission in the host country [42]. General and mental health status also affected self-perception of having good teeth. Those who brushed their teeth every day reported significantly better general and mental health and vice versa, a finding confirmed by the literature [43,44]. Parental status was further expressed as a predictor of oral health perception. The results of our study show that childless adults have a higher perception of their dental health compared with those who had children. The literature provides mixed findings about this phenomenon. Some researchers argue that immigrant parents engage in healthy dental practices to maintain their and their children's dental health, while others argue that parental acculturation is unrelated to their child's dmft/DMFT level [45,46]. Finally, feeling discriminated negatively affected self-perceived good oral health in our study, as confirmed elsewhere [37,47,48].

Migrants who had never visited the dentist reported higher odds of having good dental conditions, compared with migrants who had visited the dentist <1 year ago or had visited the dentist >1 year ago. This is an expected outcome, given that those who have never visited the dentist may have never needed treatment. Migrants with the perception that one has the chance of visiting a dentist when needed augmented self-rated oral health status and the perception of having good teeth in our study. Access to dental services is usually limited, especially for undocumented migrants who also report lower oral health status in other studies [49]. These disparities in dental health outcomes and access within the migrant population are an urgent concern for the public health community, specifically the European public health community, due to its self-ascribed priorities. As part of Health 21, the World Health Organization's EU-focused policy framework to carry out its "health-for-all policy for the twenty-first century" platform, oral health care is prioritized for its impact on quality of life, disease prevention, and maintenance of good health, alongside its ease of prevention. In fact, the WHO notes that oral healthcare is the singular area of public health where "such a major problem can be so easily prevented through very simple methods" [50]. However, oral health services have become marked by a growing dental health divide. The WHO has raised concern that oral health services are not sufficiently used and accessible to those who face housing and diet insecurity, disability, ethnic or racial discrimination [27]. Meanwhile, a 2014 study by Tchicaya and Lorentz revealed "considerable socioeconomic inequalities" in the use of dental care in Europe [51]. Our study highlights this dental divide through the found prevalence of inadequate dental health among migrants with heightened disparities within migrant groups, according to age, sex, and education status, for example. It seems that while obtaining access to dental health services is an entitlement of asylum status, migrants continue to suffer from poor/fair dental health and dental caries, as well as a lack of service uptake.

A future program that addresses the combined issues of diet and dental prevention and hygiene information should be tested on these vulnerable population groups. As is already known, the current diet-print, especially for migrants and refugees from low-income countries in Europe, usually corresponds to low-calorie, fast-food type diets with ultra-processed foods, such as burgers, pizzas, French fries, chips, cakes, biscuits, and sweetened breakfast cereals [13,52], which are cheap, rich in fat, sugar, and other refined carbohydrates [53], but can obviously lead to a high prevalence of obesity and dental caries. [17,18,54–56] This dietary acculturation issues could be fulfilled by the implementation of a voluntary dental network, offering information on diet and dental health hygiene measures, as well as simple treatments, as a temporary solution. [49] However, the voluntary nature of dental care can result in a fragmented provision of oral health care, especially among undocumented migrants. [42] To reduce inequalities in oral health in the long term, systemic barriers in access to oral health care need to be addressed to understand that demographic factors act as risk factors for dental caries perception.

5. Strengths, Limitations, and Future Research Directions

To the best of our knowledge, this is among the first studies to assess demographic determinants of status and access to dental health care among multicultural migrant populations in several EU member states. Other studies [57] have addressed and shown how SES status affects oral health mainly in the general population, but this study has focused specifically on migrants from war-torn or poor countries characterized by specific vulnerabilities, such as feelings of discrimination, when accessing the health system. Another strength includes the breadth of our questionnaire, which investigated numerous demographic factors alongside assessments of physical and mental health. As a result, our study allowed for numerous comparative analyses between demographic and dental health factors. However, this study had also several limitations. Due to the nature of the design, the cross-sectional study provided a snapshot of the respondents' dental condition, and temporal constraints reduced the viability of assessing causal relationships between dental health determinants and outcomes. Furthermore, the self-administered questionnaire depended on self-reported information and assessments of health, which may introduce reporting bias. However, this study did not aim to clinically assess the dental health of newly arrived migrants, but rather to assess demographic factors and health determinants that may have an influence on it, in planning preventive interventions and to potentially predict the dental health status of populations in future migrant waves. In addition, the sample consisted of an overwhelmingly large proportion of male respondents, which may be a result of cultural barriers to female respondents. Finally, self-perception of oral health was not related to clinical epidemiological oral health indicators or dento-occlusal aesthetic indicators, except for untreated decayed teeth. Although the dmft Index, DMFT Index and dmft/DMFT Index have been used intensively in clinical settings to assess dental caries prevalence, as well as dental treatment needs among populations [58], the use of these scores was not part of the study to check the association with self-perceived oral health, as was the case in other studies. [59] Since the proximal consequence of dental decay is pain [60], it is likely that the contribution of decayed untreated teeth to self-reported oral health is viewed by migrants through their subjective measures [61].

For the basis of prospective studies, our results could be used to further investigate the dental caries prevalence and oral health status in different migrant settings (urban/non-urban) within documented or non-documented people and access among subsets of migrant groups, as divided by the identified demographic determinants.

6. Conclusions

This study showed that many migrants report poor dental health and that the risk factors for poor dental health are numerous. Overall, higher education levels, legal permission status, better general and mental health, lower age, parental status, lower sense of discrimination, and better access to dentists are positive predictors of having good teeth (teeth with no dental caries), and thus generally good oral health. The barriers to oral healthcare-seeking behaviors of immigrants, the change in dietary habits due to immigration, and the limited access to dental services in remote settling areas pose public health problems in the host countries. These findings act as important baseline indicators upon which oral health improvement policy for migrants can be set and monitored in the future.

Author Contributions: Conceptualization, A.G.-S.; methodology, P.K., E.R. and A.G.-S.; formal analysis, K.K. and E.R.; resources, D.V.D. and M.S.R. and M.A.; data curation, K.K. and E.R.; writing—original draft preparation, D.V.D., M.S.R. and M.A.; writing—review and editing, P.K., E.R., M.A., A.G.-S. and I.G.; supervision, A.L.; project administration, P.K., A.G.-S. and E.R. All authors have read and agreed to the published version of the manuscript.

Funding: This research was funded partially by the European Union's Health program (2014–2020) Consumers, Health, Agriculture and Food Executive Agency (CHAFEA), Grant Agreement 738186, Project Acronym Mig-HealthCare. This research is a part of the EU project "Strengthen Community Based Care to minimize health inequalities and improve the integration of vulnerable migrants and refugees into local communities". Alejandro Gil-Salmerón was funded by the Training Program for Academic Staff (FPU) fellowship, from the Ministry of Education and Vocational Training, Spanish Government (ref.: FPU15/05251).

Institutional Review Board Statement: The study was conducted in accordance with the Declaration of Helsinki and approved by the Ethics Committee of both the National and Kapodistrian University of Athens, Medical School and the University of Valencia.

Informed Consent Statement: Informed consent was obtained from all subjects involved in the study.

Data Availability Statement: The limited dataset used for this analysis is available upon reasonable request from the Mig-HealthCare consortium.

Acknowledgments: The authors would like to thank the Mig-HealthCare partner consortium (www.mighealthcare.eu, accessed on 14 July 2022) for their collaboration in the project implementation. The project partners are as follows: Austria Verein Multikulturell, Bulgaria National Center of Infectious and Parasitic Diseases, Cyprus Centre for Advancement of Research and Development in Educational Technology Ltd. Cardet, Germany Ethno-Medizinisches Zentrum EV, Greece Institute of Preventive Medicine, Environmental and Occupational Health, Prolepsis, National and Kapodistrian University of Athens, Perifereia Stereas Elladas, Kentriki Enosi Dimon kai Koinotiton Ellados, Ministry of Health, France Ecole des Hautes Etudes en Sante Publique, Italy Oxfam Italia Onlus, Malta Koperazzjoni Internazzjonali Malta (Kopin) Association, Spain Universitat de Valencia and Sweden Uppsala Universitet.

Conflicts of Interest: The authors declare no conflict of interest.

References

1. Clayton, J.; Holland, H. *Over One Million Sea Arrivals Reach Europe in 2015*; UNHCR: Geneva, Switzerland, 2015; Available online: https://www.unhcr.org/news/latest/2015/12/5683d0b56/million-sea-arrivals-reach-europe-2015.html (accessed on 28 December 2021).
2. UNHCR. *Global Report 2020*; UNHCR, The UN Refugee Agency: Geneva, Switzerland, 2020; Available online: https://reporting.unhcr.org/sites/default/files/gr2020/pdf/GR2020_English_Full_lowres.pdf#_ga=2.252375049.1635139560.1640694849-877351995.1640694849 (accessed on 28 December 2021).
3. UNICEF. *Refugee and Migrant Crisis in Europe*; UNICEF: Geneva, Switzerland, 2021; Available online: https://reliefweb.int/sites/reliefweb.int/files/resources/2021-HAC-Refugee-migrant-response-Europe-July-Revision.pdf (accessed on 28 December 2021).
4. Venturi, E.; Vallianatou, A.I. Ukraine Exposes Europe's Double Standards for Refugees. 22 March 2022. Available online: https://www.chathamhouse.org/2022/03/ukraine-exposes-europes-double-standards-refugees?gclid=EAIaIQobChMIn4-7677r-AIVdgIGAB2QAAySEAAYAyAAEgKdv_D_BwE (accessed on 9 June 2022).
5. Dastyari, A.; Ghezelbash, D. Asylum at Sea: The Legality of Shipboard Refugee Status Determination Procedures. *Int. J. Refug. Law* **2020**, *32*, 1–27. [CrossRef]
6. Frattini, T. L'intégration des immigrés dans les pays d'accueil-Ce que nous savons et ce qui marche. *Revue d'économie du développement* **2017**, *1*, 105–134. [CrossRef]
7. Gil-Salmerón, A.; Katsas, K.; Riza, E.; Karnaki, P.; Linos, A. Access to Healthcare for Migrant Patients in Europe: Healthcare Discrimination and Translation Services. *Int. J. Environ. Res. Public Health* **2021**, *18*, 7901. [CrossRef] [PubMed]
8. Lebano, A.; Hamed, S.; Bradby, H.; Gil-Salmerón, A.; Durá-Ferrandis, E.; Garcés-Ferrer, J.; Azzedine, F.; Riza, E.; Karnaki, P.; Zota, D.; et al. Migrants' and refugees' health status and healthcare in Europe: A scoping literature review. *BMC Public Health* **2020**, *20*, 1039. [CrossRef] [PubMed]
9. Riza, E.; Karnaki, P.; Gil-Salmerón, A.; Zota, K.; Ho, M.; Petropoulou, M.; Katsas, K.; Garces-Ferrer, J.; Linos, A. Determinants of Refugee and Migrant Health Status in 10 European Countries: The Mig-HealthCare Project. *Int. J. Environ. Res. Public Health* **2020**, *17*, 6353. [CrossRef]
10. Paisi, M.; Baines, R.; Burns, L.; Plessas, A.; Radford, P.; Shawe, J.; Witton, R. Barriers and facilitators to dental care access among asylum seekers and refugees in highly developed countries: A systematic review. *BMC Oral Health* **2020**, *20*, 337. [CrossRef]
11. Fratzke, S.; Kainz, L. *Preparing for the Unknown: Designing Effective Predeparture Orientation for Resettling Refugees*; MPI (Migration Policy Institute): Washington, DC, USA, 2019; Available online: https://www.migrationpolicy.org/research/designing-effective-predeparture-orientation-resettling-refugees (accessed on 10 June 2022).
12. Salim, N.A.; Maayta, W.A.; Hassona, Y.; Hammad, M. Oral health status and risk determinants in adult Syrian refugees in Jordan. *Community Dent. Health* **2021**, *38*, 53–58.
13. Gsir, S. Social Interactions between Immigrants and Host Country Populations: A Country-of-Origin Perspective, Migration Policy Centre, INTERACT, 2014/02, Position Paper. Retrieved from Cadmus, EUI Research Repository. Available online: http://hdl.handle.net/1814/31243 (accessed on 10 June 2022).
14. Batra, M.; Gupta, S.; Erbas, B. Oral Health Beliefs, Attitudes, and Practices of South Asian Migrants: A Systematic Review. *Int. J. Environ. Res. Public Health* **2019**, *16*, 1952. [CrossRef]
15. Keboa, M.T.; Hiles, N.; Macdonald, M.E. The oral health of refugees and asylum seekers: A scoping review. *Glob. Health* **2016**, *12*, 59. [CrossRef]
16. Scholten, P.; Entzinger, H.; Penninx, R. Research-Policy Dialogues on Migrant Integration in Europe: A Conceptual Framework and Key Questions. In *Integrating Immigrants in Europe: Research-Policy Dialogues*; Scholten, P., Entzinger, H., Penninx, R., Verbeek, S., Eds.; IMISCOE Research Series; Springer International Publishing: Cham, Switzerland, 2015; pp. 1–16. [CrossRef]
17. Antoniadou, M.; Varzakas, T. Diet and Oral Health Coaching Methods and Models for the Independent Elderly. *Appl. Sci.* **2020**, *10*, 4021. [CrossRef]
18. Antoniadou, M.; Varzakas, T. Breaking the vicious circle of diet, malnutrition and oral health for the independent elderly. *Crit. Rev. Food Sci. Nutr.* **2021**, *61*, 3233–3255. [CrossRef] [PubMed]
19. Sischo, L.; Broder, H.L. Oral health-related quality of life: What, why, how, and future implications. *J. Dent. Res.* **2011**, *90*, 1264–1270. [CrossRef] [PubMed]
20. Paula, J.S.; Sarracini, K.L.M.; Meneghim, M.C.; Pereira, A.C.; Ortega, E.M.M.; Martins, N.S.; Mialhe, F.L. Longitudinal evaluation of the impact of dental caries treatment on oral health-related quality of life among schoolchildren. *Eur. J. Oral Sci.* **2015**, *123*, 173–178. [CrossRef]
21. Banu, A.; Șerban, C.; Pricop, M.; Urechescu, H.; Vlaicu, B. Dental health between self-perception, clinical evaluation and body image dissatisfaction—A cross-sectional study in mixed dentition pre-pubertal children. *BMC Oral Health* **2018**, *18*, 74. [CrossRef] [PubMed]
22. Chalmers, N.I.; Wislar, J.S.; Boynes, S.G.; Doherty, M.; Nový, B.B. Improving health in the United States: Oral health is key to overall health. *J. Am. Dent. Assoc.* **2017**, *148*, 477–480. [CrossRef] [PubMed]
23. Gilbert, P.A.; Khokhar, S. Changing dietary habits of ethnic groups in Europe and implications for health. *Nutr. Rev.* **2008**, *66*, 203–215. [CrossRef] [PubMed]
24. Leal, S.C.; Bronkhorst, E.M.; Fan, M.; Frencken, J.E. Untreated Cavitated dentine lesions: Impact on Children's quality of life. *Caries Res.* **2012**, *46*, 102–106. [CrossRef]

25. Costa, S.M.; Vasconcelos, M.; Haddad, J.P.A.; Abreu, M.H.N. The severity of dental caries in adults aged 35 to 44 years residing in the metropolitan area of a large city in Brazil: A cross-sectional study. *BMC Oral Health* **2012**, *12*, 25. [CrossRef]
26. Hoover, J.; Vatanparast, H.; Uswak, G. Risk Determinants of Dental Caries and Oral Hygiene Status in 3–15-Year-Old Recent Immigrant and Refugee Children in Saskatchewan, Canada: A Pilot Study. *J. Immigr. Minority Health* **2017**, *19*, 1315–1321. [CrossRef]
27. Northridge, M.E.; Kumar, A.; Kaur, R. Disparities in Access to Oral Health Care. *Annu. Rev. Public Health* **2020**, *41*, 513–535. [CrossRef]
28. WHO. Oral Health. 2021. Available online: https://www.who.int/news-room/fact-sheets/detail/oral-health (accessed on 29 December 2021).
29. Wickramage, K.; Vearey, J.; Zwi, A.B.; Robinson, C.; Knipper, M. Migration and health: A global public health research priority. *BMC Public Health* **2018**, *18*, 987. [CrossRef] [PubMed]
30. Riza, E.; Lazarou, A.; Karnaki, P.; Zota, D.; Nassi, M.; Kantzanou, M.; Linos, A. Using an IT-Based Algorithm for Health Promotion in Temporary Settlements to Improve Migrant and Refugee Health. *Healthcare* **2021**, *9*, 1284. [CrossRef] [PubMed]
31. Ponce-Gonzalez, I.; Cheadle, A.; Aisenberg, G.; Cantrell, L.F. Improving oral health in migrant and underserved populations: Evaluation of an interactive, community-based oral health education program in Washington state. *BMC Oral Health* **2019**, *19*, 30. [CrossRef]
32. Mig-HealthCare. The Mg-HealthCare Project. Available online: https://www.mighealthcare.eu/ (accessed on 9 June 2022).
33. Peek, M.E.; Nunez-Smith, M.; Drum, M.; Lewis, T.T. Adapting the Everyday Discrimination Scale to Medical Settings: Reliability and Validity Testing in a Sample of African American Patients. *Ethn. Dis.* **2011**, *21*, 502–509. [PubMed]
34. Ware, J.E., Jr.; Snow, K.K.; Kosinski, M.; Gandek, B.; New England Medical Center, The Health Institute. *SF-36 Health Survey: Manual and Interpretation Guide*; The Health Institute, New England Medical Center: Boston, MA, USA, 1997.
35. Zinah, E.; Al-Ibrahim, H.M. Oral health problems facing refugees in Europe: A scoping review. *BMC Public Health* **2021**, *21*, 1207. [CrossRef] [PubMed]
36. Lauritano, D.; Moreo, G.; Carinci, F.; Campanella, V.; Della Vella, F.; Petruzzi, M. Oral Health Status among Migrants from Middle- and Low-Income Countries to Europe: A Systematic Review. *Int. J. Environ. Res. Public Health* **2021**, *18*, 2203. [CrossRef] [PubMed]
37. Rodriguez-Alvarez, E.; Borrell, L.N.; Marañon, E.; Lanborena, N. Immigrant Status and Ethnic Inequities in Dental Caries in Children: Bilbao, Spain. *Int. J. Environ. Res. Public Health* **2022**, *19*, 4487. [CrossRef]
38. Pabbla, A.; Duijster, D.; Grasveld, A.; Sekundo, C.; Agyemang, C.; van der Heijden, G. Oral Health Status, Oral Health Behaviours and Oral Health Care Utilisation Among Migrants Residing in Europe: A Systematic Review. *J. Immigr. Minority Health* **2021**, *23*, 373–388. [CrossRef]
39. Crespo, E. The Importance of Oral Health in Immigrant and Refugee Children. *Children* **2019**, *6*, 102. [CrossRef]
40. Azodo, C.C.; Unamatokpa, B. Gender difference in oral health perception and practices among Medical House Officers. *Russ. Open Med. J.* **2012**, *1*, 0208. [CrossRef]
41. Valdez, R.; Spinler, K.; Kofahl, C.; Seedorf, U.; Heydecke, G.; Reissmann, D.R.; Lieske, B.; Dingoyan, D.; Aarabi, G. Oral Health Literacy in Migrant and Ethnic Minority Populations: A Systematic Review. *J. Immigr. Minority Health* **2022**, *24*, 1061–1080. [CrossRef] [PubMed]
42. Wilson, F.A.; Wang, Y.; Borrell, L.N.; Bae, S.; Stimpson, J.P. Disparities in oral health by immigration status in the United States. *J. Am. Dent. Assoc.* **2018**, *149*, 414–421.e3. [CrossRef] [PubMed]
43. Luo, H.; Hybels, C.; Wu, B. Acculturation, depression and oral health of immigrants in the USA. *Int. Dent. J.* **2018**, *68*, 245–252. [CrossRef] [PubMed]
44. Phlypo, I.; Palmers, E.; Janssens, L.; Marks, L.; Jacquet, W.; Declerck, D. The perception of oral health and oral care needs, barriers and current practices as perceived by managers and caregivers in organizations for people with disabilities in Flanders, Belgium. *Clin. Oral Invest.* **2020**, *24*, 2061–2070. [CrossRef]
45. Puthiyapurayil, J.; Kumar, A.; Syriac, G.; Maneesha, R.; Najmunnisa, R. Parental perception of oral health related quality of life and barriers to access dental care among children with intellectual needs in Kottayam, central Kerala-A cross sectional study. *Spec. Care Dent.* **2022**, *42*, 177–186. [CrossRef]
46. Dahlan, R.; Bohlouli, B.; Salami, B.; Saltaji, H.; Amin, M. Parental acculturation and oral health of children among immigrants. *J. Public Health Dent.* **2021**. [CrossRef]
47. Schuch, H.S.; Haag, D.G.; Bastos, J.L.; Paradies, Y.; Jamieson, L.M. Intersectionality, racial discrimination and oral health in Australia. *Community Dent. Oral Epidemiol.* **2021**, *49*, 87–94. [CrossRef]
48. Ramos-Gomez, F.; Kinsler, J.J. Addressing social determinants of oral health, structural racism and discrimination and intersectionality among immigrant and non-English speaking Hispanics in the United States. *J. Public Health Dent.* **2022**, *82* (Suppl. 1), 133–139. [CrossRef]
49. van Midde, M.; Hesse, I.; van der Heijden, G.J.; Duijster, D.; van Elteren, M.; Kroesen, M.; Agyemang, C.; Beune, E. Access to oral health care for undocumented migrants: Perspectives of actors involved in a voluntary dental network in the Netherlands. *Community Dent. Oral Epidemiol.* **2021**, *49*, 330–336. [CrossRef]
50. World Health Organization (Ed.) *Health21: The Health for All Policy Framework for the WHO European Region*; European Health for All Series; World Health Organization, Regional Office for Europe: Copenhagen, Denmark, 1999; 224p.
51. Tchicaya, A.; Lorentz, N. Socioeconomic inequalities in the non-use of dental care in Europe. *Int. J. Equity Health* **2014**, *13*, 7. [CrossRef]

52. Mellin-Olsen, T.; Wandel, M. Changes in food habits among Pakistani immigrant women in Oslo, Norway. *Ethn. Health* **2005**, *10*, 311–339. [CrossRef] [PubMed]

53. Popkin, B.; Adair, L.; Wen Ng, S. Global nutrition transition and the pandemic of obesity in developing countries. *Nutr. Rev.* **2011**, *70*, 3–21. [CrossRef] [PubMed]

54. Holmboe-Ottesen, G.; Wandel, M. Changes in dietary habits after migration and consequences for health: A focus on South Asians in Europe. *Food Nutr. Res.* **2012**, *56*, 18891. [CrossRef] [PubMed]

55. Sheiham, A.; James, W.P.T. First Diet and Dental Caries: The Pivotal Role of Free Sugars Reemphasized. *J. Dent. Res.* **2015**, *94*, 1341–1347. [CrossRef] [PubMed]

56. Tjomsland, A. US Immigrants Adopt Native Food Habits after Five Years. 28 September 2020. NIBIO—Norwegian Institute of Bioeconomy Research. Available online: https://partner.sciencenorway.no/food-nibio-nutrition/us-immigrants-adopt-native-food-habits-after-five-years/1748302 (accessed on 10 June 2022).

57. Hakeberg, M.; Wide Boman, U. Self-reported oral and general health in relation to socioeconomic position. *BMC Public Health* **2018**, *18*, 63. [CrossRef] [PubMed]

58. Broadbent, J.M.; Thomson, W.M. For debate: Problems with the DMF index pertinent to dental caries data analysis. *Community Dent. Oral Epidemiol.* **2005**, *33*, 400–409. [CrossRef]

59. Pattussi, M.P.; Anselmo Olinto, M.T.; Hardy, R.; Sheiham, A. Clinical, social and psychosocial factors associated with self-rated oral health in Brazilian adolescents. *Community Dent. Oral Epidemiol.* **2007**, *35*, 377–386. [CrossRef]

60. Ferraz, N.K.; Nogueira, L.C.; Pinheiro, M.L.; Marques, L.S.; Ramos-Jorge, M.L.; Ramos-Jorge, J. Clinical consequences of untreated dental caries and toothache in preschool children. *Pediatr. Dent.* **2014**, *36*, 389–392.

61. Benyamini, Y.; Leventhal, H.; Leventhal, E.A. Self-rated oral health as an independent predictor of self-rated general health, self-esteem and life satisfaction. *Soc. Sci. Med.* **2004**, *59*, 1109–1116. [CrossRef]

applied sciences

MDPI

Review

Mushrooms as Functional Foods for Ménière's Disease

Victoria Bell [1] and Tito Horácio Fernandes [2,*]

[1] Faculty of Pharmacy, University of Coimbra, Pólo das Ciências da Saúde, Azinhaga de Santa Comba, 3000-548 Coimbra, Portugal; victoriabell@ff.uc.pt
[2] CIISA, Faculty of Veterinary Medicine, University of Lisbon, 1300-477 Lisbon, Portugal
* Correspondence: profcattitofernandes@gmail.com

Abstract: Food, not nutrients, is the fundamental unit in nutrition, and edible mushrooms are fungi that supply unique biological bioactive compounds, different from plant or animal origin, which significantly impact human health status. However, to date all these concepts are interpreted in different ways, with rapidly increasing knowledge on nutrition, medicine, molecular biology, and plant biotechnology changing the concepts of food, health, and agriculture. The bioactive elements conveyed by foodstuffs as nutrients or non-nutrients interfere with human metabolism and have influence on health, aging, and well-being. The influence of edible mushrooms on medicinal interventions has been known and studied for many years and their latest role in neurodegenerative disorders has been recently investigated, while their significance on many other diseases has been well demonstrated. Despite considerable research, the etiology and pathogenesis of Ménière's disease remains controversial and undefined, although usually associated with allergic, genetic, or trauma sources, and with viral infections and/or immune system-mediated mechanisms. With treatment still unknown, our attention is towards the eventual impact of complementary dietary interventions, synthesizing the recent knowledge of some edible mushrooms and preparations on Ménière's disease, which is a lifelong condition that can develop at any age, but most commonly emerges between 40 and 60 years of age. It is demonstrated that the oral administration of a biomass preparation, with 3 g/day of the mushroom *Coriolus versicolor* for 2 to 6 months, on some 40 human Ménière's disease patients reduced systemic oxidative stress and cellular stress response, decreased the number of crises and their duration, and the frequency of symptoms, improving the clinical grading of tinnitus severity.

Keywords: vertigo; tinnitus; functional foods; macrofungi; neuronal diseases

Citation: Bell, V.; Fernandes, T.H. Mushrooms as Functional Foods for Ménière's Disease. *Appl. Sci.* **2023**, *13*, 12348. https://doi.org/10.3390/app132212348

Academic Editor: Monica Gallo

Received: 3 October 2023
Revised: 6 November 2023
Accepted: 13 November 2023
Published: 15 November 2023

1. Introduction

Foods are generally designated as functional if they encompass a bioactive element. These biologically active dietary elements are extrinsic non-nutritional substances that can regulate biochemical metabolic activities, leading to health promotion [1].

The term nutraceutical from nutrition and pharmaceutical was first coined in 1989 and next interchangeably used with the term pharmanutrient or functional food. However, functional foods are edibles that provide health benefits or disease risk reduction beyond their nutritional value, whereas nutraceuticals, different from functional foods, are commodities that may be considered a food or part of a food, but are supplied in different medicinal oral forms [2].

With treatment still unknown, our attention is towards the eventual impact of complementary dietary interventions, synthesizing the recent knowledge of some edible mushrooms and preparations on Ménière's disease. Mushrooms are edible fungi widely used as medicaments in Asia for ages. They are valuable macro-fungi that exist as an integral and vital component of the ecosystem as major decomposers. The unique composition of mushrooms, namely on specific enzymes, contributes to biodiversity, to traditional herbal medicines, and the supply of useful bioactive compounds [3]. Although some of these hold promising preventive and therapeutic opportunities, there is no universal definition and harmonized regulatory framework among countries [4]. There are many thousands of mushroom species but just a handful are edible or defined as functional food, and there is as yet scarce clinical evidence for their efficacy, safety, and effectiveness [5].

We have recently reviewed how the enormous potential of the bioactive elements present in mushrooms complement the human diet, with various active molecules undetected or insufficient in common foodstuffs of plant and animal origin, being considered a functional food for health benefits or the prevention of several human diseases [6].

Edible mushrooms represent not only a huge storehouse of vitamins, minerals, and dietary fiber, but they are also an important source of bioactive components such as polysaccharides, terpenes, steroids, anthraquinone, phenolic acid, and benzoic acid, while primary metabolites contain proteins, oxalic acid, and peptides [7].

Without a complete understanding of the influence of mushroom bioactive constituents and their mode of action as nutraceuticals, it is challenging to effectively understand the role of mushrooms as dietary interventions in malfunctions and diseases [8]. The structural diversity of various mushroom bioactive secondary metabolites (e.g., terpenoids, acids, alkaloids, sesquiterpenes, polyphenolic compounds, lactones, sterols, nucleotide analogues, vitamins, and metal chelating agents), as well as their specific potency as a therapeutic prospect and/or antioxidants, have been widely investigated [9–11].

In general, mushrooms contain large amounts of chitin, mannans, galactans, xylans, glucans, krestin, lentinan, and hemicelluloses; therefore, they perform as potential candidates for prebiotics [12]. The prebiotic activity of mushrooms beneficially affects gut homeostasis performance and the balance of gut microbiota is enhanced [13].

Edible mushrooms (e.g., *Lentinula edodes*, *Pleurotus* spp., *Agaricus* spp., and *Ganoderma* spp.) are valuable sources of protein for both food and medicine, containing more protein than vegetables, fruits, and grains. Mushrooms contain bioactive proteins and peptides known to have antihypertensive, immunomodulatory, antifungal, antibiotic, antibacterial activities, anticancer, antiviral, and antioxidant properties [14,15].

Bioactive components reported in different edible mushrooms include β-glucans, lentinan, peptidoglycan, ergosterol, cordycepin, tocopherols, quercetin, catechin, lovastatin, eritadenine, hericenones, erinacines, among many others [16]. Furthermore, mushroom bioactive elements include ribosome inactivating proteins, proteases, antifungal proteins, and lectins, present namely in *Ganoderma lucidum*, *Agaricus bisporus*, and *Boletus satanus* [17,18].

Protein from mushrooms contrasts from animal, vegetable, and microbial proteins, usually forming cytotoxic enzymes (efficient tools to combat cancer), and include fungal immunomodulatory proteins (FIPs), Ribosome Inactivating Proteins (RIPs), nucleases, ubiquitin-like proteins, and proteins possessing enzymatic activity such as ribonucleases and laccases [19–21].

Mushrooms contain ergothioneine, which humans are unable to synthesize, a unique sulfur-containing amino acid, an antioxidant, cytoprotective, and anti-inflammatory element with therapeutic potential, approved by world food agencies. The novel food, synthetic L-ergothioneine, has also been approved by the FDA and EFSA at a recommended dosage of 30 mg/day for adults [22–24].

Ergothioneine is a thiourea derivative of histidine synthesized by few bacteria and fungi and exclusively acquired by animals and plants from exogenous sources, representing a rich source of nutrients for microbiota in the host environment. Ergothioneine is a natural powerful antioxidant activity, derived from microorganisms, especially in edible mushrooms, and a close relationship with various oxidative stress-related diseases. Besides the antioxidant properties, it has a powerful cytoprotective role in some important cells and tissues [25,26]. Mushrooms have been proven to be the highest dietary source of ergothioneine, accounting for about 95% of dietary intake [27].

Ergothioneine has a protective role in chronic inflammatory disorders. The organic cation transporter OCTN1 has been presumed to carry organic cations across the plasma membrane, but the key substrate of this transporter is in fact ergothioneine [28]. The antioxidant activity of ergothioneine has many advantages over other antioxidants such as glutathione and ascorbic acid due to its specific transportation of OCTN1. The molecular and genetic defects and the pathophysiology behind Ménière's disease, as well as the dysregulation of these ion transporters, can result in severe defects in hearing or even deafness, enhancing the dietary role of ergothioneine from mushrooms [29].

Ergothioneine is a chief amino acid but an under-recognized dietary nutrient known to avert several inflammatory and cardiovascular diseases, diabetes, and liver and neurodegenerative diseases and has been suggested as a "vitamin" and to have nutraceutical uses [26] (Figure 1). It is also considered that ergothioneine deficiency may be related to neuropathy and aging, which may increase the risk of aging-related oxidative stress diseases in the elderly. Ergothioneine ameliorates the deterioration of sleep quality caused by psychological stress, possibly through anti-inflammatory and antioxidant mechanisms in the central and peripheral nervous system (Figure 1) [30].

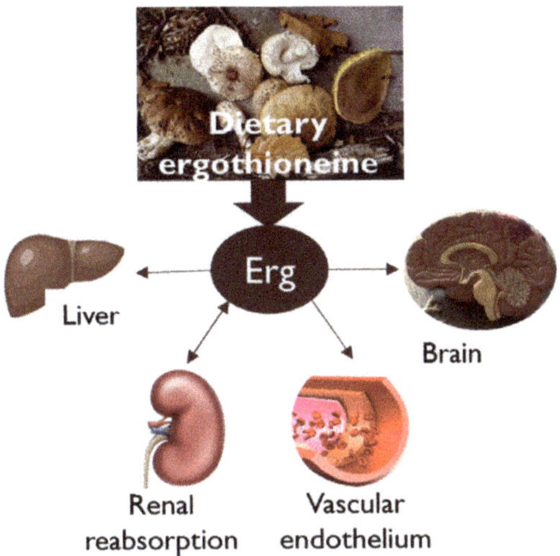

Figure 1. Ergothioneine, an antioxidant present in significant amounts in mushrooms, is an essential amino acid for humans, capable of clearing up hydroxyl radicals and maintaining bioenergetic homeostasis.

Ergothioneine extracted from *Pleurotus ostreatus*, the second most cultivated edible mushroom worldwide, was shown to exhibit strong antioxidant activity; hence, it could be a functional food for the prevention and treatment of ulcerative colitis [31].

The *Pleurotus ostreatus* (oyster) and *Ganoderma lucidum* (reishi) bioactive compounds have antimicrobial and prebiotic properties which are distributed in the mushroom mycelium and fruiting body. These mushrooms are rich in non-digestible carbohydrates (e.g., chitin and glucan) and secondary metabolites (phenolic compounds, terpenoids, and lectins), which act as prebiotics and support the growth and activity of beneficial gut microbiota, thereby maintaining a healthy balance in the gut microbiome and reducing the risk of antibiotic resistance [32].

Approved for many years in Asian countries, *Ganoderma lucidum* food components or supplements were recently given the Generally Recognized as Safe (GRAS) status and approved by the Food and Drug Administration (FDA) [33].

It is well-documented that the microbiota in the gastrointestinal tract is continuously reshaped by multiple environmental factors, especially diet, and it is decisive in human health and its balance is immunomodulated by mushroom bioactive components [34]. Nevertheless, and surprisingly, mushrooms are rarely included in most food guidelines [6].

Indeed, indigestible (by human enzymes) mushroom polysaccharides provide exceptional origin of prebiotics and curtail the reproduction of opportunistic pathogens, promoting the growth of beneficial probiotic bacteria and restoring the bacterial imbalance in the gastrointestinal tract [35,36]. Mushrooms polysaccharides have also been shown to enhance the antioxidant status, exercising anti-diabetic activity by lowering glycaemia and improving insulin resistance, bearing anti-intestinal inflammation and antineoplastic effects by regulating gut microbiota through improving microbiome diversity in the gut [37].

Nutrition for the increasingly susceptible aging brain has been vastly studied, and it is now well established that many dietary components play an important role in early prevention against neurodegenerative diseases. In the growing elderly population, the progressive loss of structure or function of brain neurons negatively influences the beginning, seriousness, and span of neurodegenerative diseases.

Several studies on neurological impact and contributions to the growth of nerves and brain cells indicated that the presence of polyphenols in edible mushrooms demonstrated protective effects against neurodegenerative disorders and aging [38].

Very recently, mushroom components, through multiple mechanisms, were shown to have a protective role in redox homeostasis and modulated effects by hormetic nutrients in complex neurodevelopment disorders such as autism spectrum disorders [39].

Presently, the medicinal role of mushrooms in nutrient balancing, in strengthening the human immune system, in enhancing natural body resistance, and in lowering proneness to disease is well established [40].

Coriolus versicolor is a type of white rot fungus found primarily on dead logs and the fungal mycelium secretes a variety of compounds into its substrate, altering its chemical composition [41]. Some mushroom bioactive elements, enzymes, and secondary metabolites have been identified and used for medicinal purposes in a purified form (Table 1) [42].

Table 1. Bioactive compounds of some edible mushrooms and their health benefits [5,6,40,42].

Edible Mushrooms	Bioactive Compounds	Health Benefits
Agaricus bisporus *Pleurotus ostreatus* *Coriolus versicolor* *Lentinula edodes* *Flammulina velutipes* *Ganoderma lucidum* *Cordyceps sinensis* *Auricularia auricular* *Pleurotus sajor-caju* *Hericium erinaceus* *Grifola frondosa*	Polyphenols Dietary fibres Lectins Terpenoids Antioxidants Flavonoids Peptidoglycans B-glucan Phytosterois Funcional proteins	Anti-diabetic Anti-inflamatory Anti-carcinogenic Anti-microbial Anti-oxidative Anti-proliferative Cholesterol-lowering Anti-viral Immuno-modulatory Osteoporosis Pre-biotic Anti-hypertensive Obesity Anti-cataractogenic Anti-viral Anti-ageing Gastrointestinal health

These enzymes from *Coriolus versicolor* prevent oxidative stress (i.e., superoxide dismutase-SOD; catalase; glutathione-GSH peroxidase), inhibit cell growth (i.e., protease, glucose-2-oxidase), and play a role in detoxification (i.e., peroxidases, cytochrome P450) (Table 2).

Table 2. *Coriolus versicolor* (CV) in vitro enzyme activity (analyzed per tablet of 500 mg CV plus 225 mg excipient). The presence of pepsin or trypsin was performed to evaluate eventual gastric degradation [43].

	In the Absence of Proteolytic Enzymes	In the Presence of	
		Pepsin	Trypsin
Cytochrome P-450	0.51 nmoles	0.49 nmoles	0.52 nmoles
Cytochrome P-450 reductase	11.9 mU	9.52 mU	11.1 mU
Glucoamylase/Beta-glucanase activity	6.9 U	ND	ND
Glucose 2-oxidase activity	49.5 mU	ND	ND
Laccase activity	521.5 mU	522.6 mU	ND
Peroxidase activity	67.2 mU	60.4 mU	52.6 mU
Protease activity	5.9 U	5.0 U	5.7 U
Protein content	17.3 mg	15.7 mg	16.6 mg
Protein-bound polysaccharide	91.5 mg	80.5 mg	78.1 mg
Reducing sugars	14.8 mg	14.5 mg	261 mg *
Secondary metabolites (Thrombin inhibitors)	59%	54.20%	52%
Superoxide dismutase (SOD) activity	77.1 mU	61.2 U	68.5 U

* The presence of reducing sugars is due to the use of maltodextrin in the manufacturing process. ND-non-determined.

2. Ménière's Disease

Ménière's disease (MD) was first described by Prosper Ménière in 1861 when investigating migraine headaches [44]. MD represents a non-communicable disorder of the inner ear with a high clinical heterogeneity but characterized by episodes of spontaneous vertigo, tinnitus, fluctuating sensorineural hearing loss, and a sensation of the ear being full that affects one or both ears [45]. Hearing loss is the most common sensory defect and affects 450 million people worldwide in a disabling form.

Thus far, the etiology of MD remains largely unknown, mainly caused by a buildup of fluid in the chambers in the inner ear. There is growing evidence implying that oxidative stress and neuroinflammation, involving proinflammatory cytokines, may be fundamental to the occurrence of abnormal fluctuations of primary endolymphatic hydrops in the inner ear's labyrinth [46] (Figure 2).

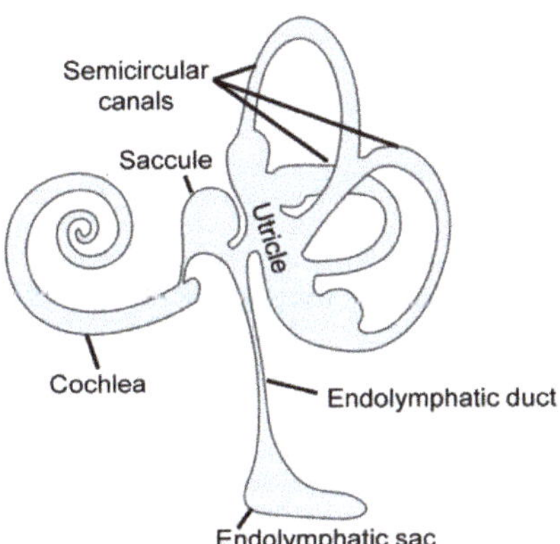

Figure 2. Diagram of the inner ear anatomy [47] showing the endolymphatic sac where the hydrostatic pressure and endolymph homeostasis is maintained.

Symptoms of Meniere's disease include: (a) Tinnitus: ringing, buzzing, roaring, whistling, or hissing sound in the ear. (b) Vertigo: regular dizzy bouts, spinning feeling starting and stopping unexpectedly, without warning. It usually lasts 20 min to 12 h, but not more than 24 h; serious vertigo can cause nausea. (c) Hearing disorder. Hearing loss may come and go, especially early on. Over time, hearing loss can be long-lasting and eventually can be permanent. (d) Aural fullness: feeling of fullness in the ear, feeling pressure in the ear.

Complaints can range from mild to severe dizziness and nausea and vomiting. Symptoms can last for days, weeks, months, and may recur even after years [48]. Due to aging, otholitic degeneration and displacement in the reuniting duct and stockpile of debris in the semicircular canals may cause an otolithic crisis [49].

Although the evidence of a causal association between allergy and MD is inconclusive, the inclusion of allergy control as part of the treatment plan for MD is low risk and has been suggested by multiple authors [50,51].

Several researchers using different methodologies have been investigating candidate genes related to MD and the prevalence of autoimmune diseases diagnosed along with MD among different populations [52–55].

As our understanding concerning the etiology and medical intervention of the disease expands, the argument encircling the pathogenesis of MD deepens. Our aim here is not to describe this disorder or treatment but to outline the impact of diet and mushroom nutrition on MD, as already demonstrated in other neurodegenerative diseases.

Further to the ongoing genetic research, multiple studies have concentrated on the pharmacology and usefulness of bioactive elements and metabolites, as well as gut microbiome mediation, as a novel relevant procedure to address a number of human diseases, but mainly those related to neurological degeneration [56].

Treatment of Ménière's Disease

Ménière's disease does not have a cure yet, but there are some recommendations to help in coping with the condition and lifestyle modifications and the following treatments can help affected people cope with the symptoms [57].

However, none are considered effective by the scientific community. They include:

Medications. The most debilitating symptom of an outbreak of MD is dizziness. Prescription drugs such as anxiolytic diazepam (e.g., valium®), anticholinergic glycopyrrolate (e.g., robinul®), antihistamine meclizine (e.g., verticalm®) to control nausea, vomiting, and dizziness, and sedative lorazepam (e.g., ativan®) can help relieve dizziness and shorten the onset [58].

Holistic medicine. Some alternative treatments are occasionally used for MD but nothing indicates the advantages of therapies such as acupuncture or acupressure, martial arts, or herbal remedies such as gingko biloba, B3 niacin, or ginger rhizome.

Restriction of salt and use of diuretics. Low-salt or a salt-free diet and taking diuretics (e.g., diamox®-acetazolamide) may help some control dizziness by reducing the fluid retention in the body and lowering fluid volume and pressure in the inner ear.

Cognitive Processing Therapy. A type of psychotherapy talk may help people focus on how they interpret and react to life experiences, reducing symptoms of various mental health conditions, primarily depression and anxiety [59].

Infusions. Injecting antibiotics (e.g., intratympanic gentamycin) into the middle ear or a corticosteroid often helps reduce dizziness and has no risk of hearing loss [60].

Positive pressure therapy. The U.S. Food and Drug Administration (FDA) approved a portable minimally invasive but costly device (Meniett) for Ménière's disease that delivers intermittent air pressure pulses to the middle ear and fits into the outer ear. The air pressure pulses appear to act on endolymph fluid to prevent dizziness [61]. However, this treatment has been recently claimed to be very uncertain [62].

Surgical procedures. Surgery may be recommended when all other treatments have failed to relieve dizziness. A number of surgical modalities, of varying levels of invasiveness, have been developed on the endolymphatic sac to decompress it. Another possible surgery is to cut the vestibular nerve or labyrinthectomy, although this occurs less frequently [63].

Dietary control and attitude shifts. Eliminating caffeine, chocolate, and alcohol may reduce symptoms; hence, the need to avoid or limit them in their diet. Not smoking and reducing stress also may help lessen the symptoms [64]. SPC (special processed cereals, in flakes) show a significant reduction in vertigo spells and a positive effect on tinnitus severity [65].

Triple semicircular canal plugging (TSCP). A study with a total of 116 MD patients revealed that this technique was comparable to labyrinthectomy for the control of vertigo in intractable MD, representing an effective therapy [66].

3. Ménière's Disease and Brain Function

Understandably, available research on cognitive function in Ménière's disease (MD) human patients is quite limited. Cognitive function, hearing thresholds, emotional stress, and speech discrimination scores were recently studied in patients using detailed neuropsychological tests to measure how well a person's brain is working [67].

No gold standard diagnostic test for Ménière's disease exists. MD is distinguished by discontinuous events of vertigo, fluctuating sensorineural hearing loss, tinnitus, and the feeling of fullness and stuffiness in the ear (aural pressure), while some consider that it could be a migraine-related phenomenon [68].

Vestibular dysfunction is considered a likely adjustable risk factor for cognitive decline [69] on occasions combined with "brain fog" in the form of dullness, difficulty concentrating, poor memory, or confusion [70]. In MD patients, anxiety, depression, panic, and dyssomnia may arbitrate the link between vestibule dysfunction and cognition [71].

This is observed namely in Ménière's Disease chronic vestibular syndromes and migraine-associated dizziness, when compared to acute spinning vertigo such as benign paroxysmal positional vertigo (BPPV) [72]. The same vulnerability to derangements in homeostasis may also explain the common triggering factors of both MD attacks and migraine headaches, including stress, weather, and diet. Therefore, different foods and nutrients may assist in maintaining a metabolic balance, which is essential to both maintain energy homeostasis and prevent MD and neurological disorders [73]. The outlook was further complicated when hearing loss was recognized as a risk factor for dementia, widening the spectrum of negative consequences [74].

There are still no studies aimed at matching cognitive performance and multiple clinical features of these MD patients, or on whether mental personality changes could be protected or improved after successful treatment [75]. Recent studies with some 500 patients confirm a possible link between late-onset MD and an increased incidence of all-cause dementia, where the loss of hearing and vestibular function emerged as an important risk factor for senile mental illness [76].

Despite few claims [76], to date, the relationship between MD and dementia, including Alzheimer's disease and vascular dementia, has not yet been clarified. The absence of research and data in this field is justified by the lack of neurotologist experts and work being performed by general neurologists [77].

The clinical grading of tinnitus severity (ringing, rustling, or buzzing sound) has brought renewed hope to the treatment or prevention of auditory neurodegeneration. However, a correct diagnosis of the underlying vestibular disorder is necessary and new techniques to diagnose these disorders have been developed, where machine learning methods have the potential to perform better than clinical scores [78,79].

The goal of this review is to provide clues about how mushroom nutritional studies may help to better understand the tight relationship between food, metabolic balance, MD, brain activity, and aging.

4. Five Way Interactions and Ménière's Disease

The five way event interactions between Environment–Host–Drug–Microbiota–Nutrient reflect the importance of gut metabolites in energy metabolism, cell communication, and host immunity, and on the mediation of several physiological activities [80] (Figure 3).

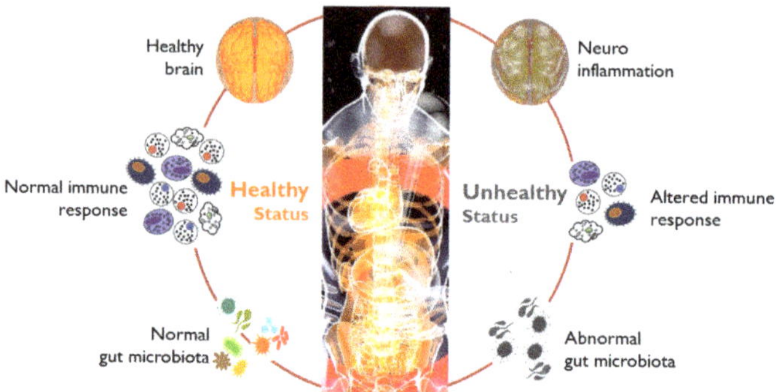

Figure 3. The multiple role of gut microbiome in neuroinflammation. Adapted from [81].

The correlation between gut microbiota neurotransmitters (e.g., dopamine, serotonin, norepinephrine, and δ-amino butyric acids-GABA) and brain functions has been well described under the designation of the "gut-brain axis" [82,83].

Numerous studies have enhanced the existing interaction between the microbiota and human health and disease [84]. Although the composition of gut microbiota remains quite stable since birth and in eubiosis, possible changes may induce dysbiosis and damages to the intestinal cell wall villi which may induce inflammatory reactions [85].

Considering that 99% of the genetic information in the human body is from the oral and gut microbiome, since microbiota (e.g., Firmicutes and Bacteriodetes) have an estimated 3.3 million genes compared to a human's 23,000 [86], this is extremely critical for regulating gut metabolism, which is also important for the human immune system [87].

The multidirectional communication between loose nutrients in the lumen, gut microbiota, and the brain and other tissues is related to several neurological disorders [88], but specific crosstalk between the microbiome, gut, and host has not been clearly elucidated in relation to diet [39,89].

5. Nutrition in Idiopathic Endolymphatic Hydrops

Diet is a key element in healthy living and adequate nutrition impacts metabolic balance as a key component of healthy aging; however, little is known about the mechanisms involved in the beneficial effects of diets on cognitive function and general health [90].

It is important to remember that studies on healthy populations are infrequent and there has been a limited number of epidemiological studies on Ménière's disease, characterized by spontaneous vertigo attacks (each lasting 20 min to 12 h). The disease is presumed to have a frequency of some 0.2% in western countries, with 50,000 people diagnosed annually [40], namely in people aged >65 years [91].

An array of potential etiopathologies have been pinpointed during the course of MD. Perhaps the most entrenched is the finding of endolymphatic hydrops (ELH). These are considered the main etiology of Ménière's disease; however, it seems that supplementary intermediate cofactors (e.g., allergies, viral infections) are necessary to develop the usual symptoms [92]. The role of the endolymphatic sac as an immune mediator for the middle ear and hormonal mechanisms presents other modes by which MD could be provoked, with endolymph in excess [93].

To prevent permanent effects of the symptoms on the hearing and balance system, changes in lifestyle are promoted. Dietary recommendations for the control of MD include abundant water intake, a low salt diet, moderate alcohol and caffeine consumption, and a gluten-free diet [94], together with diuretics, vasodilator corticosteroids, and intra-tympanic steroids [95]. Taking extra vitamin D plus calcium is considered to cut the odds of getting a debilitating form of benign paroxysmal positional vertigo [96]. However, the daily requirement of vitamin D has been recently under debate as most guidelines state 600–800 IU daily when other authors claim that these levels are low and should be around 8000 IU for young adults and thereafter, or even 20,000 IU once every two weeks [97,98].

However, although some research has revealed an improvement with these first line dietary mediations, namely in the control of relapsing, presently there is no constant agreement on their effectiveness [99].

6. Mushroom Nutrition in Neurodegenerative Diseases

Ménière's disease has a degenerative course that often results in permanent sensorineural hearing loss with different "clinical phenotypes", but its etiology remains elusive [100]. The pivotal role of mitochondria in redox regulation and oxidative stress has a critical performance in the development of several age-related conditions and several chronic diseases, but it can also be considered as a healing perspective to certain clinical conditions [101].

The brain and nervous tissues have a large potential oxidative capacity but a limited ability to counteract oxidative stress [102]. The administration of mushroom nutritional supplements has been the subject of research in several diseases, mostly associated with enhancement of antioxidant factors against oxidative stress and free-radical-induced cell damage [103].

Mushroom supplementation represents a valid support in health-promoting strategies and has shown effective prophylactic and therapeutic antioxidant intervention to maintain the wholeness and persistence of neurons and to oppose age-related neurodegenerative pathologies [104,105].

The possible mechanisms of action of edible mushrooms on preventing several age-based neuronal diseases are still undisclosed, but it is advocated that they could contribute through a reduction in oxidative stress, neuroinflammation, and on the modulation of acetylcholinesterase activity, protecting neurons or stimulation, and regulating neurotrophins synthesis, on the rough endoplasmic reticulum [106,107].

The tripeptide glutathione, commonplace in every cell, is a reliable biomarker for the redox balance, being reduced in neurodegenerative disorders such as stroke, and Alzheimer's, Huntington's, and Parkinson's disease. Many edible mushroom species (e.g., *Hericium erinaceus, Ganoderma lucidum, Agaricus bisporus, Grifola frondosa, Pleurotus ostreatus, Lentinula edodes*) are reliable sources of glutathione and thus are good nutritional supporters of the regulation of homeostasis and metabolism in the nervous system [108].

Vitagenes are genes implicated in cellular homeostasis by perceiving the intracellular nutrient and energy status, the functional state of mitochondria, and the concentration of ROS produced in mitochondria [109].

These vitagenes encode for heat shock proteins, the small ubiquitous redox proteins, and the sirtuin family of signaling protein systems, which are significant in longevity processes [110]. Dietary antioxidants from exogenous nutritional approaches, such as mushrooms, have recently been demonstrated to be neuroprotective through the activation of hormetic pathways, including vitagenes [111].

Brain neuroinflammation has been linked to chronic neurodegenerative disorders, including: Amyotrophic Lateral Sclerosis (ALS), Multiple Sclerosis (MS), Parkinson´s disease (PD), Alzheimer´s disease (AD), Dementia with Lewy bodies (DLB), depression and stress, psychosis, cognitive functions, and aging [110].

Heat shock proteins (HSPs) are one of the major groups of proteins which help respond to and mitigate stresses. To cope with stress, organisms, including mushrooms, express Hsps or chaperons to stabilize client proteins involved in various cell functions in fungi [112]. The HSP mushroom-derived lipoxin A4 (LXA4) is a short-lived endogenous bioactive lipid eicosanoid (oxidized derivatives of arachidonic acid) able to promote the resolution of inflammation, acting as an endogenous "braking signal" in the inflammatory process [113,114].

Lipoxin A4 may serve as biomarker and play a significant role in several auto-immune diseases [115]. *Hericium erinaceus* and *Coriolus versicolor* mushrooms administered to mice were found to be neuroprotectors through their ability to increase levels of the anti-inflammatory mediator lipoxin A4 [114,116].

7. Targeting Neurogenesis with Mushroom Nutraceuticals

Neurogenesis, or formation of neurons de novo, is the process by which new neurons are formed in the brain even late throughout one's lifespan. The mature brain has many specialized areas of function and neurons that differ in structure and connections. The hippocampus alone, which is a brain region that plays an important role in memory function and spatial navigation, has at least 122 different types of neurons [117,118].

We have previously reviewed this subject showing that ongoing neurogenesis does decline with age, which is possibly linked to compromised neurocognitive-psychological human resilience. Hippocampal neurogenesis drops sharply during the early stages of Alzheimer's disease, while older individuals have less angiogenesis and neuroplasticity and a smaller quiescent neural stem cell pool [119,120].

Ménière's disease patients, exposed to chronic stress, also have significantly decreased hippocampal volume, interrelated with memory and key clinical, vascular, and genetic risk factors, which is consistent with severity hyperacusis and vestibular balance disorders of the affected side [121,122].

We have previously evaluated in mice the safety and toxicity of *Coriolus versicolor* based on EU guidelines [123]. Other in vivo trials with mice fed an edible mushroom, *Coriolus versicolor*, revealed no change in the dentate gyrus volume or proliferation in newly generated neurons. It was found that mice treated with this mushroom biomass supplementation had a significant increase in the complexity of the long and short immature neurons (increase in dendritic complexity) [118].

This indicated that *Coriolus versicolor* biomass promoted hippocampal neurogenic reserve in mice by increasing levels of β-catenin in the nucleus and cytoplasm of newly developed neurons, which may translate into enhanced cognitive reserves which are essential for learning and memory [124].

Although mushroom bioactive compounds elicit beneficial health outcomes, exercised via numerous approaches [125], little consideration is given to how their elements may generate internal mechanisms of safeguarding immunity by modulating cellular signaling, processes such as key transcription factors, regulating the pathways and cellular responses against reactive electrophilic and oxygen species stresses [126].

8. Mushroom Nutrition in Ménière's Disease

Dizzy spells and vertigo may be caused by different factors and may cause nausea and vomiting. Diet and dehydration can also cause blood pressure to drop, which can lead to dizzy spells. Prolonged episodes of whirling vertigo along with hearing problems in one ear could be Ménière's, while frequent bouts of dizziness and vertigo can also indicate B12 deficiency [127].

Immune system dysregulation is increasingly being attributed to the development of a multitude of neurodegenerative diseases [128]. It is admissible that MD, as a systemic oxidant disorder involved in its pathogenesis and by the neurodegenerative nature of the inner ear cochlear spiral ganglion neurons, can be considered a neurodegenerative disorder [129].

Many studies have reported that *Coriolus versicolor* has several well researched effects, namely antioxidant, hypoglycemic, and immune-enhancing outcomes (Figure 4).

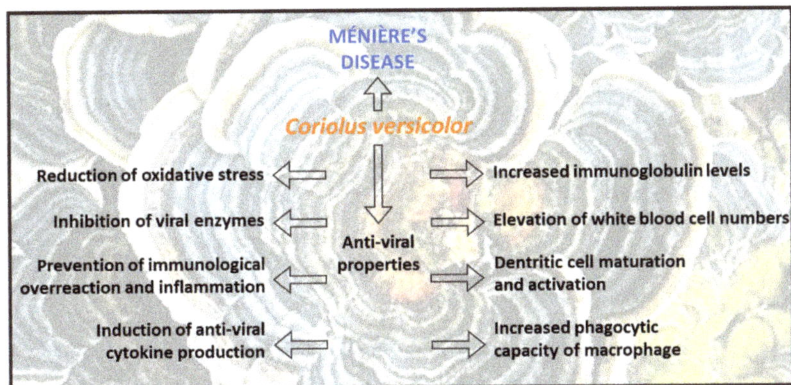

Figure 4. Some of the multiple roles of *Coriolus versicolor*.

One emerging and complementary strategy in tackling MD is the nutritional supplementation with mushrooms. In a specific and pioneer human trial, conducted by the Italian team of Professor V. Calabrese on 40 MD patients, it was evaluated the neurotoxic insult as a critical primary mediator operating in MD pathogenesis, exhibited by quantitative changes in biomarkers of oxidative stress and cellular stress response in the peripheral blood of MD patients (Figure 5) [130].

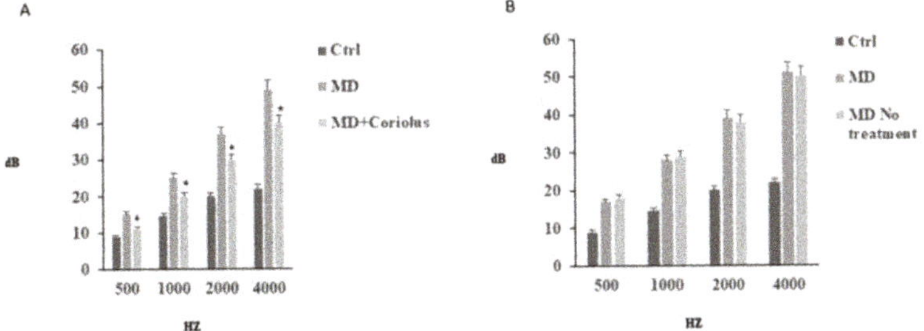

Figure 5. Tonal audiometry analysis. Tonal interest was centered on medium–high frequencies, with an average intensity of 55 dB loss. All subjects reported in both T0 (**B**) and T1 (**A**) phases no significant changes, either in the frequency range, or in the average loss in dB. Speech audiometry analysis, revealed in subjects of group A, who received mushrooms, a significant improvement of intellection threshold [130]. * Significantly different vs. untreated MD patients.

Comprehensive oxidative stress and a wide range of molecular changes in the cells of MD patients were investigated in the absence or in the presence of oral administration with a mushroom (*Coriolus versicolor*) biomass of commercial preparation (Manufactured by Mycology Research Laboratories Ltd., Luton, UK. *Coriolus versicolor* containing both mycelium and primordia biomass).

In a controlled study in 40 individuals with Meniere's disease (MD), the effects of 2 months of supplementation with 3 g/day of a biomass of commercial preparation from *Coriolus versicolor* (3 tablets of 500 mg every 12 h) on their peripheral blood antioxidant levels were measured to evaluate systemic oxidative stress and cellular stress response. This study has been replicated for a longer period (6 months) of administration with auspicious outcomes, but is presently still under publication [131].

With *Coriolus versicolor* treatment, it was observed in the plasma a significant stimulation of vitagenes (e.g., lipoxin A4, heat shock proteins 70, heme oxygenase-1, sirtuin-1, thioredoxin, and γ-GC ligase) and a significant increase in ratio-reduced glutathione vs. oxidized glutathione. This ratio is used as an indicator of cellular health [132,133].

This study also underlined the advantage of researching MD as a suitable facsimile of cochlear neuropathy spectrum disorder. Auditory neuropathy spectrum disorder (ANSD), designated as a spectrum since it effects each person in different ways, with symptoms ranging from mild to severe, a relatively rare form of sensorineural impairments and deafness, is characterized by a range of hearing impairments, namely compromised acoustic transmission from the inner ear to the brain due to defective synaptic function or neural conduction [134,135].

ANSD, a possible consequence of cranial nerve VIII degeneration with clinical profiles that are largely heterogeneous, can result from syndromic and non-syndromic (a partial or total loss of hearing) genetic abnormalities, as well as environmental causes (e.g., lack of oxygen, noise-pollution, chemotherapy drugs) and aging [136]. A rare degenerative ANSD familial disorder is caused by a riboflavin transporter deficiency, reflecting the need for high-dose oral supplementation of riboflavin (vitamin B_2) between 10 mg and 50 mg/kg/day, thus indicating the importance of diets [137,138].

Sensorineural hearing rehabilitation depends on the many varieties of etiologies and is currently being treated by genetic therapies which are in development. By looking for state-of-the-art activators of the vitagene system, the development of new pharmacological strategies will be possible, resulting in enhanced defense against energy and stress-resistant homeostasis disruption [139].

By incrementing the inherent pool of sensitive neurons, such as retinal ganglion cells, boosting anti-degenerative feedback, and through the study of major neurological biomarkers of brain disorders, it will be possible to deliver neurohealing, neurorescue, neuroregeneration, and neurorestauration [140].

Therefore, patients affected by Ménière´s disease are considered as being under conditions of systemic oxidative stress, and the induction of vitagenes by mushroom supplementation indicates a sustained response to counteract intracellular compounds that initiate, facilitate, or accelerate lipid oxidation [130].

9. Conclusions

The benefits and risks of different lifestyles and dietary changes for Ménière's disease are currently unclear and the efficacy of different dietary interventions for preventing vertigo attacks and their associated symptoms is still vague. However, it is clear from a study in 40 human MD patients that oral consumption of a biomass preparation, with 3 g/day from mushroom *Coriolus versicolor* for 2 or 6 months, reduced systemic oxidative stress and cellular stress response, decreasing the number of crises and their duration and the frequency of symptoms.

Novel techniques are being explored for the extraction of bioactive components from edible mushrooms and/or the use of complete biomass, and the nutraceutical potential of mushrooms needs to be investigated in clinical trials. Only a limited number of clinical trials have been carried out so far, mainly due to ethical reasons.

An understanding of the key drivers of the functional food market alongside a consistent and well-defined regulatory framework will provide further opportunities for growth, expansion, and segmentation of different applications of mushrooms.

Author Contributions: Conceptualization, T.H.F. and V.B.; investigation, T.H.F.; writing—original draft preparation, T.H.F.; writing—review and editing, T.H.F. and V.B. All authors have read and agreed to the published version of the manuscript.

Funding: This research received no external funding.

Conflicts of Interest: The authors declare no conflict of interest.

References

1. Galanakis, C.M. Functionality of Food Components and Emerging Technologies. *Foods* **2021**, *10*, 128. [CrossRef]
2. Shahidi, F. Nutraceuticals and Functional Foods: Whole versus Processed Foods. *Trends Food Sci. Technol.* **2009**, *20*, 376–387. [CrossRef]
3. Łysakowska, P.; Sobota, A.; Wirkijowska, A. Medicinal Mushrooms: Their Bioactive Components, Nutritional Value and Application in Functional Food Production—A Review. *Molecules* **2023**, *28*, 5393. [CrossRef]
4. Coppens, P.; Da Silva, M.F.; Pettman, S. European Regulations on Nutraceuticals, Dietary Supplements and Functional Foods: A Framework Based on Safety. *Toxicology* **2006**, *221*, 59–74. [CrossRef] [PubMed]
5. Rani, M.; Mondal, S.M.; Kundu, P.; Thakur, A.; Chaudhary, A.; Vashistt, J.; Shankar, J. Edible mushroom: Occurrence, management and health benefits. *Food Mater. Res.* **2023**, *3*, 21. [CrossRef]
6. Bell, V.; Silva, C.R.P.G.; Guina, J.; Fernandes, T.H. Mushrooms as Future Generation Healthy Foods. *Front. Nutr.* **2022**, *9*, 1050099. [CrossRef]
7. Yadav, D.; Negi, P.S. Bioactive Components of Mushrooms: Processing Effects and Health Benefits. *Food Res. Int.* **2021**, *148*, 110599. [CrossRef]
8. Chugh, R.M.; Mittal, P.; MP, N.; Arora, T.; Bhattacharya, T.; Chopra, H.; Cavalu, S.; Gautam, R.K. Fungal Mushrooms: A Natural Compound With Therapeutic Applications. *Front. Pharmacol.* **2022**, *13*, 925387. [CrossRef]
9. Dávila Giraldo, L.R.; Pérez Jaramillo, C.C.; Méndez Arteaga, J.J.; Murillo-Arango, W. Nutritional Value and Antioxidant, Antimicrobial and Cytotoxic Activity of Wild Macrofungi. *Microorganisms* **2023**, *11*, 1158. [CrossRef]
10. Galappaththi, M.C.A.; Patabendige, N.M.; Premarathne, B.M.; Hapuarachchi, K.K.; Tibpromma, S.; Dai, D.Q.; Suwannarach, N.; Rapior, S.; Karunarathna, S.C. A Review of Ganoderma Triterpenoids and Their Bioactivities. *Biomolecules* **2023**, *13*, 24. [CrossRef]
11. Bhambri, A.; Srivastava, M.; Mahale, V.G.; Mahale, S.; Karn, S.K. Mushrooms as Potential Sources of Active Metabolites and Medicines. *Front. Microbiol.* **2022**, *13*, 837266. [CrossRef] [PubMed]
12. Fernandes, A.; Nair, A.; Kulkarni, N.; Todewale, N.; Jobby, R. Exploring Mushroom Polysaccharides for the Development of Novel Prebiotics: A Review. *Int. J. Med. Mushrooms* **2023**, *25*, 1–10. [CrossRef]
13. Kerezoudi, E.N.; Mitsou, E.K.; Gioti, K.; Terzi, E.; Avgousti, I.; Panagiotou, A.; Koutrotsios, G.; Zervakis, G.I.; Mountzouris, K.C.; Tenta, R.; et al. Fermentation of *Pleurotus Ostreatus* and *Ganoderma Lucidum* Mushrooms and Their Extracts by the Gut Microbiota of Healthy and Osteopenic Women: Potential Prebiotic Effect and Impact of Mushroom Fermentation Products on Human Osteoblasts. *Food Funct.* **2021**, *12*, 1529–1546. [CrossRef] [PubMed]
14. Ayimbila, F.; Keawsompong, S. Nutritional Quality and Biological Application of Mushroom Protein as a Novel Protein Alternative. *Curr. Nutr. Rep.* **2023**, *12*, 290–307. [CrossRef] [PubMed]
15. González, A.; Cruz, M.; Losoya, C.; Nobre, C.; Loredo, A.; Rodríguez, R.; Contreras, J.; Belmares, R. Edible Mushrooms as a Novel Protein Source for Functional Foods. *Food Funct.* **2020**, *11*, 7400–7414. [CrossRef]
16. Rauf, A.; Joshi, P.B.; Ahmad, Z.; Hemeg, H.A.; Olatunde, A.; Naz, S.; Hafeez, N.; Simal-Gandara, J. Edible Mushrooms as Potential Functional Foods in Amelioration of Hypertension. *Phytother. Res.* **2023**, *37*, 2644–2660. [CrossRef]
17. Singh, R.S.; Walia, A.K.; Kennedy, J.F. Mushroom Lectins in Biomedical Research and Development. *Int. J. Biol. Macromol.* **2020**, *151*, 1340–1350. [CrossRef]
18. Guillamón, E.; García-Lafuente, A.; Lozano, M.; D'arrigo, M.; Rostagno, M.A.; Villares, A.; Martínez, J.A. Edible Mushrooms: Role in the Prevention of Cardiovascular Diseases. *Fitoterapia* **2010**, *81*, 715–723. [CrossRef]
19. Sousa, A.S.; Araújo-Rodrigues, H.; Pintado, M.E. The Health-Promoting Potential of Edible Mushroom Proteins. *Curr. Pharm. Des.* **2022**, *29*, 804–823. [CrossRef]
20. El-Maradny, Y.A.; El-Fakharany, E.M.; Abu-Serie, M.M.; Hashish, M.H.; Selim, H.S. Lectins Purified from Medicinal and Edible Mushrooms: Insights into Their Antiviral Activity against Pathogenic Viruses. *Int. J. Biol. Macromol.* **2021**, *179*, 239–258. [CrossRef]
21. Zhou, R.; Liu, Z.K.; Zhang, Y.N.; Wong, J.H.; Ng, T.B.; Liu, F. Research Progress of Bioactive Proteins from the Edible and Medicinal Mushrooms. *Curr. Protein Pept. Sci.* **2018**, *20*, 196–219. [CrossRef]
22. Tian, X.; Thorne, J.L.; Moore, J.B. Ergothioneine: An Underrecognised Dietary Micronutrient Required for Healthy Ageing? *Br. J. Nutr.* **2023**, *129*, 104–114. [CrossRef]
23. Turck, D.; Bresson, J.; Burlingame, B.; Dean, T.; Fairweather-Tait, S.; Heinonen, M.; Hirsch-Ernst, K.I.; Mangelsdorf, I.; McArdle, H.J.; Naska, A.; et al. Safety of Synthetic L-ergothioneine (Ergoneine®) as a Novel Food Pursuant to Regulation (EC) No 258/97. *EFSA J.* **2016**, *14*, e04629. [CrossRef]
24. Turck, D.; Bresson, J.L.; Burlingame, B.; Dean, T.; Fairweather-Tait, S.; Heinonen, M.; Hirsch-Ernst, K.I.; Mangelsdorf, I.; McArdle, H.J.; Naska, A.; et al. Statement on the Safety of Synthetic L-Ergothioneine as a Novel Food—Supplementary Dietary Exposure and Safety Assessment for Infants and Young Children, Pregnant and Breastfeeding Women. *EFSA J.* **2017**, *15*, e05060. [CrossRef]
25. Halliwell, B.; Cheah, I.K.; Tang, R.M.Y. Ergothioneine—A Diet-Derived Antioxidant with Therapeutic Potential. *FEBS Lett.* **2018**, *592*, 3357–3366. [CrossRef]
26. Fu, T.T.; Shen, L. Ergothioneine as a Natural Antioxidant Against Oxidative Stress-Related Diseases. *Front. Pharmacol.* **2022**, *13*, 850813. [CrossRef] [PubMed]
27. Kalaras, M.D.; Richie, J.P.; Calcagnotto, A.; Beelman, R.B. Mushrooms: A Rich Source of the Antioxidants Ergothioneine and Glutathione. *Food Chem.* **2017**, *233*, 429–433. [CrossRef] [PubMed]

28. Gründemann, D.; Harlfinger, S.; Golz, S.; Geerts, A.; Lazar, A.; Berkels, R.; Jung, N.; Rubbert, A.; Schömig, E. Discovery of the Ergothioneine Transporter. *Proc. Natl. Acad. Sci. USA* **2005**, *102*, 5256–5261. [CrossRef]
29. Mittal, R.; Aranke, M.; Debs, L.H.; Nguyen, D.; Patel, A.P.; Grati, M.; Mittal, J.; Yan, D.; Chapagain, P.; Eshraghi, A.A.; et al. Indispensable Role of Ion Channels and Transporters in the Auditory System. *J. Cell. Physiol.* **2017**, *232*, 743–758. [CrossRef] [PubMed]
30. Katsube, M.; Watanabe, H.; Suzuki, K.; Ishimoto, T.; Tatebayashi, Y.; Kato, Y.; Murayama, N. Food-Derived Antioxidant Ergothioneine Improves Sleep Difficulties in Humans. *J. Funct. Foods* **2022**, *95*, 105165. [CrossRef]
31. Pang, L.; Wang, T.; Liao, Q.; Cheng, Y.; Wang, D.; Li, J.; Fu, C.; Zhang, C.; Zhang, J. Protective Role of Ergothioneine Isolated from Pleurotus Ostreatus against Dextran Sulfate Sodium-Induced Ulcerative Colitis in Rat Model. *J. Food Sci.* **2022**, *87*, 415–426. [CrossRef] [PubMed]
32. Törős, G.; El-Ramady, H.; Prokisch, J.; Velasco, F.; Llanaj, X.; Nguyen, D.H.H.; Peles, F. Modulation of the Gut Microbiota with Prebiotics and Antimicrobial Agents from Pleurotus Ostreatus Mushroom. *Foods* **2023**, *12*, 2010. [CrossRef] [PubMed]
33. Viceconte, F.R.; Diaz, M.L.; Soresi, D.S.; Lencinas, I.B.; Carrera, A.; Prat, M.I.; Gurovic, M.S.V. Ganoderma Sessile Is a Fast Polysaccharide Producer among Ganoderma Species. *Mycologia* **2021**, *113*, 513–524. [CrossRef] [PubMed]
34. Li, M.; Yu, L.; Zhao, J.; Zhang, H.; Chen, W.; Zhai, Q.; Tian, F. Role of Dietary Edible Mushrooms in the Modulation of Gut Microbiota. *J. Funct. Foods* **2021**, *83*, 104538. [CrossRef]
35. Yadav, D.; Negi, P.S. Role of Mushroom Polysaccharides in Improving Gut Health and Associated Diseases. In *Microbiome, Immunity, Digestive Health and Nutrition: Epidemiology, Pathophysiology, Prevention and Treatment*; Bagchi, D., Downs, B.W., Eds.; Academic Press: Cambridge, MA, USA, 2022; pp. 431–448, ISBN 9780128222386.
36. Ma, T.; Shen, X.; Shi, X.; Sakandar, H.A.; Quan, K.; Li, Y.; Jin, H.; Kwok, L.Y.; Zhang, H.; Sun, Z. Targeting Gut Microbiota and Metabolism as the Major Probiotic Mechanism—An Evidence-Based Review. *Trends Food Sci. Technol.* **2023**, *138*, 178–198. [CrossRef]
37. Zhao, J.; Hu, Y.; Qian, C.; Hussain, M.; Liu, S.; Zhang, A.; He, R.; Sun, P. The Interaction between Mushroom Polysaccharides and Gut Microbiota and Their Effect on Human Health: A Review. *Biology* **2023**, *12*, 122. [CrossRef]
38. Trovato-Salinaro, A.; Siracusa, R.; Di Paola, R.; Scuto, M.; Fronte, V.; Koverech, G.; Luca, M.; Serra, A.; Toscano, M.A.; Petralia, A.; et al. Redox Modulation of Cellular Stress Response and Lipoxin A4 Expression by Coriolus Versicolor in Rat Brain: Relevance to Alzheimer's Disease Pathogenesis. *Neurotoxicology* **2016**, *53*, 350–358. [CrossRef]
39. Sergio, M.; Gabriella, L.; Mario, T.; Francesco, R.; Marialaura, O.; Maria, S.; Angela, T.S.; Antonio, A.; Daniela, A.C.; Maria, L.; et al. Antioxidants, Hormetic Nutrition, and Autism. *Curr. Neuropharmacol.* **2023**, *21*. [CrossRef]
40. Kumar, K.; Mehra, R.; Guiné, R.P.F.; Lima, M.J.; Kumar, N.; Kaushik, R.; Ahmed, N.; Yadav, A.N.; Kumar, H. Edible Mushrooms: A Comprehensive Review on Bioactive Compounds with Health Benefits and Processing Aspects. *Foods* **2021**, *10*, 2996. [CrossRef]
41. Benson, K.F.; Stamets, P.; Davis, R.; Nally, R.; Taylor, A.; Slater, S.; Jensen, G.S. The Mycelium of the *Trametes versicolor* (Turkey Tail) Mushroom and Its Fermented Substrate Each Show Potent and Complementary Immune Activating Properties in Vitro. *BMC Complement. Altern. Med.* **2019**, *19*, 342. [CrossRef]
42. Kıvrak, İ.; Kıvrak, Ş.; Karababa, E. Assessment of Bioactive Compounds and Antioxidant Activity of Turkey Tail Medicinal Mushroom *Trametes versicolor* (Agaricomycetes). *Int. J. Med. Mushrooms* **2020**, *22*, 559–571. [CrossRef]
43. Karmali, A.; Bugalho, A.; Fernandes, T. Coriolus Versicolor Supplementation in CIN-1 (LSIL) HPV Infection: Mode of Action. *Clin. J. Mycol.* **2007**, *2*, 6–10.
44. Moshtaghi, O.; Sahyouni, R.; Lin, H.W.; Ghavami, Y.; Djalilian, H.R. A Historical Recount: Discovering Menière's Disease and Its Association with Migraine Headaches. *Otol. Neurotol.* **2016**, *37*, 1199–1203. [CrossRef]
45. Nakashima, T.; Pyykkö, I.; Arroll, M.A.; Casselbrant, M.L.; Foster, C.A.; Manzoor, N.F.; Megerian, C.A.; Naganawa, S.; Young, Y.H. Meniere's Disease. *Nat. Rev. Dis. Primers* **2016**, *2*, 16028. [CrossRef] [PubMed]
46. Huang, C.; Wang, Q.; Pan, X.; Li, W.; Liu, W.; Jiang, W.; Huang, L.; Peng, A.; Zhang, Z. Up-Regulated Expression of Interferon-Gamma, Interleukin-6 and Tumor Necrosis Factor-Alpha in the Endolymphatic Sac of Meniere's Disease Suggesting the Local Inflammatory Response Underlies the Mechanism of This Disease. *Front. Neurol.* **2022**, *13*, 781031. [CrossRef]
47. Møller, M.N.; Kirkeby, S.; Caye-Thomasen, P. Innate Immune Defense in the Inner Ear—Mucines Are Expressed by the Human Endolymphatic Sac. *J. Anat.* **2017**, *230*, 297–302. [CrossRef]
48. Koenen, L.; Andaloro, C. *Meniere Disease*; StatPearls Publishing: Treasure Island, FL, USA, 2023.
49. Koç, A. Benign Paroxysmal Positional Vertigo: Is It Really an Otolith Disease? *J. Int. Adv. Otol.* **2022**, *18*, 62–70. [CrossRef]
50. Weinreich, H.M.; Agrawal, Y. The Link Between Allergy and Menière's Disease. *Curr. Opin. Otolaryngol. Head Neck Surg.* **2014**, *22*, 227. [CrossRef]
51. Pan, T.; Zhao, Y.; Ding, Y.J.; Lu, Z.Y.; Ma, F.R. The Pilot Study of Type I Allergic Reaction in Meniere's Disease Patients. *Chin. J. Otorhinolaryngol. Head Neck Surg.* **2017**, *52*, 89–92. [CrossRef]
52. Dai, Q.; Wang, D.; Zheng, H. The Polymorphic Analysis of the Human Potassium Channel Kcne Gene Family in Meniere's Disease. A Preliminary Study. *J. Int. Adv. Otol.* **2019**, *15*, 130–134. [CrossRef]
53. Martín-Sierra, C.; Gallego-Martinez, A.; Requena, T.; Frejo, L.; Batuecas-Caletrió, A.; Lopez-Escamez, J.A. Variable Expressivity and Genetic Heterogeneity Involving DPT and SEMA3D Genes in Autosomal Dominant Familial Meniere's Disease. *Eur. J. Hum. Genet.* **2016**, *25*, 200–207. [CrossRef] [PubMed]

54. Lopes, K.D.C.; Sartorato, E.L.; Da Silva-Costa, S.M.; De Macedo Adamov, N.S.; Ganança, F.F. Ménière's Disease: Molecular Analysis of Aquaporins 2, 3 and Potassium Channel KCNE1 Genes in Brazilian Patients. *Otol. Neurotol.* **2016**, *37*, 1117–1121. [CrossRef] [PubMed]
55. Doi, K.; Sato, T.; Kuramasu, T.; Hibino, H.; Kitahara, T.; Horii, A.; Matsushiro, N.; Fuse, Y.; Kubo, T. Ménière's Disease Is Associated with Single Nucleotide Polymorphisms in the Human Potassium Channel Genes, KCNE1 and KCNE3. *ORL* **2005**, *67*, 289–293. [CrossRef] [PubMed]
56. Brandalise, F.; Roda, E.; Ratto, D.; Goppa, L.; Gargano, M.L.; Cirlincione, F.; Priori, E.C.; Venuti, M.T.; Pastorelli, E.; Savino, E.; et al. Hericium Erinaceus in Neurodegenerative Diseases: From Bench to Bedside and Beyond, How Far from the Shoreline? *J. Fungi* **2023**, *9*, 551. [CrossRef]
57. NIH. Ménière's Disease. Available online: https://www.nidcd.nih.gov/health/menieres-disease (accessed on 20 March 2023).
58. Shih, R.D.; Walsh, B.; Eskin, B.; Allegra, J.; Fiesseler, F.W.; Salo, D.; Silverman, M. Diazepam and Meclizine Are Equally Effective in the Treatment of Vertigo: An Emergency Department Randomized Double-Blind Placebo-Controlled Trial. *J. Emerg. Med.* **2017**, *52*, 23–27. [CrossRef]
59. Watkins, L.E.; Sprang, K.R.; Rothbaum, B.O. Treating PTSD: A Review of Evidence-Based Psychotherapy Interventions. *Front. Behav. Neurosci.* **2018**, *12*, 400414. [CrossRef] [PubMed]
60. de Cates, C.; Winters, R. *Intratympanic Steroid Injection*; StatPearls Publishing: Treasure Island, FL, USA, 2023.
61. Russo, F.Y.; Nguyen, Y.; De Seta, D.; Bouccara, D.; Sterkers, O.; Ferrary, E.; Bernardeschi, D. Meniett Device in Meniere Disease: Randomized, Double-Blind, Placebo-Controlled Multicenter Trial. *Laryngoscope* **2017**, *127*, 470–475. [CrossRef] [PubMed]
62. Webster, K.E.; George, B.; Galbraith, K.; Harrington-Benton, N.A.; Judd, O.; Kaski, D.; Maarsingh, O.R.; MacKeith, S.; Ray, J.; Van Vugt, V.A.; et al. Positive pressure therapy for Ménière's disease. *Cochrane Database Syst. Rev.* **2023**, *2*, CD015248. [CrossRef]
63. Pullens, B.; Verschuur, H.P.; van Benthem, P.P. Surgery for Ménière's disease. *Cochrane Database Syst. Rev.* **2013**, *2*, CD005395. [CrossRef]
64. Chiarella, G.; Marcianò, G.; Viola, P.; Palleria, C.; Pisani, D.; Rania, V.; Casarella, A.; Astorina, A.; Scarpa, A.; Esposito, M.; et al. Nutraceuticals for Peripheral Vestibular Pathology: Properties, Usefulness, Future Perspectives and Medico-Legal Aspects. *Nutrients* **2021**, *13*, 3646. [CrossRef]
65. Viola, P.; Pisani, D.; Scarpa, A.; Cassandro, C.; Laria, C.; Aragona, T.; Ciriolo, M.; Spadera, L.; Ralli, M.; Cavaliere, M.; et al. The Role of Endogenous Antisecretory Factor (AF) in the Treatment of Ménière's Disease: A Two-Year Follow-up Study. Preliminary Results. *Am. J. Otolaryngol.* **2020**, *41*, 102673. [CrossRef] [PubMed]
66. Li, X.; Lyu, Y.; Li, Y.; Jian, H.; Wang, J.; Song, Y.; Kong, L.; Fan, Z.; Wang, H.; Zhang, D. Triple Semicircular Canal Plugging versus Labyrinthectomy for Meniere Disease: A Retrospective Study. *Laryngoscope* **2023**, *133*, 3178–3184. [CrossRef] [PubMed]
67. Eraslan Boz, H.; Kırkım, G.; Koçoğlu, K.; Çakır Çetin, A.; Akkoyun, M.; Güneri, E.A.; Akdal, G. Cognitive Function in Meniere's Disease. *Psychol. Health Med.* **2023**, *28*, 1076–1086. [CrossRef] [PubMed]
68. Sarna, B.; Abouzari, M.; Lin, H.W.; Djalilian, H.R. A Hypothetical Proposal for Association between Migraine and Meniere's Disease. *Med. Hypotheses* **2020**, *134*, 109430. [CrossRef]
69. Zhong, J.; Li, X.; Xu, J.; Chen, W.; Gao, J.; Lu, X.; Liang, S.; Guo, Z.; Lu, M.; Li, Y.; et al. Analysis of Cognitive Function and Its Related Factors after Treatment in Meniere's Disease. *Front. Neurosci.* **2023**, *17*, 1137734. [CrossRef]
70. Chari, D.A.; Liu, Y.H.; Chung, J.J.; Rauch, S.D. Subjective Cognitive Symptoms and Dizziness Handicap Inventory (DHI) Performance in Patients With Vestibular Migraine and Ménière's Disease. *Otol. Neurotol.* **2021**, *42*, 883–889. [CrossRef]
71. Xie, D.; Welgampola, M.S.; Miller, L.A.; Young, A.S.; D'Souza, M.; Breen, N.; Rosengren, S.M. Subjective Cognitive Dysfunction in Patients with Dizziness and Vertigo. *Audiol. Neurotol.* **2022**, *27*, 122–132. [CrossRef]
72. Dornhoffer, J.R.; Liu, Y.F.; Zhao, E.E.; Rizk, H.G. Does Cognitive Dysfunction Correlate with Dizziness Severity in Ménière's Disease Patients. *Otol. Neurotol.* **2021**, *42*, E323–E331. [CrossRef]
73. Cornelius, C.; Perrotta, R.; Graziano, A.; Calabrese, E.J.; Calabrese, V. Stress Responses, Vitagenes and Hormesis as Critical Determinants in Aging and Longevity: Mitochondria as a "Chi". *Immun. Ageing* **2013**, *10*, 15. [CrossRef]
74. Livingston, G.; Huntley, J.; Sommerlad, A.; Ames, D.; Ballard, C.; Banerjee, S.; Brayne, C.; Burns, A.; Cohen-Mansfield, J.; Cooper, C.; et al. Dementia Prevention, Intervention, and Care: 2020 Report of the Lancet Commission. *Lancet* **2020**, *396*, 413–446. [CrossRef]
75. Jia, X.; Wang, Z.; Huang, F.; Su, C.; Du, W.; Jiang, H.; Wang, H.; Wang, J.; Wang, F.; Su, W.; et al. A Comparison of the Mini-Mental State Examination (MMSE) with the Montreal Cognitive Assessment (MoCA) for Mild Cognitive Impairment Screening in Chinese Middle-Aged and Older Population: A Cross-Sectional Study. *BMC Psychiatry* **2021**, *21*, 485. [CrossRef]
76. Lee, I.H.; Yu, H.; Ha, S.S.; Son, G.M.; Park, K.J.; Lee, J.J.; Kim, D.K. Association between Late-Onset Ménière's Disease and the Risk of Incident All-Cause Dementia. *J. Pers. Med.* **2021**, *12*, 19. [CrossRef] [PubMed]
77. Halmágyi, G.M.; Akdal, G.; Welgampola, M.S.; Wang, C. Neurological Update: Neuro-Otology 2023. *J. Neurol.* **2023**, *270*, 6170–6192. [CrossRef] [PubMed]
78. Kabade, V.; Hooda, R.; Raj, C.; Awan, Z.; Young, A.S.; Welgampola, M.S.; Prasad, M. Machine Learning Techniques for Differential Diagnosis of Vertigo and Dizziness: A Review. *Sensors* **2021**, *21*, 7565. [CrossRef] [PubMed]
79. Ahmadi, S.A.; Vivar, G.; Navab, N.; Möhwald, K.; Maier, A.; Hadzhikolev, H.; Brandt, T.; Grill, E.; Dieterich, M.; Jahn, K.; et al. Modern Machine-Learning Can Support Diagnostic Differentiation of Central and Peripheral Acute Vestibular Disorders. *J. Neurol.* **2020**, *267*, 143–152. [CrossRef] [PubMed]

80. Zhang, Y.; Chen, R.; Zhang, D.D.; Qi, S.; Liu, Y. Metabolite Interactions between Host and Microbiota during Health and Disease: Which Feeds the Other? *Biomed. Pharmacother.* **2023**, *160*, 114295. [CrossRef]

81. Farooq, R.K.; Alamoudi, W.; Alhibshi, A.; Rehman, S.; Sharma, A.R.; Abdulla, F.A. Varied Composition and Underlying Mechanisms of Gut Microbiome in Neuroinflammation. *Microorganisms* **2022**, *10*, 705. [CrossRef]

82. Fung, T.C.; Olson, C.A.; Hsiao, E.Y. Interactions between the Microbiota, Immune and Nervous Systems in Health and Disease. *Nat. Neurosci.* **2017**, *20*, 145–155. [CrossRef]

83. Strandwitz, P. Neurotransmitter Modulation by the Gut Microbiota. *Brain Res.* **2018**, *1693*, 128–133. [CrossRef]

84. Altveş, S.; Yildiz, H.K.; Vural, H.C. Interaction of the Microbiota with the Human Body in Health and Diseases. *Biosci. Microbiota Food Health* **2020**, *39*, 23–32. [CrossRef]

85. Fakharian, F.; Thirugnanam, S.; Welsh, D.A.; Kim, W.K.; Rappaport, J.; Bittinger, K.; Rout, N. The Role of Gut Dysbiosis in the Loss of Intestinal Immune Cell Functions and Viral Pathogenesis. *Microorganisms* **2023**, *11*, 1849. [CrossRef] [PubMed]

86. Zhu, B.; Wang, X.; Li, L. Human Gut Microbiome: The Second Genome of Human Body. *Protein Cell* **2010**, *1*, 718–725. [CrossRef] [PubMed]

87. Elzayat, H.; Mesto, G.; Al-Marzooq, F. Unraveling the Impact of Gut and Oral Microbiome on Gut Health in Inflammatory Bowel Diseases. *Nutrients* **2023**, *15*, 3377. [CrossRef]

88. Sittipo, P.; Choi, J.; Lee, S.; Lee, Y.K. The Function of Gut Microbiota in Immune-Related Neurological Disorders: A Review. *J. Neuroinflamm.* **2022**, *19*, 154. [CrossRef] [PubMed]

89. Beane, K.E.; Redding, M.C.; Wang, X.; Pan, J.H.; Le, B.; Cicalo, C.; Jeon, S.; Kim, Y.J.; Lee, J.H.; Shin, E.C.; et al. Effects of Dietary Fibers, Micronutrients, and Phytonutrients on Gut Microbiome: A Review. *Appl. Biol. Chem.* **2021**, *64*, 36. [CrossRef]

90. Carneiro, L.; Pellerin, L. Nutritional Impact on Metabolic Homeostasis and Brain Health. *Front. Neurosci.* **2022**, *15*, 767405. [CrossRef]

91. Jeng, Y.; Young, Y.H. Evolution of Geriatric Meniere's Disease during the Past Two Decades. *J. Formos. Med. Assoc.* **2023**, *122*, 65–72. [CrossRef]

92. Basura, G.J.; Adams, M.E.; Monfared, A.; Schwartz, S.R.; Antonelli, P.J.; Burkard, R.; Bush, M.L.; Bykowski, J.; Colandrea, M.; Derebery, J.; et al. Clinical Practice Guideline: Ménière's Disease. *Otolaryngol. Neck Surg.* **2020**, *162*, S1–S55. [CrossRef]

93. Oberman, B.S.; Patel, V.A.; Cureoglu, S.; Isildak, H. The Aetiopathologies of Ménière's Disease: A Contemporary Review L'eziopatogenesi Della Sindrome Di Ménière: Stato Dell'arte. *Aggiorn. Acta Otorhinolaryngol. Ital.* **2017**, *37*, 250–263. [CrossRef]

94. De Luca, P.; Cassandro, C.; Ralli, M.; Gioacchini, F.M.; Turchetta, R.; Orlando, M.P.; Iaccarino, I.; Cavaliere, M.; Cassandro, E.; Scarpa, A. Dietary Restriction for The Treatment of Meniere's Disease. *Transl. Med. UniSa* **2020**, *22*, 5–9.

95. Oğuz, E.; Cebeci, A.; Geçici, C.R. The Relationship between Nutrition and Ménière's Disease. *Auris Nasus Larynx* **2021**, *48*, 803–808. [CrossRef] [PubMed]

96. Jeong, S.H.; Kim, J.S.; Kim, H.J.; Choi, J.Y.; Koo, J.W.; Choi, K.D.; Park, J.Y.; Lee, S.H.; Choi, S.Y.; Oh, S.Y.; et al. Prevention of Benign Paroxysmal Positional Vertigo with Vitamin D Supplementation. *Neurology* **2020**, *95*, e1117–e1125. [CrossRef] [PubMed]

97. McCullough, P.; Amend, J. Results of Daily Oral Dosing with up to 60,000 International Units (Iu) of Vitamin D3 for 2 to 6 Years in 3 Adult Males. *J. Steroid Biochem. Mol. Biol.* **2017**, *173*, 308–312. [CrossRef] [PubMed]

98. Papadimitriou, D.T. The Big Vitamin D Mistake. *J. Prev. Med. Public Health* **2017**, *50*, 278–281. [CrossRef] [PubMed]

99. Hussain, K.; Murdin, L.; Schilder, A.G. Restriction of salt, caffeine and alcohol intake for the treatment of Ménière's disease or syndrome. *Cochrane Database Syst. Rev.* **2018**, *12*, CD012173. [CrossRef] [PubMed]

100. Eckhard, A.H.; Zhu, M.Y.; O'Malley, J.T.; Williams, G.H.; Loffing, J.; Rauch, S.D.; Nadol, J.B.; Liberman, M.C.; Adams, J.C. Inner Ear Pathologies Impair Sodium-Regulated Ion Transport in Meniere's Disease. *Acta Neuropathol.* **2019**, *137*, 343–357. [CrossRef]

101. Atayik, M.C.; Çakatay, U. Redox Signaling in Impaired Cascades of Wound Healing: Promising Approach. *Mol. Biol. Rep.* **2023**, *50*, 6927–6936. [CrossRef]

102. Poon, H.F.; Calabrese, V.; Scapagnini, G.; Butterfield, D.A. Free Radicals: Key to Brain Aging and Heme Oxygenase as a Cellular Response to Oxidative Stress. *J. Gerontol. Ser. A* **2004**, *59*, M478–M493. [CrossRef]

103. Liuzzi, G.M.; Petraglia, T.; Latronico, T.; Crescenzi, A.; Rossano, R. Antioxidant Compounds from Edible Mushrooms as Potential Candidates for Treating Age-Related Neurodegenerative Diseases. *Nutrients* **2023**, *15*, 1913. [CrossRef]

104. Fekete, M.; Szarvas, Z.; Fazekas-Pongor, V.; Feher, A.; Csipo, T.; Forrai, J.; Dosa, N.; Peterfi, A.; Lehoczki, A.; Tarantini, S.; et al. Nutrition Strategies Promoting Healthy Aging: From Improvement of Cardiovascular and Brain Health to Prevention of Age-Associated Diseases. *Nutrients* **2022**, *15*, 47. [CrossRef]

105. Rai, S.N.; Mishra, D.; Singh, P.; Vamanu, E.; Singh, M.P. Therapeutic Applications of Mushrooms and Their Biomolecules along with a Glimpse of in Silico Approach in Neurodegenerative Diseases. *Biomed. Pharmacother.* **2021**, *137*, 111377. [CrossRef]

106. Jiang, X.; Li, S.; Feng, X.; Li, L.; Hao, J.; Wang, D.; Wang, Q. Mushroom Polysaccharides as Potential Candidates for Alleviating Neurodegenerative Diseases. *Nutrients* **2022**, *14*, 4833. [CrossRef]

107. Al-Qudah, M.A.; Al-Dwairi, A. Mechanisms and Regulation of Neurotrophin Synthesis and Secretion. *Neurosci. J.* **2016**, *21*, 306–313. [CrossRef]

108. Iskusnykh, I.Y.; Zakharova, A.A.; Pathak, D. Glutathione in Brain Disorders and Aging. *Molecules* **2022**, *27*, 324. [CrossRef]

109. Calabrese, V.; Cornelius, C.; Dinkova-Kostova, A.T.; Calabrese, E.J. Vitagenes, Cellular Stress Response, and Acetylcarnitine: Relevance to Hormesis. *BioFactors* **2009**, *35*, 146–160. [CrossRef] [PubMed]

110. Calabrese, V.; Ontario, M. Mushroom Nutrition In Neurodegenerative Diseases. *Clin. J. Mycol.* **2022**, *6*. [CrossRef]

111. Trovato Salinaro, A.; Pennisi, M.; Di Paola, R.; Scuto, M.; Crupi, R.; Cambria, M.T.; Ontario, M.L.; Tomasello, M.; Uva, M.; Maiolino, L.; et al. Neuroinflammation and Neurohormesis in the Pathogenesis of Alzheimer's Disease and Alzheimer-Linked Pathologies: Modulation by Nutritional Mushrooms. *Immun. Ageing* **2018**, *15*, 8. [CrossRef] [PubMed]

112. Tiwari, S.; Thakur, R.; Shankar, J. Role of Heat-Shock Proteins in Cellular Function and in the Biology of Fungi. *Biotechnol. Res. Int.* **2015**, *2015*, 132635. [CrossRef]

113. Zhao, X.; Yin, K.; Feng, R.; Miao, R.; Lin, J.; Cao, L.; Ni, Y.; Li, W.; Zhang, Q. Genome-Wide Identification and Analysis of the Heat-Shock Protein Gene in *L. Edodes* and Expression Pattern Analysis under Heat Shock. *Curr. Issues Mol. Biol.* **2023**, *45*, 614–627. [CrossRef]

114. Trovato, A.; Siracusa, R.; Di Paola, R.; Scuto, M.; Ontario, M.L.; Bua, O.; Di Mauro, P.; Toscano, M.A.; Petralia, C.C.T.; Maiolino, L.; et al. Redox Modulation of Cellular Stress Response and Lipoxin A4 Expression by Hericium Erinaceus in Rat Brain: Relevance to Alzheimer's Disease Pathogenesis. *Immun. Ageing* **2016**, *13*, 23. [CrossRef]

115. Das, U.N. Lipoxins as Biomarkers of Lupus and Other Inflammatory Conditions. *Lipids Health Dis.* **2011**, *10*, 76. [CrossRef]

116. Cordaro, M.; Modafferi, S.; D'Amico, R.; Fusco, R.; Genovese, T.; Peritore, A.F.; Gugliandolo, E.; Crupi, R.; Interdonato, L.; Di Paola, D.; et al. Natural Compounds Such as Hericium Erinaceus and Coriolus Versicolor Modulate Neuroinflammation, Oxidative Stress and Lipoxin A4 Expression in Rotenone-Induced Parkinson's Disease in Mice. *Biomedicines* **2022**, *10*, 2505. [CrossRef]

117. Ólafsdóttir, H.F.; Bush, D.; Barry, C. The Role of Hippocampal Replay in Memory and Planning. *Curr. Biol.* **2018**, *28*, R37–R50. [CrossRef] [PubMed]

118. Wheeler, D.W.; White, C.M.; Rees, C.L.; Komendantov, A.O.; Hamilton, D.J.; Ascoli, G.A. Hippocampome.Org: A Knowledge Base of Neuron Types in the Rodent Hippocampus. *Elife* **2015**, *4*, e09960. [CrossRef] [PubMed]

119. Ferreiro, E.; Fernandes, T. Targeting Neurogenesis with Mushroom Nutrition: A Mini Review. *Clin. J. Mycol.* **2022**, *6*. [CrossRef]

120. Audesse, A.J.; Webb, A.E. Mechanisms of Enhanced Quiescence in Neural Stem Cell Aging. *Mech. Ageing Dev.* **2020**, *191*, 111323. [CrossRef]

121. Ruan, J.; Hu, X.; Liu, Y.; Han, Z.; Ruan, Q. Vulnerability to Chronic Stress and the Phenotypic Heterogeneity of Presbycusis with Subjective Tinnitus. *Front. Neurosci.* **2022**, *16*, 1046095. [CrossRef] [PubMed]

122. Seo, Y.J.; Kim, J.; Kim, S.H. The Change of Hippocampal Volume and Its Relevance with Inner Ear Function in Meniere's Disease Patients. *Auris Nasus Larynx* **2016**, *43*, 620–625. [CrossRef]

123. Barros, A.B.; Ferrão, J.; Fernandes, T. A Safety Assessment of Coriolus Versicolor Biomass as a Food Supplement. *Food Nutr. Res.* **2016**, *60*, 29953. [CrossRef]

124. Piatti, V.C.; Ewe, L.A.; Leutgeb, J.K. Neurogenesis in the Dentate Gyrus: Carrying the Message or Dictating the Tone. *Front. Neurosci.* **2013**, *7*, 45461. [CrossRef]

125. Uffelman, C.N.; Chan, N.I.; Davis, E.M.; Wang, Y.; McGowan, B.S.; Campbell, W.W. An Assessment of Mushroom Consumption on Cardiometabolic Disease Risk Factors and Morbidities in Humans: A Systematic Review. *Nutrients* **2023**, *15*, 1079. [CrossRef]

126. He, F.; Ru, X.; Wen, T. NRF2, a Transcription Factor for Stress Response and Beyond. *Int. J. Mol. Sci.* **2020**, *21*, 4777. [CrossRef]

127. Özdemir, D.; Mehel, D.M.; Küçüköner, Ö.; Ağrı, İ.; Yemiş, T.; Akgül, G.; Özgür, A. Vestibular Evoked Myogenic Potentials in Patients With Low Vitamin B12 Levels. *Ear Nose Throat J.* **2021**, *100*, NP231–NP235. [CrossRef] [PubMed]

128. Jorfi, M.; Maaser-Hecker, A.; Tanzi, R.E. The Neuroimmune Axis of Alzheimer's Disease. *Genome Med.* **2023**, *15*, 6. [CrossRef] [PubMed]

129. Kishimoto-Urata, M.; Urata, S.; Fujimoto, C.; Yamasoba, T. Role of Oxidative Stress and Antioxidants in Acquired Inner Ear Disorders. *Antioxidants* **2022**, *11*, 1469. [CrossRef]

130. Scuto, M.; Di Mauro, P.; Ontario, M.L.; Amato, C.; Modafferi, S.; Ciavardelli, D.; Salinaro, A.T.; Maiolino, L.; Calabrese, V. Nutritional Mushroom Treatment in Meniere's Disease with Coriolus Versicolor: A Rationale for Therapeutic Intervention in Neuroinflammation and Antineurodegeneration. *Int. J. Mol. Sci.* **2019**, *21*, 284. [CrossRef]

131. Di Paola, R.; Siracusa, R.; Fusco, R.; Ontario, M.; Cammilleri, G.; Pantano, L.; Scuto, M.; Tomasello, M.; Spano, S.; Salinaro, A.T.; et al. Redox Modulation of Meniere Disease by Coriolus Versicolor Treatment, a Nutritional Mushroom Approach with Neuroprotective Potential. *Curr. Neuropharmacol.* **2023**, *in press.* CN-2023-0107.R2-MS.

132. Owen, J.B.; Allan Butterfiel, D. Measurement of Oxidized/Reduced Glutathione Ratio. *Methods Mol. Biol.* **2010**, *648*, 269–277.

133. Habtemariam, S. Trametes versicolor (Synn. *Coriolus versicolor*) Polysaccharides in Cancer Therapy: Targets and Efficacy. *Biomedicines* **2020**, *8*, 135. [CrossRef] [PubMed]

134. Saidia, A.R.; Ruel, J.; Bahloul, A.; Chaix, B.; Venail, F.; Wang, J. Current Advances in Gene Therapies of Genetic Auditory Neuropathy Spectrum Disorder. *J. Clin. Med.* **2023**, *12*, 738. [CrossRef]

135. Shearer, A.E.; Hansen, M.R. Auditory Synaptopathy, Auditory Neuropathy, and Cochlear Implantation. *Laryngoscope Investig. Otolaryngol.* **2019**, *4*, 429–440. [CrossRef]

136. Moser, T.; Starr, A. Auditory Neuropathy—Neural and Synaptic Mechanisms. *Nat. Rev. Neurol.* **2016**, *12*, 135–149. [CrossRef]

137. Cali, E.; Dominik, N.; Manole, A.; Houlden, H. Riboflavin Transporter Deficiency. In *GeneReviews*; Adam, M.P., Feldman, J., Mirzaa, G.M., Pagon, R.A., Wallace, S.E., Bean, L.J.H., Gripp, K.W., Amemiya, A., Eds.; University of Washington: Seattle, WA, USA, 2021.

138. De Siati, R.D.; Rosenzweig, F.; Gersdorff, G.; Gregoire, A.; Rombaux, P.; Deggouj, N. Auditory Neuropathy Spectrum Disorders: From Diagnosis to Treatment: Literature Review and Case Reports. *J. Clin. Med.* **2020**, *9*, 1074. [CrossRef] [PubMed]

139. Alberio, T.; Brughera, M.; Lualdi, M. Current Insights on Neurodegeneration by the Italian Proteomics Community. *Biomedicines* **2022**, *10*, 2297. [CrossRef] [PubMed]
140. Reddy, D.S.; Abeygunaratne, H.N. Experimental and Clinical Biomarkers for Progressive Evaluation of Neuropathology and Therapeutic Interventions for Acute and Chronic Neurological Disorders. *Int. J. Mol. Sci.* **2022**, *23*, 11734. [CrossRef] [PubMed]

applied sciences

MDPI

Review

Vitamin D and Vitamin D Receptor Polymorphisms Relationship to Risk Level of Dental Caries

Marios Peponis [1], Maria Antoniadou [1,*], Eftychia Pappa [1], Christos Rahiotis [1] and Theodoros Varzakas [2,*]

[1] Department of Dentistry, School of Health Sciences, National and Kapodistrian University of Athens, 11527 Athens, Greece; mariospep@gmail.com (M.P.); effiepappa@dent.uoa.gr (E.P.); craxioti@dent.uoa.gr (C.R.)
[2] Department of Food Science and Technology, University of the Peloponnese, 24100 Kalamata, Greece
* Correspondence: mantonia@dent.uoa.gr (M.A.); t.varzakas@uop.gr (T.V.)

Featured Application: Factors influencing Vitamin D and its receptor polymorphisms on risk level of dental caries.

Abstract: Dental caries is a multifactorial disease with multiple risk factors. Vitamin D levels (VDLs) and vitamin D receptor polymorphisms (VDRPs) have been investigated for this reason. The aim of this narrative review is to investigate the relation and the factors affecting vitamin D deficiency (VDD), VDRP, Early Childhood Caries (ECC) and Severe Early Childhood Caries (S-ECC) in children (primary and mixed dentition) and dental caries risk in adults (permanent dentition). Additionally, we present a model incorporating factors and interactions that address this relationship. Methods: Three databases (PubMed/MEDLINE, Web of Science, Cochrane Library) were comprehensively searched until 17 January 2023 using the following keywords: "vitamin D", "vitamin D receptor polymorphism", "dental caries", and "dental caries risk", finding 341 articles. Two reviewers searched, screened, and extracted information from the selected articles. All pooled analyses were based on random-effects models. Eligibility criteria were articles using dmft/DMFT diagnostic criteria with calibrated examiners, probability sampling, and sample sizes. We excluded studies conducted on institutionalized patients. A total of 32 studies were finally used. Results: In most studies, *TaqI*, *FokI*, and *BsmI* polymorphisms affected the prevalence of dental caries. A strong correlation between ECC, S-ECC, and the prevalence of dental caries was reported in association with VDD and maternal intake of VD in primary dentition. Regarding the influence in mixed dentition, the results were found to be inconclusive. A slight positive influence was reported for permanent dentition. Conclusions: Factors affecting caries risk were maternal intake, socioeconomic factors, and level of VD. There is a certain need for more well-conducted studies that will investigate the association between VDR gene polymorphisms and the prevalence of dental caries in mixed and permanent dentition, specifically in adult patients.

Keywords: vitamin D receptor; polymorphism; dental caries; dentistry; prevalence; prevention; dmft/DMFT index

Citation: Peponis, M.; Antoniadou, M.; Pappa, E.; Rahiotis, C.; Varzakas, T. Vitamin D and Vitamin D Receptor Polymorphisms Relationship to Risk Level of Dental Caries. *Appl. Sci.* 2023, 13, 6014. https://doi.org/10.3390/app13106014

Academic Editors: Luca Testarelli and Andrea Scribante

Received: 27 March 2023
Revised: 10 May 2023
Accepted: 11 May 2023
Published: 13 May 2023

1. Introduction

Dental caries is a multifactorial disease with multiple risk factors [1]. Vitamin D levels (VDLs) and vitamin D receptor polymorphisms (VDRPs) have been investigated for this reason [2]. Vitamin D is a fat-soluble steroid hormone which regulates calcium and phosphorus levels through the intestine [3]. This vitamin can be found in the human body as vitamin D3 (cholecalciferol) and vitamin D2 [2]. Both are converted to 25-hydroxyvitamin D (25(OH)D), which acts as a biological marker for vitamin D levels in serum. When 25(OH)D reaches the kidneys, it is converted to calcitriol (1,25(OH)2D), the most active form of vitamin D, with a shorter half-life [3].

Vitamin D deficiency (VDD) is defined when levels of 25(OH)D are below 20 ng/mL (50 nmol/L) and there is an insufficiency between 21–29 ng/mL (50–75 nmol/L) [4]. VDD has, therefore, abnormality in calcium, phosphorus, and bone metabolism [4] and is associated with increased risk of neoplastic, metabolic, and immune disorders [5,6]. Tooth mineralization co-occurs to skeletal mineralization, so disturbances in mineral metabolism affect bone and teeth. As VD has a crucial role in tooth and bone mineralization, low levels of VD may lead to a defective and hypomineralized tooth [6]. The main mechanism related to severe VDD causes hypophosphatemia and hypocalcemia with secondary hyperparathyroidism [7]. Hyperparathyroidism, in turn, stimulates renal production of 1,25(OH)2D and absorption of calcium in intestines. It increases bone turnover and may lead to increased serum levels of calcium ions and decreased serum levels of inorganic phosphate [8]. Therefore, proper mineralization of teeth is inhibited due to the loss of VD signaling pathways in tooth cells as concentrations of calcium ions and phosphate ions are low [9–11]. The biological activity of VD is modulated and modified by the vitamin D receptor (VDR) protein. VDR is responsible for the expression of many genes involved in cellular proliferation and differentiation, calcium–phosphate homeostasis, and immune response [12]. VDR is regulated by the VDR gene, whose polymorphisms affect the function of the VDR protein. Polymorphisms are genetic variations in non-coding parts of the gene (introns) or in the exonic parts of the DNA that influence at least 1% of the population [13]. Changes in introns do not influence the protein product but can modify the degree of gene expression. On the other hand, changes in exonic parts of the gene affect the protein sequence, except for synonymous polymorphisms, which are alterations in exonic parts that do not affect the protein structure [13]. These variations are responsible for the creation or removal of restriction enzyme sites in DNA, which, in turn, create DNA fragments with various lengths [14]. The most common polymorphisms (VDRP) of the VDR gene, which have been related to oral and systemic conditions, are *BsmI (rs1544410)*, *TaqI (rs731236)*, *BglI (rs739837)*, *ApaI (rs7975232)*, *FokI (rs10735810)*, and *FokI (rs2228570)* [14,15].

VDD has been associated with Early Childhood Caries (ECC), and Severe Early Childhood Caries (S-ECC) [10]. ECC is a global health problem, affecting almost half of preschool children, defined as the presence of one or more decayed, missing, or filled primary teeth in children aged 71 months (5 years) or younger [11]. ECC has a multi-factorial etiology, including susceptible teeth due to colonization with high levels of cariogenic bacteria, such as *S. mutans*, enamel hypoplasia, and sugar metabolism, and by bacteria that produce acid, which, in turn, demineralizes tooth structure [10]. A subtype of ECC is Severe Early Childhood Caries (S-ECC). S-ECC is defined as any sign of smooth-surface caries in a child under the age of three and, from ages three to five, one or more cavitated, missing, or filled smooth surfaces in primary maxillary anterior teeth or a decayed, missing, or filled score greater than or equal to 4 (for 3 years of age), 5 (for 4 years of age), or 6 (for 5 years of age) [11].

As awareness of VDD has increased among patients and the health community, many authors have conducted clinical trials to study a potential association between VDD and dental caries [12–14]. In addition, there is a systematic review investigating vitamin D receptor (VDR) gene polymorphisms in relation to increased risk of dental caries only in children [15–17]. The aim of this study was to further investigate the relation between VDD, VDRP, ECC, and S-ECC in children (primary and mixed dentition) and dental caries risk in adults (permanent dentition). Additionally, we present a model incorporating factors and interactions that address this relationship.

2. Materials and Methods

For this study, an extended review of the relevant literature was performed. The algorithm of the search was "vitamin D and Polymorpisms and dental caries and abstract and vitamin D receptor". Records have been selected from three databases (PubMed/Medline, Web of Science, Cochrane Library). Two reviewers searched, screened, and extracted information from the selected articles. All pooled analyses were based on random-effects models. Eligibility criteria were articles using dmft/DMFT diagnostic criteria with calibrated examiners, probability sampling, and sample sizes. We excluded studies conducted on institutionalized patients. A total of 341 articles were reported in the initial search, of which 104 were removed as duplicates. A total of 40 articles were excluded for other reasons, such as inadequate methodology or irrelevance to our search content through abstract evaluation. In the second phase of the study, from the remaining 197 articles, 145 were excluded due to inadequate methodology. In the third phase of the screening, 32 studies were finally included in the review, as seen in detail in the following Prisma flow chart (Figure 1).

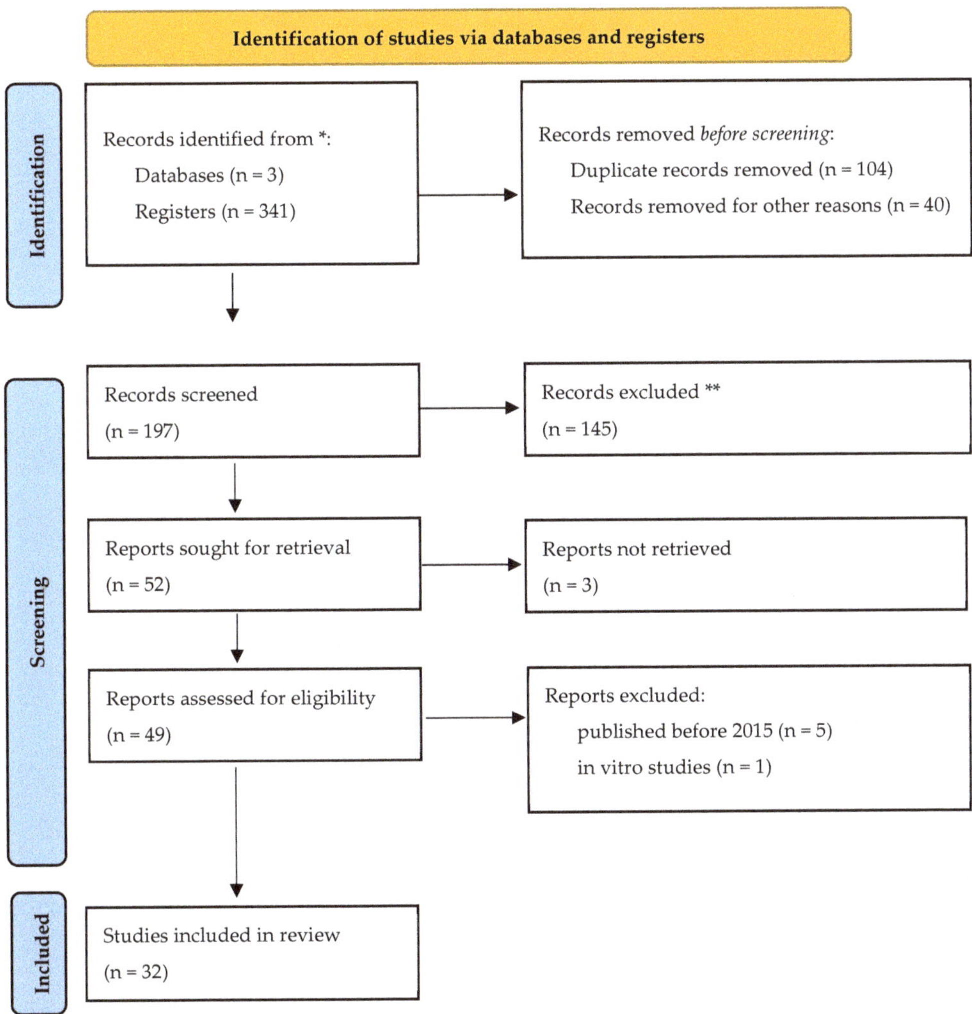

Figure 1. Prisma flow chart of the present study. * Eligibility criteria were articles using DMFT diagnostic criteria with calibrated examiners, probability sampling, and sample sizes.** We excluded studies conducted on institutionalized patients.

3. Results

Vitamin D Receptor Gene Polymorphisms

To discuss possible relations between VDPP and caries, we needed to report on the possible alleles of the polymorphisms of the VDR gene, presented in Table 1.

Table 1. Different types of alleles and genotypes of the VDR gene polymorphisms.

Genotype	*ApaI* (rs7975232)	*TaqI* (rs731236)	*FokI* (rs10735810)	*FokI* (rs2228570)	*BglI* (rs739837)	*BsmI* (rs1544410)
Homozygous dominant	AA	TT	FF	FF	BB	BB
Heterozygous	Aa	Tt	Ff	Ff	Bb	Bb
Homozygous recessive	Aa	tt	Ff	Ff	bb	bb

From our findings we can support the fact that all relevant studies are differentiated in both methodology and sample size, thus revealing differences in the hypothesis that certain polymorphisms may play a specific role in caries risk assessment (Table 2).

For example, Kong et al. found a correlation between the *BsmI* polymorphism containing the Bb genotype and increased risk of caries in deciduous teeth of Chinese children [18]. On the other hand, two studies in the Chinese population did not show a statistically significant difference between the caries-free and caries-active groups [19,20]. In numerous studies, no statistically significant difference was found for a higher caries prevalence in the population with *ApaI* polymorphism [18–21]. In the case of the *TaqI* polymorphism, one study suggested the allele 't' as a possible genetic factor for determining dental caries in Chinese individuals because the 't' allele was more frequent in the caries-active group [22]. Similar results were found by Cogulu et al. [19] concerning the *TaqI* polymorphism, showing a statistically significant difference in *TaqI* genotypes between the case and control groups. On the other hand, many other case–control studies did not report significant differences in allele and genotype frequency of the *TaqI* polymorphism between individuals with and without caries [18–20,23,24]. Further, the *BglI* polymorphism has been studied by two research groups in the Brazilian population. The authors concluded that the *BglI* polymorphism was not associated with an increased risk of dental caries in the respective population [25,26]. From all polymorphisms studied, the *FokI* polymorphism, distinguished in *rs10735810* and *rs2228570*, is reported to play a significant role in the issue studied here. Although two studies supported that different alleles and genotypes of *rs10735810* were not associated with tooth decay in Chinese and Turkish populations [18,19], two other studies found a statistically significant difference when the *FokI rs10735810* genotype was present in Chinese and Brazilian populations [20,25]. In terms of the *rs2228570* polymorphism, no correlation was found regarding a higher dental caries risk in the Brazilian population [25,26]. A systematic review and meta-analysis compared the influence of the *ApaI*, *FokI (rs10735810)*, *TaqI*, *BsmI*, *FokI (rs2228570)*, and *BglI* polymorphisms of the VDR gene on dental caries risk [15]. Based on nine studies, the meta-analysis reported an association between the *FokI (rs10735810)* polymorphism and dental caries risk in children, with the f allele and the ff genotype reported as having a protective role. Additionally, authors found that there was no association between the *ApaI (rs7975232)*, *TaqI (rs731236)*, *BsmI (rs1544410)*, *FokI (rs2228570)*, and *BglI (rs739837)* polymorphisms of the VDR gene and the prevalence of dental caries in children [15]. The authors claimed that the protective role of the f allele and ff genotype might be due to their interactions with co-transcription factors and location [15]. It is then concluded that there is a lack of consistency among the case–control studies reported here, their results related to the different VDR gene polymorphisms, and their influence on dental caries risk [27–29]. This fact is attributed to the statistical heterogeneity among studies, the small number of existing relative studies, and the small sample sizes. Thus, there is a significant need for additional, well-conducted research that will investigate the association between VDR gene polymorphisms and the prevalence of dental caries in mixed and permanent dentition, specifically in adult patients.

Table 2. Studies investigating correlation between VDR polymorphisms and prevalence of dental caries.

Authors Publication Year	ApaI rs7975232	TaqI rs731236	FokI rs10735810	FokI rs2228570	Cdx2 rs11568820	BglI rs739837	BsmI rs1544410	Type of Study/Country	Results
Hu et al., 2015 [22]		+						Case (264)–control (219) study/China Permanent dentition	• 't' allele more frequent in case group (7%) than control (2.1%) ($p = 0.0003$) • allele 't' may be a genetic factor for the determination of dental caries individuals in Chinese population
Holla et al., 2017 [23]		+						Case (235)–control (153) study/Czech Permanent dentition	No significant differences in allele and genotype frequency of *TaqI* between case and control group
Cogulu et al., 2016 [19]	+	+	+		+			Case (57)–control (38) study/Turkey Primary dentition	• statistically significant difference between *TaqI* genotypes between control and case group ($p = 0.029$) • no statistically significant difference for genotypes of *ApaI, FokI, Cdx2*
Kong et al., 2017 [18]	+	+	+				+	Case (249)–control (131) study/China Primary dentition	• increased risk of deciduous tooth decay in Bb genotype of *BsmI* • *ApaI, TaqI, FokI* polymorphisms are not associated with deciduous tooth decay
Yu et al., 2017 [20]	+	+	+				+	Case (200)–control (200) study/China Permanent dentition	• C allele frequency of the FokI VDR polymorphism was significantly increased in the case group ($p < 0.001$) • *BsmI, TaqI, ApaI* showed no statistically significant difference between case and control group • *FokI* gene polymorphism might be related with susceptibility to permanent decayed teeth in Chinese adolescent

Table 2. *Cont.*

Authors Publication Year	ApaI rs7975232	TaqI rs731236	FokI rs10735810	FokI rs2228570	Cdx2 rs11568820	BglI rs739837	BsmI rs1544410	Type of Study/Country	Results
Qin et al., 2019 [21]	+	+	+		+		+	Case (304)–control study (245)/China Primary dentition	• *FokI* genotype was statistically greater in the high caries risk group than moderate risk and control groups ($p = 0.028$) • The use of VDR polymorphisms as markers for increased risk of dental caries in Chinese children is not reliable
Barbosa et al., 2020 [25]				+		+		Case–control study/Brazil Permanent dentition	*FokI, BgII* VDR gene polymorphisms were not associated with dental caries
Aribam et al., 2020 [24]		+						Case (60)–control (60) study/India Permanent dentition	• No statistically significant differences between 'TT', 'Tt', 'tt' genotypes among case and control group • Individuals with 't' allele and 'tt', 'Tt' genotypes might be susceptible to dental caries
Fatturi et al., 2020 [26]				+		+		Case–control study/Brazil Permanent dentition	• No relation between MIH, HPSM, and dental caries with *BbII, FokI* VDR gene polymorphisms • Higher prevalence of MIH was observed when an individual carried at least one 'G' allele ($p = 0.03$)

Table 2. *Cont.*

Authors Publication Year	ApaI rs7975232	TaqI rs731236	FokI rs10735810	FokI rs2228570	Cdx2 rs11568820	BglI rs739837	BsmI rs1544410	Type of Study/Country	Results
Sadeghi et al., 2021 [15]	+	+	+	+		+	+	Systematic review, meta-analysis Primary and permanent dentition	• Out of nine studies included in the meta-analysis, there was no association between *TaqI, Apal, Bsml, FokI (rs2228570), BglI* VDR polymorphisms and dental caries risk • Protective role of f allele and ff genotype *Fokl (rs10735810)* VDR polymorphism in relation to dental caries
Nireeksha et al., 2022 [7]				+				Case (239)–control (138) study/India Permanent dentition	• Salivary vitamin D levels were higher in the control group (caries-free) ($p < 0.001$) • T allele of Fokl VDR polymorphism significantly associated with having active caries, while C allele with being caries-free • 2.814-fold increased possibility of TC genotype of *rs2228570* to be caries-active and 3.116-fold increased possibility of TT genotype to be caries-active

The effect of vitamin D serum levels on the prevalence of dental caries has also been studied in many scientific papers among different ages, populations, and dentitions. Table 3 presents studies correlating vitamin D serum levels with the prevalence of dental caries.

From the studies presented in Table 3, we can report on 10 studies addressing the issue on primary dentition, 8 on mixed, 3 on permanent, and one on mixed and permanent dentition. Results from the above-mentioned 10 studies on primary dentition indicate an association between prenatal levels of 25(OH)D and caries risk in primary dentition. It is mentioned that maternal 25(OH)D, during pregnancy, diffuses across the placental barrier, while 25(OH)D cord blood concentrations are 75–90% of maternal concentrations at delivery [30]. Thus, in four out of ten studies, there was an inverse relation between maternal intake of VD and dental caries in children. In favor of this correlation, we refer to the study of Tanaka et al., which examined 1210 Japanese mother–child pairs and agreed on the issue [31]. This inverse relation was also verified by another study, in which the authors found that insufficient levels of 25(OH)D (<50 nmol/L) during the third semester of pregnancy is associated with higher caries experience in primary teeth by the age of six [32]. In a third study, data were extracted from an extensive cohort survey in Austria on VDLs during the 12th week of gestation and after gestation at 4 and 8 years of age, while a dental examination was held at 6 and 10 years of age. Although sugar consumption and incorrect brushing were the main factors associated with dental caries in children participating in the study, low levels of 25(OH)D (<20 ng/mL) during pregnancy magnified the effect of cariogenic parameters. For this reason, the authors claimed that supplementing VD for pregnant women and children could be an option [33]. These findings agree with a 6-year follow-up randomized clinical trial, which concluded that supplementation with a high dose of VD during pregnancy leads to a lower probability of enamel defects in the offspring [34]. On the contrary, Schroth et al. did not find a statistically significant difference when they administered prenatally two oral doses of 50,000 (IU) of VD in the prevalence of Early Childhood Caries [35].

Furthermore, nine studies have investigated the relationship between VD deficiency and ECC or S-ECC. More specifically, there was a negative correlation between VDD and increased risk of ECC [36–45]; for example, Singleton et al. [41] stated that deficient cord blood 25(OH)D levels (<30 nmol/L) had a significant impact on the dmft in the follow-up period between 12 and 35 months of infants' age, with dmft scores that were double in relation to nondeficient concentrations. In the same study, no significant association between insufficient levels of 25(OH)D (<50 nmol/L) and ECC was reported. In another study, with a significant sample size of 1510 Chinese children, VD deficiency and insufficiency led to an increased risk of ECC [38]. Chhonkar et al. suggested supplementation of VD to prevent dental caries because VDD is a significant risk factor for dental caries in children [39]. In addition to VDD, children with S-ECC showed significantly lower levels of Ca and serum albumin and higher levels of PTH [40]. Deficient and insufficient levels of VD led to a higher odds ratio for S-ECC in children compared to optimal VD levels, according to two studies that have been conducted in Canada and North America [43,44]. Only one case–control study reported levels of VD in children without caries comparatively higher than in children with severe caries at a mean age of 40.82 months, even if the mean levels of VD were close to optimal [42]. A cross-sectional study with a sample of 1638 children from Poland concluded that the prevalence of ECC and S-ECC was significantly lower in children receiving VD supplementation. The authors also reported that children and mothers of higher education received VD supplementation and had fewer caries, highlighting the socioeconomic background of ECC [45]. In six out of nine studies addressing the issue, it is reported that socioeconomic factors, and especially the mothers' educational background, have a negative impact on caries risk and a positive effect on levels of VD.

Furthermore, many studies have investigated the relation between 25(OH)D levels and dental caries risk in mixed dentition, of which eight are included in this study. In all studies included here, there seems to be a weak indication that improving children's VD status might be an additional preventive measure against caries during the period of mixed dentition. In more detail, a cross-sectional study with a sample of 1017 Canadian children from 6 to 11 years of age stated that optimal levels of 25(OH)D (>75 nmol/L) had a significantly lower odds ratio for caries [46]. Further, Navarro et al. reported a weak association between the risk of dental caries and 25(OH)D serum, and the authors did not recommend supplementation of VD as a preventive measure for managing dental caries [47]. A weak inverse correlation between VD and dental caries risk in mixed dentition was found also in another study with a sample of 206 Swedish children [48]. Supplementation of VD and fluoride during the entire first year of a child's life led to a lower probability of caries-related restorations at the age of 10 in primary teeth in relation to children who received the same supplementation for less than 6 months [49]. A cohort study with a sample of 335 Portuguese children and a mean age of 7 years reported that 25(OH)D levels less than 30 ng/mL were associated with advanced dental caries in permanent teeth, while no statistically significant difference was found in terms of 25(OH)D levels and dental caries risk in mixed dentition [50]. The same results, regarding mixed dentition, were reported in a study with a sample of 121 Polish children with growth hormone deficiency. There was no statistically significant impact of VD on the mean DMFT index, but the authors recommended VD as a potentially effective agent in reducing dental caries, especially for growth hormone deficiency patients [51].

Apart from primary and mixed dentitions, a possible correlation between levels of 25(OH)D and dental caries risk has been investigated for permanent dentition. In our study, we report on three studies regarding permanent dentition and one with both mixed and permanent ones. From our findings in all four relevant studies, it seems that permanent dentition presents a higher correlation with VD deficiency than the mixed one. To support this, we can report that in a cross-sectional study with a sample of 2579 American adolescents, VD deficiency in permanent dentition has a limited but statistically non-significant association with caries in permanent dentition of adolescents [52]. A cross-sectional study on 1688 children from Korea found that children with serum 25(OH)D levels lower than 50 nmol/L were 1.29 times more likely to develop dental caries and that 25(OH)D concentration and DMFT were negatively correlated. However, the authors reported that there is difficulty in confirming the association between dental caries experience and 25(OH)D levels [53]. Another study that has investigated permanent dentition and 25(OH)D levels was conducted in the USA with 4244 participants with a mean age of 51.22 years. This study stated that the severely deficient group was strongly correlated with dental caries and that 25(OH)D is significantly associated with the prevalence of dental caries among US adults [54].

Concerning the VD supplementation for caries prevention in permanent dentition, we found data in all three relevant studies. Firstly, a systematic review of control clinical trials studied the influence of VD supplementation in two different forms (vitamin D2, D3) on dental caries prevention [55]. The results showed that VD supplementation reduced the risk of dental caries by about 47%, but with low certainty [55]. In two different mendelian randomization studies, the results failed to find a strong and statistically significant causal relationship between vitamin D concentrations and risk of dental caries [56,57].

Driven by the analysis performed in this study, we present a relevant model of the VD, VDRP, and dental caries equation. The model was designed with the Vensim program (Vensim PLE 8.1.0), showing factors influencing the equation. Vensim, developed by Ventana Systems (a simulation system employing causal tracing; US Patent Application EP19910909851, 26 February 1991), is precision simulation software. It primarily shows continuous simulation (system dynamics), with agent-based modeling capabilities. This modeling language supports arrays (subscripts) and permits mapping among factors. The built-in allocation functions satisfy, in our case, constraints that may not be met by conventional approaches (Figure 2).

Table 3. Studies investigating the relation between 25(OH)D levels and dental caries.

Authors Publication Year	Type of Study	Population	Age	Dentition	Results	Conclusions
Schroth et al., 2013 [40]	Case–control study	• 144 Canadian children with S-ECC • 122 Canadian children caries-free	• 42 +/− 11.9 (case group) • 39.4 +/− 16.3 (control group)	Primary	• More children receiving vitamin D drops were in caries-free group ($p < 0.001$) • 25(OH)D levels were significantly lower in S-ECC group ($p < 0.001$) • Significantly more children with S-ECC had 25(OH)D < 75 nmol/L compared to control group ($p = 0.006$)	• Significantly lower vitamin D in children with S-ECC compared to caries-free control group • Children with S-ECC show significantly lower levels of Ca, serum albumin, and higher levels of PTH compared to caries-free control group
Tanaka et al., 2015 [31]	Prospective study	1210 Japanese mother–child pairs	36–46 months postnatal for the evaluation	Primary	• Linear relationship between vitamin D intake and log-odds of dental caries • Inverse relationship between maternal intake of vitamin D and dental caries in children	Association between higher maternal vitamin D intake during pregnancy and a reduced risk of dental caries
Schroth et al., 2016 [46]	Cross-sectional study	1017 Canadian children	6 to 11 years	Mixed	• 25(OH)D levels > 75 nmol/L in children had significantly lower odds ratio for caries (OR = 0.57, 95% CI 0.39 to 0.82) • Levels >50 nmol/L had lower odd ratio for caries (OR = 0.56, 95% CI 0.39 to 0.80) • Levels of 25(OH)D > 50 nmol/L were significantly and independently related to lower adjusted odds for caries (OR = 0.46, 95% CI 0.26 to 0.83) • Concentrations > 75 nmol/L were significantly and independently related to lower dmft/DMFT scores	Association between caries and lower serum vitamin D in a sample of Canadian children

Table 3. *Cont.*

Authors Publication Year	Type of Study	Population	Age	Dentition	Results	Conclusions
Kühnisch et al., 2017 [49]	Clinical trial	• 406 children	• Enrollment at the 1st year of age • Dental examination at the age of 10	Mixed	• Supplementation of vitamin D and fluoride during the entire first year of life led to significantly lower probability of having caries-related restorations in primary teeth in relation to those who received supplementation for less than 6 months • No statistically significant difference between supplementation and prevalence of MIH	• Preventive effect of fluoride/vitamin D supplementation over the first year of life in the primary dentition
Chhonkar et al., 2018 [39]	Case–control study	• 60 Indian children	• 4.4 +/− 0.89 years (case group) • 4.5 +/− 1.1 years (control group)	Primary	• Statistically significant difference in mean levels of 25(OH) vitamin D levels between two groups ($p < 0.0001$) • Statistically significant inverse correlation between vitamin D levels and S-ECC ($p < 0.0001$)	• VDD is an important risk factor for incidence and severity of dental caries in children • Prevention of dental caries in children by supplementing vitamin D and by preventing VDD
Deane et al., 2018 [44]	Case–control study	• 266 children in Canada	• Mean age 40.8 +/− 14.1 months	Primary	• Association between 25(OH)D levels below 75 nmol/L and S-ECC ($p = 0.007$) • Combined deficiency of hemoglobin and 25(OH)D showed higher odds for dental caries ($p < 0.001$) • Combined deficiency of iron or iron deficiency anemia with 25(OH)D levels < 75 nmol/L showed higher odds for dental caries ($p < 0.001$, $p = 0.004$, respectively)	• Due to the low frequency of children with S-ECC and combined deficiencies, authors do not suggest laboratory investigation • Dietary history and dietary advice may have a helpful role in children with S-ECC, especially those of lower income

Table 3. *Cont.*

Authors Publication Year	Type of Study	Population	Age	Dentition	Results	Conclusions
Gyll et al., 2018 [48]	Intervention study	• 206 children from Sweden	• 6 years old at the baseline • 8 years old for examination	Mixed	• Weak inverse association between vitamin D levels at 6 years of age and caries 2 years later (OR = 0.96, p = 0.024) • Multivariate projection regression showed insufficient vitamin D concentration correlated with caries, while higher levels of 25(OH)D were correlated with being caries-free • Vitamin D positively correlated with the levels of LL37 in saliva	• Vitamin D was not related to enamel defects on permanent incisors and molars • Association between 25(OH)D and LL37 levels • Negative correlation between vitamin D and caries; however, a small study group and weak association
Wójcik et al., 2019 [51]		• 121 polish children and adolescents with growth hormone deficiency	• 6 to 18 years of age	Mixed	• Statistically significant impact of vitamin D levels on the average DMFT index in children and adolescents from rural areas (p = 0.049) • No statistically significant impact of vitamin D levels on mean DMFT index (p = 0.73)	• Decrease in dental caries when vitamin D levels are increased in children who live in rural areas and are treated for growth hormone deficiency • In urban areas, vitamin D supplementation and intensification of dental care is needed in children with growth hormone deficiency • Promotion of vitamin D as a potentially effective agent in reducing dental caries in patients with growth hormone deficiency
Akinkugbe et al., 2018 [52]	Cross-sectional	• 2579 American adolescents	Two age groups: • 12–14 • 15–19	Permanent	Participants with insufficiency and deficiency of vitamin D had non-statistically significant adjusted estimates of 1.02 (0.72, 1.44) and 1.23 (0.7, 2.16), respectively, for caries experience	Deficiency of vitamin D appears to have limited but statistically non-significant association with adolescent caries

Table 3. *Cont.*

Authors Publication Year	Type of Study	Population	Age	Dentition	Results	Conclusions
Kim et al., 2018 [53]	Cross-sectional study	• 1688 Korean children	10–12 years of age	Permanent	• 25(OH)D levels lower than 50 nmol/L had a higher incidence with respect to dental caries in the permanent dentition and permanent first molar ($p = 0.012$, $p = 0.006$, respectively) • 25(OH)D concentration and DMFT were negatively correlated ($p < 0.01$)	• There is a difficulty in confirming the association between dental caries experience and 25(OH)D levels • Insufficiency of 25(OH)D might be associated with dental caries
Singleton et al., 2019 [41]	Cohort study	• 76 prenatal Alaskan mother–infant pairs with prenatal blood were examined • 57 Alaskan infants were examined by measuring cord blood	Two periods for follow-up of the infants: • 12 to 35 months • 36 to 59 months	Primary	• Significant difference in 12–35 months age group in DMFT score with deficient cord blood 25(OH)D ($p = 0.002$) • Negative correlation between cord blood 25(OH)D levels and DMFT in the 12 to 35 months age group (R = -0.37, $p = 0.016$) • No significant difference in mean DMFT for age groups 12–35, 36–59 months of age with prenatal 25(OH)D levels below or above 50 nmol/L	• Prenatal vitamin D levels may have an impact on the primary dentition and the risk of developing ECC • Improving vitamin D levels in pregnant women may affect ECC rates in their infants
Schroth et al., 2020 [35]	Prospective cohort	• 283 mothers in Canada • 175 for the follow-up		Primary	• No significant difference between intervention group (2 oral prenatal doses of 50,000 (IU) vitamin D) and control group in prevalence of ECC ($p = 0.21$) • Inverse correlation between dt scores and 25(OH)D ($p = 0.001$) • Age, socioeconomic factor index, and enamel hypoplasia were significantly and independently associated with dt, while vitamin D supplementation was not	• Significant inverse relation between the number of teeth with caries and the levels of 25(OH)D • Supplementation with vitamin D during pregnancy did not influence the prevalence of ECC

Table 3. *Cont.*

Authors Publication Year	Type of Study	Population	Age	Dentition	Results	Conclusions
Zhou et al., 2020 [54]	Cross-sectional study	• 4244 participants in the USA	• 20–80 years of age, mean age 51.22 +/− 17.86	Permanent	• In three different models with different covariates (crude model, model 1, model 2) there was a negative correlation between 25(OH) serum levels and dental caries • Severely deficient group was strongly associated with dental caries	• 25(OH)D concentration is significantly associated with prevalence of dental caries among US adults
Jha et al., 2021 [42]	Case–control study	266 children from India	Mean age 40.82 +/− 14.09 months	Primary	Children with severe caries had significantly lower vitamin D_3 in very young childhood (68.87 +/− 28.04 vs. 82.89 +/− 31.12 nmol/L, $p < 0.001$)	Levels of vitamin D_3 in children without caries were comparatively higher than in children with severe caries
Navarro et al., 2021 [47]	Prospective cohort study	5257 multi-ethnic children	Mean age 6.1 (4.8–9.1)	Mixed	• Severe deficiencies of 25(OH)D (<25 nmol/L) prenatally and in early childhood associated with higher prevalence of dental caries than children with sufficient concentrations (>75 nmol/L) • Longitudinal association between low early childhood 25(OH)D serum concentrations and caries at 6 years of age • Children with genetically predisposed low serum 25(OH)D concentrations do not show higher risk of developing caries in primary dentition	• Weak association between risk of dental caries and 25(OH)D serum concentrations • Authors do not suggest vitamin D supplementation as a preventive measure for managing dental caries

Table 3. *Cont.*

Authors Publication Year	Type of Study	Population	Age	Dentition	Results	Conclusions
Silva et al., 2021 [50]	Cohort study	335 Portuguese children	Mean age 7 years	Mixed and permanent	• Levels of 25(OH)D below 30 ng/mL lead to more dental caries in permanent teeth ($p = 0.016$, $p = 0.034$ for advanced dental caries) • No statistically significant difference in terms of 25(OH)D levels and dental caries risk in mixed dentition ($p = 0.288$)	• 25(OH)D levels < 30 ng/mL are associated with advanced dental caries in 7-year-old children's permanent teeth • In mixed dentition, social and behavioral factors influence the prevalence of dental caries in the examined Portuguese sample • Sufficiency of vitamin D might have an additionally protective role in terms of dental caries in permanent teeth
Williams et al., 2021 [43]	Case–control study	144 caries-free children 200 children with S-ECC from North America	42.1 +/− 14.6 months	Primary	• Children with S-ECC had significantly lower mean levels of 25(OH)D than the control group ($p < 0.001$) • Children with deficient and adequate concentrations of 25(OH)D were significantly more likely to have S-ECC than children with optimal levels (OR = 10.1, $p < 0.001$ and OR = 1.8, $p = 0.01$, respectively)	• Significant and independent association between caries and 25(OH)D levels • Children with optimal levels of 25(OH)D had lower odds for S-ECC
Olczak-Kowalczyk et al., 2021 [45]	Cross-sectional study	1638 children from Poland	3 years of age	Primary	• Significantly lower prevalence of ECC/S-ECC in children receiving vitamin D supplementation ($p < 0.05$) • After controlling confounding variables, only dt/ds associated with supplementation of vitamin D	• Lower caries incidence in those who received vitamin D supplementation • Children and mothers of higher education received vitamin D supplementation and had less caries • During periods of significant growth and development, children should take supplements

Table 3. *Cont.*

Authors Publication Year	Type of Study	Population	Age	Dentition	Results	Conclusions
Suárez-Calleja et al., 2021 [33]	Cohort study	188 children	Dental examination 6–10 years of age	Mixed	• Levels of 25(OH)D below 20 ng/mL in mother and at 8 years of age is a risk factor for dental caries in children • The risk of caries tripled when 25(OH)D were below 20 ng/ml	• An incorrect brushing technique and sugar consumption were the main reasons for caries in children • Low levels of 25(OH)D during pregnancy have magnified the effect of cariogenic factors • Supplementation of vitamin D for pregnant women and children could be an option
Chen et al., 2021 [38]	Cross-sectional	1510 Chinese children	44 +/− 8.2 months	Primary	• VDI and VDD leads to higher prevalence of ECC; 25.95%, 29.65%, respectively ($p = 0.016$) • Negative correlation between 25(OH)D and number of caries (r = −0.103, $p < 0.001$)	• VDD, VDI increased the risk of ECC in preschool children
Beckett et al., 2022 [32]	Observational study	81 children from New Zealand	Mean age of 6.6 years +/− 0.6	Mixed	• Maternal vitamin D insufficiency (<50 nmol/L) during third trimester of pregnancy was associated with increased dental caries risk by 6 years of age (IRR = 3.55, $p < 0.05$) • Vitamin D insufficiency was not related to enamel defect prevalence at any timepoint	• Maternal insufficiency of vitamin D during third semester of pregnancy is associated with higher caries experience in primary teeth by the age of 6 • Recommendation for vitamin D supplementation during pregnancy and early life of the infant

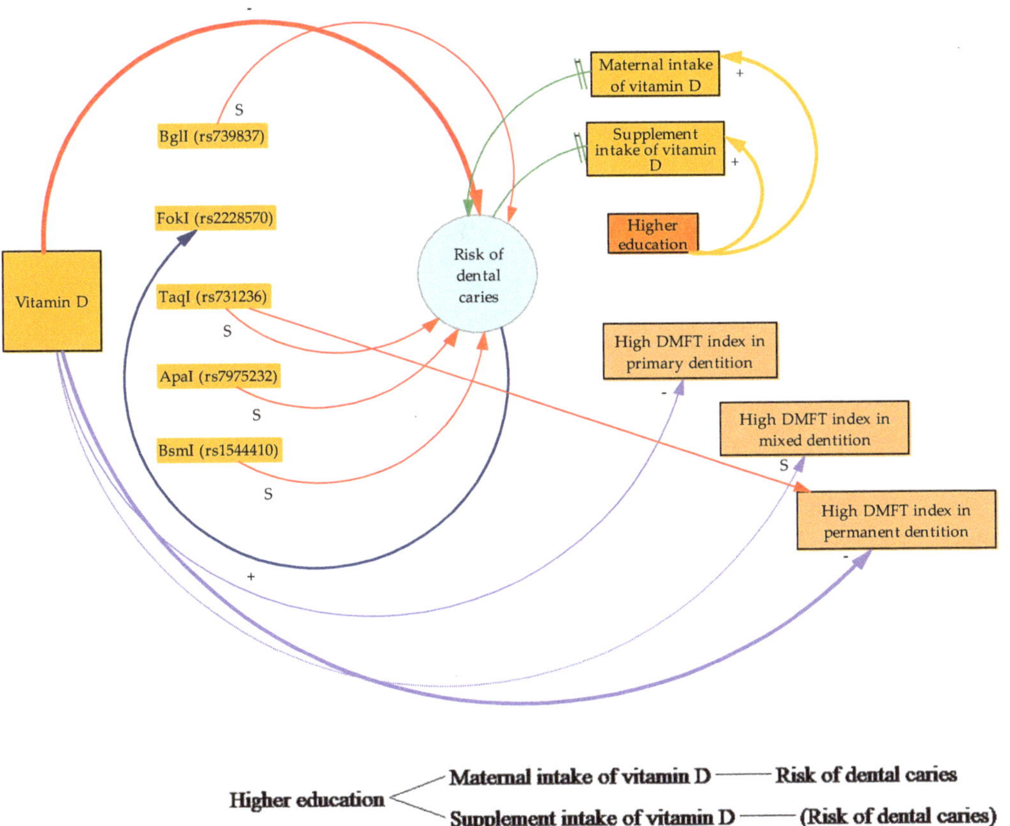

Figure 2. A Vensim diagram provides a graphical modeling interface with basic VDRPs and risk of dental caries. The diagram presents flows and causal loops among the basic factors affecting VD serum levels and risk of caries reported in our study. It also reveals a text-based system of equations in a declarative programming language. (+) positive or in tandem influence, (−) negative or inverse influence, (S) neutral relationship, (=) delayed response among factors, (…) border lines with dots support lack of sufficient data. Border lines support a quantitatively stronger relationship among factors. Red causal loops show the VDRP relationship with risk of dental caries; purple causal loops show VD serum levels and type of dentition; green ones show the way specific factors relate to risk of dental caries; orange ones show the interconnection between higher education and factors affecting dental caries; and blue is a reverse loop showing the interconnection of *FokI (rs2228579)* and risk of dental caries.

4. Discussion

Dental caries incidence can derive from a host of factors that may be related to the structure of dental enamel, immunologic response to cariogenic bacteria, or the composition of saliva. Dental caries causes a modern problem even in developed countries, leading to a decrease in the quality of life of the affected individuals and high economic costs for both individuals and society, with disparities related to well-known issues of socioeconomics, immigration, lack of preventive efforts, and dietary changes [58]. The burden of dental caries in children is incredibly high, causing pain that can affect school attendance, eating and speaking, and impair growth and development [21]. More data on the correlation of dental caries with VDD and polymorphisms were derived from studies conducted in children. However, it is not only in children that this problem could cause dysfunction in social life and diet. Related problems can occur in young or middle-aged adults due to the cumulative and chronic nature of the disease [59]. Our study reveals the positive correlation of the DMFT index in permanent dentition with VDLs both in maternal and early birth stages. Our model also shows the importance of the *FokI (rs2228570)* polymorphism, in conjunction with the risk of dental caries, thus suggesting that relevant blood and genetic examinations can arise awareness of ECC, S-ECC, and high dental caries risk in permanent dentition, as already supported elsewhere [15].

As a higher risk of carious lesions has been associated with lower socioeconomic levels [60,61], we could expect further reductions in preventive care and frequency of dental visits [62]. According to the new Global Oral Health Status Report from the WHO [63], it is estimated that almost half of the world's population (45%, or 3.5 billion people) suffer from oral diseases such as caries, with 3 out of every 4 affected people living in low- and middle-income countries. Global cases of dental caries have increased by 1 billion over the last 30 years. This clearly indicates that many people do not have access to prevention measures and treatment of oral diseases such as dental caries. Socioeconomic factors should then be considered when planning preventive programs for caries. Higher education is a positive factor for controlling caries risk and an important part of the relevant model.

Thus, new ways of identifying high-risk individuals are future tools in controlling caries. As observed in our study, genetic variation in host factors such as VDPs may contribute to increased risks of the disease. Thus, it seems imperative that VDR gene polymorphisms can be proposed as a marker for identifying patients with high caries risk [21]. As mentioned by Di Spigna et al. [64], the genetic screening of VDRPs could be a valuable tool for the early identification of other health problems too, such as osteoporosis in female patients with rheumatoid arthritis (RA). Commercial kits, based on the Restriction Fragment Length Polymorphism (RFLP) method for VDR polymorphisms detection, could then be used in patients at high risk for dental caries too. It is proposed, then, that "the clinician and the lab manager may join to evaluate costs and availability, of the appropriate methods to setting molecular diagnostics of VDR Genotyping" [64]. Soon, genetic tests, now performed either by academic ultra-specialized labs or custom service laboratories using certified commercial kits [65], could be performed widely, in relation to demand in dental settings, under a protected insurance policy that could provide low-income citizens with cost-effective and accurate diagnostic and preventive tools for dental caries. In these cases, future studies should focus on the importance of the application of fluorides [66] or other remineralizing biomimetic materials [67] for prevention purposes.

This review highlights the need for more detailed and extensive studies to establish the cost and effectiveness of genotyping, introducing it more in everyday dental clinical practice in terms of caries prevalence. The limitations of the present study include the risk of misinterpreting data from clinical studies that have different methodology due to inadequate explanations and the possible incomplete retrieval of identified relevant research. In the future, with new genetic markers being identified and validated, dentists will have new ways and means to tailor specific dental therapy to individual caries risk genetic profiles [65]. Overall, this would reduce the final functional treatment costs within national health systems while further enhancing the economic status of lower- and middle-class individuals. Consequently, pharmaceutical and biotechnology companies must join their future investments to develop accurate and low-cost genetic tests for routine dental diagnostics.

5. Conclusions

This study strengthens the relationship of dental caries with VD levels, particularly in primary dentition and less in mixed and permanent dentition. VD levels in children are directly related to the maternal intake and socioeconomic factors and education of the mother.

There is an inconsistency among the case–control studies and their results regarding the different VDR gene polymorphisms and their influence on ECC and S-ECC for primary and mixed dentition and dental caries risk in permanent dentition. This fact is attributed to the statistical heterogeneity between studies, the small number of existing relative studies, and the small sample sizes. There is a need for more well-conducted studies that will investigate other factors possibly influencing the current suggested model as well as the association between VDR gene polymorphisms and the risk of dental caries. VDR gene polymorphisms could be a marker for identifying not only children but also adult patients with high caries risk.

Author Contributions: Conceptualization, M.A., E.P. and C.R.; methodology, M.P. and M.A.; validation, M.P. and M.A.; formal analysis, M.P. and M.A.; investigation, M.P.; writing—original draft preparation, M.P. and M.A.; writing—review and editing, M.P., M.A., E.P., C.R. and T.V.; visualization, M.P., M.A. and E.P.; supervision, M.A., C.R. and T.V.; project administration, M.A. and T.V. All authors have read and agreed to the published version of the manuscript.

Funding: This research received no external funding.

Institutional Review Board Statement: Not applicable.

Informed Consent Statement: Not applicable.

Data Availability Statement: Not applicable.

Conflicts of Interest: The authors declare no conflict of interest.

Abbreviations

Vitamin D (VD), vitamin D deficiency (VDD), vitamin D receptor (VDR), vitamin D receptor polymorphism (VDRP), Restriction Fragment Length Polymorphism (RFLP), Early Childhood Caries (ECC), Severe Early Childhood Caries (S-ECC), Decayed, Missing, and Filled Teeth (DMFT) index.

References

1. Norman, A.W. From Vitamin D to Hormone D: Fundamentals of the Vitamin D Endocrine System Essential for Good Health. *Am. J. Clin. Nutr.* **2008**, *88*, 491S–499S. [CrossRef]
2. Borel, P.; Caillaud, D.; Cano, N.J. Vitamin D Bioavailability: State of the Art. *Crit. Rev. Food Sci. Nutr.* **2015**, *55*, 1193–1205. [CrossRef] [PubMed]
3. Christakos, S.; Ajibade, D.V.; Dhawan, P.; Fechner, A.J.; Mady, L.J. Vitamin D: Metabolism. *Endocrinol. Metab. Clin. N. Am.* **2010**, *39*, 243–253. [CrossRef]
4. Holick, M.F.; Binkley, N.C.; Bischoff-Ferrari, H.A.; Gordon, C.M.; Hanley, D.A.; Heaney, R.P.; Murad, M.H.; Weaver, C.M. Evaluation, Treatment, and Prevention of Vitamin D Deficiency: An Endocrine Society Clinical Practice Guideline. *J. Clin. Endocrinol. Metab.* **2011**, *96*, 1911–1930. [CrossRef] [PubMed]
5. Pfotenhauer, K.M.; Shubrook, J.H. Vitamin D Deficiency, Its Role in Health and Disease, and Current Supplementation Recommendations. *J. Osteopath. Med.* **2017**, *117*, 301–305. [CrossRef]
6. D'Ortenzio, L.; Kahlon, B.; Peacock, T.; Salahuddin, H.; Brickley, M. The Rachitic Tooth: Refining the Use of Interglobular Dentine in Diagnosing Vitamin D Deficiency. *Int. J. Paleopathol.* **2018**, *22*, 101–108. [CrossRef] [PubMed]
7. Nireeksha, N.; Hegde, M.N.; Shetty, S.S.; Kumari, S.N. FOK 1 Vitamin D Receptor Gene Polymorphism and Risk of Dental Caries: A Case-Control Study. *Int. J. Dent.* **2022**, *2022*, 6601566. [CrossRef] [PubMed]
8. Dudding, T.; Thomas, S.J.; Duncan, K.; Lawlor, D.A.; Timpson, N.J. Re-Examining the Association between Vitamin D and Childhood Caries. *PLoS ONE* **2015**, *10*, e0143769. [CrossRef] [PubMed]
9. Bergwitz, C.; Jüppner, H. Regulation of Phosphate Homeostasis by PTH, Vitamin D, and FGF23. *Annu. Rev. Med.* **2010**, *61*, 91–104. [CrossRef]
10. AAPD. Policy on Early Childhood Caries (ECC): Classifications, consequences, and preventive strategies. *Pediatr Dent.* **2017**, *39* (Suppl. S7), 59–61.
11. Uribe, S.; Innes, N.; Maldupa, I. The global prevalence of early childhood caries: A systematic review with meta-analysis using the WHO diagnostic criteria. *Int. J. Pediatr. Dent.* **2021**, *31*, 817–830. [CrossRef] [PubMed]
12. Wang, Y.; Zhu, J.; DeLuca, H.F. Where Is the Vitamin D Receptor? *Arch. Biochem. Biophys.* **2012**, *523*, 123–133. [CrossRef] [PubMed]
13. Valdivielso, J.M.; Fernandez, E. Vitamin D Receptor Polymorphisms and Diseases. *Clin. Chim. Acta* **2006**, *371*, 1–12. [CrossRef] [PubMed]
14. Uitterlinden, A.G.; Fang, Y.; van Meurs, J.B.J.; Pols, H.A.P.; van Leeuwen, J.P.T.M. Genetics and Biology of Vitamin D Receptor Polymorphisms. *Gene* **2004**, *338*, 143–156. [CrossRef] [PubMed]

15. Sadeghi, M.; Golshah, A.; Godiny, M.; Sharifi, R.; Khavid, A.; Nikkerdar, N.; Tadakamadla, S.K. The Most Common Vitamin D Receptor Polymorphisms (ApaI, FokI, TaqI, BsmI, and BglI) in Children with Dental Caries: A Systematic Review and Meta-Analysis. *Children* **2021**, *8*, 302. [CrossRef]
16. Cooper, G.S.; Umbach, D.M. Are Vitamin D Receptor Polymorphisms Associated with Bone Mineral Density? A Meta-Analysis. *J. Bone Miner. Res.* **2010**, *11*, 1841–1849. [CrossRef]
17. Gong, G.; Stern, H.S.; Cheng, S.-C.; Fong, N.; Mordeson, J.; Deng, H.-W.; Recker, R.R. The Association of Bone Mineral Density with Vitamin D Receptor Gene Polymorphisms. *Osteoporos. Int.* **1999**, *9*, 55–64. [CrossRef]
18. Kong, Y.; Zheng, J.; Zhang, W.; Jiang, Q.; Yang, X.; Yu, M.; Zeng, S. The Relationship between Vitamin D Receptor Gene Polymorphism and Deciduous Tooth Decay in Chinese Children. *BMC Oral Health* **2017**, *17*, 111. [CrossRef]
19. Cogulu, D.; Onay, H.; Ozdemir, Y.; Aslan, G.I.; Ozkinay, F.; Eronat, C. The Role of Vitamin D Receptor Polymorphisms on Dental Caries. *J. Clin. Pediatr. Dent.* **2016**, *40*, 211–214. [CrossRef]
20. Yu, M.; Jiang, Q.-Z.; Sun, Z.-Y.; Kong, Y.-Y.; Chen, Z. Association between Single Nucleotide Polymorphisms in Vitamin D Receptor Gene Polymorphisms and Permanent Tooth Caries Susceptibility to Permanent Tooth Caries in Chinese Adolescent. *BioMed Res. Int.* **2017**, *2017*, 4096316. [CrossRef]
21. Qin, X.; Shao, L.; Zhang, L.; Ma, L.; Xiong, S. Investigation of Interaction between Vitamin D Receptor Gene Polymorphisms and Environmental Factors in Early Childhood Caries in Chinese Children. *BioMed Res. Int.* **2019**, *2019*, 4315839. [CrossRef] [PubMed]
22. Hu, X.P.; Li, Z.Q.; Zhou, J.Y.; Yu, Z.H.; Zhang, J.M.; Guo, M.L. Analysis of the Association between Polymorphisms in the Vitamin D Receptor (VDR) Gene and Dental Caries in a Chinese Population. *Genet. Mol. Res.* **2015**, *14*, 11631–11638. [CrossRef] [PubMed]
23. Holla, L.I.; Borilova Linhartova, P.; Kastovsky, J.; Bartosova, M.; Musilova, K.; Kukla, L.; Kukletova, M. Vitamin D Receptor TaqI Gene Polymorphism and Dental Caries in Czech Children. *Caries Res.* **2017**, *51*, 7–11. [CrossRef]
24. Aribam, V.; Aswath, N.; Ramanathan, A. Single-Nucleotide Polymorphism in Vitamin D Receptor Gene and Its Association with Dental Caries in Children. *J. Indian Soc. Pedod. Prev. Dent.* **2020**, *38*, 8. [CrossRef] [PubMed]
25. Barbosa, M.C.F.; Lima, D.C.; Reis, C.L.B.; Reis, A.L.M.; Rigo, D.; Segato, R.A.B.; Storrer, C.L.M.; Küchler, E.C.; Oliveira, D.S.B. Vitamin D Receptor FokI and BglI Genetic Polymorphisms, Dental Caries, and Gingivitis. *Int. J. Paediatr. Dent.* **2020**, *30*, 642–649. [CrossRef]
26. Fatturi, A.L.; Menoncin, B.L.; Reyes, M.T.; Meger, M.; Scariot, R.; Brancher, J.A.; Küchler, E.C.; Feltrin-Souza, J. The Relationship between Molar Incisor Hypomineralization, Dental Caries, Socioeconomic Factors, and Polymorphisms in the Vitamin D Receptor Gene: A Population-Based Study. *Clin. Oral Investig.* **2020**, *24*, 3971–3980. [CrossRef]
27. Schwendicke, F.; Frencken, J.E.; Bjørndal, L.; Maltz, M.; Manton, D.J.; Ricketts, D.; Van Landuyt, K.; Banerjee, A.; Campus, G.; Doméjean, S.; et al. Managing Carious Lesions: Consensus Recommendations on Carious Tissue Removal. *Adv. Dent. Res.* **2016**, *28*, 58–67. [CrossRef]
28. Kidd, E.A.M. Clinical Threshold for Carious Tissue Removal. *Dent. Clin. N. Am.* **2010**, *54*, 541–549. [CrossRef]
29. Chapple, I.L.C.; Bouchard, P.; Cagetti, M.G.; Campus, G.; Carra, M.-C.; Cocco, F.; Nibali, L.; Hujoel, P.; Laine, M.L.; Lingstrom, P.; et al. Interaction of Lifestyle, Behaviour or Systemic Diseases with Dental Caries and Periodontal Diseases: Consensus Report of Group 2 of the Joint EFP/ORCA Workshop on the Boundaries between Caries and Periodontal Diseases. *J. Clin. Periodontol.* **2017**, *44* (Suppl. S18), S39–S51. [CrossRef]
30. Keim, S.A.; Bodnar, L.M.; Klebanoff, M.A. Maternal and Cord Blood 25(OH)-Vitamin D Concentrations in Relation to Child Development and Behaviour. *Paediatr. Perinat. Epidemiol.* **2014**, *28*, 434–444. [CrossRef]
31. Tanaka, K.; Hitsumoto, S.; Miyake, Y.; Okubo, H.; Sasaki, S.; Miyatake, N.; Arakawa, M. Higher Vitamin D Intake during Pregnancy Is Associated with Reduced Risk of Dental Caries in Young Japanese Children. *Ann. Epidemiol.* **2015**, *25*, 620–625. [CrossRef] [PubMed]
32. Beckett, D.M.; Broadbent, J.M.; Loch, C.; Mahoney, E.K.; Drummond, B.K.; Wheeler, B.J. Dental Consequences of Vitamin D Deficiency during Pregnancy and Early Infancy—An Observational Study. *Int. J. Environ. Res. Public Health* **2022**, *19*, 1932. [CrossRef] [PubMed]
33. Suárez-Calleja, C.; Aza-Morera, J.; Iglesias-Cabo, T.; Tardón, A. Vitamin D, Pregnancy and Caries in Children in the INMA-Asturias Birth Cohort. *BMC Pediatr.* **2021**, *21*, 380. [CrossRef] [PubMed]
34. Nørrisgaard, P.E.; Haubek, D.; Kühnisch, J.; Chawes, B.L.; Stokholm, J.; Bønnelykke, K.; Bisgaard, H. Association of High-Dose Vitamin D Supplementation During Pregnancy with the Risk of Enamel Defects in Offspring: A 6-Year Follow-up of a Randomized Clinical Trial. *JAMA Pediatr.* **2019**, *173*, 924–930. [CrossRef]
35. Schroth, R.J.; Christensen, J.; Morris, M.; Gregory, P.; Mittermuller, B.-A.; Rockman-Greenberg, C. The Influence of Prenatal Vitamin D Supplementation on Dental Caries in Infants. *J. Can. Dent. Assoc.* **2020**, *86*, k13.
36. Tinanoff, N. Introduction to the Conference: Innovations in the Prevention and Management of Early Childhood Caries. *Pediatr. Dent.* **2015**, *37*, 198–199.
37. American Academy of Pediatric Dentistry; American Academy of Pediatrics; American Academy of Pediatric Dentistry Council on Clinical Affairs. Policy on Early Childhood Caries (ECC): Classifications, Consequences, and Preventive Strategies. *Pediatr. Dent.* **2005**, *27* (Suppl. S7), 31–33.
38. Chen, Z.; Lv, X.; Hu, W.; Qian, X.; Wu, T.; Zhu, Y. Vitamin D Status and Its Influence on the Health of Preschool Children in Hangzhou. *Front. Public Health* **2021**, *9*, 675403. [CrossRef]

39. Chhonkar, A.; Arya, V. Comparison of Vitamin D Level of Children with Severe Early Childhood Caries and Children with No Caries. *Int. J. Clin. Pediatr. Dent.* **2018**, *11*, 199–204. [CrossRef]

40. Schroth, R.J.; Levi, J.A.; Sellers, E.A.; Friel, J.; Kliewer, E.; Moffatt, M.E. Vitamin D Status of Children with Severe Early Childhood Caries: A Case–Control Study. *BMC Pediatr.* **2013**, *13*, 174. [CrossRef]

41. Singleton, R.; Day, G.; Thomas, T.; Schroth, R.; Klejka, J.; Lenaker, D.; Berner, J. Association of Maternal Vitamin D Deficiency with Early Childhood Caries. *J. Dent. Res.* **2019**, *98*, 549–555. [CrossRef]

42. Jha, A.; Jha, S.; Shree, R.; Menka, K.; Shrikaar, M. Association between Serum Ferritin, Hemoglobin, Vitamin D3, Serum Albumin, Calcium, Thyrotropin-Releasing Hormone with Early Childhood Caries: A Case–Control Study. *Int. J. Clin. Pediatr. Dent.* **2021**, *14*, 648–651. [CrossRef] [PubMed]

43. Williams, T.L.; Boyle, J.; Mittermuller, B.-A.; Carrico, C.; Schroth, R.J. Association between Vitamin D and Dental Caries in a Sample of Canadian and American Preschool-Aged Children. *Nutrients* **2021**, *13*, 4465. [CrossRef]

44. Deane, S.; Schroth, R.J.; Sharma, A.; Rodd, C. Combined Deficiencies of 25-Hydroxyvitamin D and Anemia in Preschool Children with Severe Early Childhood Caries: A Case–Control Study. *Paediatr. Child Health* **2018**, *23*, e40–e45. [CrossRef] [PubMed]

45. Olczak-Kowalczyk, D.; Kaczmarek, U.; Gozdowski, D.; Turska-Szybka, A. Association of Parental-Reported Vitamin D Supplementation with Dental Caries of 3-Year-Old Children in Poland: A Cross-Sectional Study. *Clin. Oral Investig.* **2021**, *25*, 6147–6158. [CrossRef]

46. Schroth, R.J.; Rabbani, R.; Loewen, G.; Moffatt, M.E. Vitamin D and Dental Caries in Children. *J. Dent. Res.* **2016**, *95*, 173–179. [CrossRef] [PubMed]

47. Navarro, C.L.A.; Grgic, O.; Trajanoska, K.; van der Tas, J.T.; Rivadeneira, F.; Wolvius, E.B.; Voortman, T.; Kragt, L. Associations Between Prenatal, Perinatal, and Early Childhood Vitamin D Status and Risk of Dental Caries at 6 Years. *J. Nutr.* **2021**, *151*, 1993–2000. [CrossRef]

48. Gyll, J.; Ridell, K.; Öhlund, I.; Karlsland Åkeson, P.; Johansson, I.; Lif Holgerson, P. Vitamin D Status and Dental Caries in Healthy Swedish Children. *Nutr. J.* **2018**, *17*, 11. [CrossRef]

49. Kühnisch, J.; Thiering, E.; Heinrich-Weltzien, R.; Hellwig, E.; Hickel, R.; Heinrich, J. Fluoride/Vitamin D Tablet Supplementation in Infants—Effects on Dental Health after 10 Years. *Clin. Oral Investig.* **2017**, *21*, 2283–2290. [CrossRef]

50. Carvalho Silva, C.; Gavinha, S.; Manso, M.C.; Rodrigues, R.; Martins, S.; Guimarães, J.T.; Santos, A.C.; Melo, P. Serum Levels of Vitamin D and Dental Caries in 7-Year-Old Children in Porto Metropolitan Area. *Nutrients* **2021**, *13*, 166. [CrossRef]

51. Wójcik, D.; Szalewski, L.; Pietryka-Michałowska, E.; Borowicz, J.; Pels, E.; Beń-Skowronek, I. Vitamin D 3 and Dental Caries in Children with Growth Hormone Deficiency. *Int. J. Endocrinol.* **2019**, *2019*, 2172137. [CrossRef] [PubMed]

52. Akinkugbe, A.A.; Moreno, O.; Brickhouse, T.H. Serum Cotinine, Vitamin D Exposure Levels and Dental Caries Experience in U.S. Adolescents. *Community Dent. Oral Epidemiol.* **2018**, *47*, 185–192. [CrossRef] [PubMed]

53. Kim, I.-J.; Lee, H.-S.; Ju, H.-J.; Na, J.-Y.; Oh, H.-W. A Cross-Sectional Study on the Association between Vitamin D Levels and Caries in the Permanent Dentition of Korean Children. *BMC Oral Health* **2018**, *18*, 43. [CrossRef] [PubMed]

54. Zhou, F.; Zhou, Y.; Shi, J. The Association between Serum 25-Hydroxyvitamin D Levels and Dental Caries in US Adults. *Oral Dis.* **2020**, *26*, 1537–1547. [CrossRef]

55. Hujoel, P.P. Vitamin D and Dental Caries in Controlled Clinical Trials: Systematic Review and Meta-Analysis. *Nutr. Rev.* **2013**, *71*, 88–97. [CrossRef] [PubMed]

56. Hu, Z.; Zhou, F.; Xu, H. Circulating Vitamin C and D Concentrations and Risk of Dental Caries and Periodontitis: A Mendelian Randomization Study. *J. Clin. Periodontol.* **2022**, *49*, 335–344. [CrossRef]

57. Dodhia, S.A.; West, N.X.; Thomas, S.J.; Timpson , N.J.; Johansson, I.; Lif Holgerson, P.; Dudding, T.; Haworth, S. Examining the causal association between 25-hydroxyvitamin D and caries in children and adults: A two-sample Mendelian randomization approach. *Wellcome Open Res.* **2021**, *5*, 281. [CrossRef]

58. Karnaki, P.; Katsas, K.; Diamantis, D.V.; Riza, E.; Rosen, M.S.; Antoniadou, M.; Gil-Salmerón, A.; Grabovac, I.; Linou, A. Dental Health, Caries Perception and Sense of Discrimination among Migrants and Refugees in Europe: Results from the Mig-HealthCare Project. *Appl. Sci.* **2022**, *12*, 9294. [CrossRef]

59. Bernabé, E.; Sheiham, A. Age, period and cohort trends in caries of permanent teeth in four developed countries. *Am. J. Public Health* **2014**, *104*, e115–e121. [CrossRef]

60. Schwendicke, F.; Dörfer, C.E.; Schlattmann, P.; Foster Page, L.; Thomson, W.M.; Paris, S. Socioeconomic inequality and caries: A systematic review and meta-analysis. *J. Dent. Res.* **2015**, *94*, 10–18. [CrossRef]

61. Stennett, M.; Tsakos, G. The impact of the COVID-19 pandemic on oral health inequalities and access to oral healthcare in England. *Br. Dent. J.* **2022**, *232*, 109–114. [CrossRef] [PubMed]

62. Soares, G.H.; Pereira, N.F.; Biazevic, M.G.; Braga, M.M.; Michel-Crosato, E. Dental caries in South American Indigenous peoples: A systematic review. *Community Dent. Oral Epidemiol.* **2019**, *47*, 142–152. [CrossRef] [PubMed]

63. WHO. WHO Global Oral Health Status Report. WHO Highlights Oral Health Neglect Affecting nearly Half of the World's Population. Available online: https://www.who.int/news/item/18-11-2022-who-highlights-oral-health-neglect-affecting-nearly-half-of-the-world-s-population (accessed on 25 March 2023).

64. Di Spigna, G.; Del Puente, A.; Covelli, B.; Abete, E.; Varriale, E.; Salzano, S.; Postiglione, L. Vitamin D receptor polymorphisms as tool for early screening of severe bone loss in women patients with rheumatoid arthritis. *Eur. Rev. Med. Pharmacol. Sci.* **2016**, *20*, 4664–4669. [PubMed]

65. Di Francia, R.; Amitrano, F.; De Lucia, D. Evaluation of genotyping methods and costs for VDR polymorphisms. Letter to the editor. *Eur. Rev. Med. Pharmacol. Sci.* **2017**, *21*, 1–3. Available online: https://www.europeanreview.org/wp/wp-content/uploads/1-3-Evaluation-of-genotyping-methods-and-costs-for-VDR-polymorphisms.pdf (accessed on 25 March 2023).
66. Zampetti, P.; Scribante, A. Historical and bibliometric notes on the use of fluoride in caries prevention. *Eur. J. Pediatr. Dent.* **2020**, *21*, 148–152.
67. Khanduri, N.; Kurup, D.; Mitra, M. Quantitative evaluation of remineralizing potential of three agents on artificially demineralized human enamel using scanning electron microscopy imaging and energy-dispersive analytical X-ray element analysis: An in vitro study. *Dent. Res. J.* **2020**, *17*, 366–372. [CrossRef]

applied
sciences

Review

Drug-Food Interactions with a Focus on Mediterranean Diet

Marios Spanakis *, Evridiki Patelarou and Athina Patelarou

Department of Nursing, School of Health Sciences, Hellenic Mediterranean University, GR-71410 Heraklion, Crete, Greece
* Correspondence: mspanakis@hmu.gr

Abstract: There is a growing interest among people in western countries for adoption of healthier lifestyle habits and diet behaviors with one of the most known ones to be Mediterranean diet (Med-D). Med-D is linked with daily consumption of food products such as vegetables, fruits, whole grains, seafood, beans, nuts, olive oil, low-fat food derivatives and limited consumption of meat or full fat food products. Med-D is well-known to promote well-being and lower the risk of chronic conditions such as cardiovascular diseases, diabetes, and metabolic syndrome. On the other hand bioactive constituents in foods may interfere with drugs' pharmacological mechanisms, modulating the clinical outcome leading to drug-food interactions (DFIs). This review discusses current evidence for food products that are included within the Med-Dand available scientific data suggest a potential contribution in DFIs with impact on therapeutic outcome. Most cases refer to potential modulation of drugs' absorption and metabolism such as foods' impact on drugs' carrier-mediated transport and enzymatic metabolism as well as potential synergistic or antagonistic effects that enhance or reduce the pharmacological effect for some drugs. Adherence to Med-D can improve disease management and overall well-being, but specific foods should be consumed with caution so as to not hinder therapy outcome. Proper patient education and consultation from healthcare providers is important to avoid any conflicts and side effects due to clinically significant DFIs.

Keywords: drug-food interactions; pharmacokinetic interactions; Mediterranean diet; drug metabolizing enzymes; CYPs; P-gp; DFIs

Citation: Spanakis, M.; Patelarou, E.; Patelarou, A. Drug-Food Interactions with a Focus on Mediterranean Diet. *Appl. Sci.* **2022**, *12*, 10207. https://doi.org/10.3390/app122010207

Academic Editors: Maria Antoniadou and Theodoros Varzakas

Received: 31 August 2022
Accepted: 7 October 2022
Published: 11 October 2022

Publisher's Note: MDPI stays neutral with regard to jurisdictional claims in published maps and institutional affiliations.

1. Introduction

Mediterranean diet (Med-D) represents one of the most famous and well received diet food habits, particularly in the western countries [1]. Med-D originates from the eating patterns that people around Mediterranean basin have been following since the ancient years, along with the adoption of food products that were cultivated later in the area (i.e., potatoes) [2,3]. Med-D promotes the consumption of all food types but focuses on vegetables, fruits, the nearly exclusive use of olive oil in food preparation, routine consumption of marine food, low fat white meat (chicken, turkey) along with cereals, grains, nuts, wine and lesser consumption of high-fat meat products [3]. UNESCO states regarding Med-D: *"The Mediterranean diet is characterized by a nutritional model that has remained constant over time and space, consisting mainly of olive oil, cereals, fresh or dried fruit and vegetables, a moderate amount of fish, dairy and meat, and many condiments and spices, all accompanied by wine or infusions, always respecting beliefs of each community"* [4]. Since the first observations in 1960s [5], Med-D has shown undisputable, longitudinal, and high-quality evidence to support its health benefits against specific diseases for people that remain adherent to Med-D dietary patterns [6–9]. A reduced risk of overall mortality related with metabolic syndrome, cardiovascular diseases, coronary heart disease, myocardial infarction, cancer incidence, diabetes, neurodegenerative diseases, kidney disease and arthritis are mostly identified regarding clinical data availability [10–21]. Moreover, for Med-D food products we can recognize specific active constituents (i.e., oleuropein, resveratrol, retinoids, flavonoids, terpenes, catechins, ω-3-fatty acids etc.) with pharmacological effects (i.e., antioxidant and anti-inflammatory) that are related with the health benefits from adherence in Med-D [22–31]. In this respect, the question that arises is whether there could be cases in which pharmacologically active compounds that are present in Med-D foods could potentially interfere with the pharmacological action of drugs and lead in clinically significant DFIs.

Drug interactions are an important clinical issue for optimum healthcare provision. The proper drug administration and drug combinations ensures the minimization of risks for adverse drug reactions (ADRs) due to drug-drug interactions [32–34]. Moreover, drug interactions can emerge due to co-administration of several complementary and alternative medicine (CAM) products such as herbal medicinal products (HMPs) as well as dietary supplements (DS) and food products [35–38]. A DFI describes the physical, biochemical or physiological modulation of a pharmacological process that a drug is following due to the presence of one or more constituents within a food product or other nutraceuticals [39,40]. These bioactive constituents that are present in dietary products may interfere with drugs' biological pathways and alter drugs' pharmacological action or levels in the body. Especially for food products there are well-established clinical examples of drug-food interaction (DFI) such as foods containing tyramine and monoamine-oxidase inhibitors (MAOIs), alcoholic beverages or spirits with central nervous system (CNS) drugs as well as grapefruit and its impact in systemic concentration of several drugs [38,41–43]. Overall, DFIs can occur during the habitual use of certain foods whose components can modify pharmacological processes and pathways that take place in the body for drugs. These pathways can be related with the absorption, distribution, metabolism, and elimination processes (pharmacokinetic drug-food interactions, PK-DFIs) or with biological pathways related with the main—or secondary—pharmacological action of a drug (pharmacodynamic drug-food interactions, PD-DFIs) [44,45].

The aim of this review is to summarize available knowledge regarding DFIs focusing mostly on products that are included within Med-D diet. The analysis focuses on the most common and known food products considering cultural, ethnic, agricultural variations among the regions around Mediterranean area which account for alterations in Med-D dietary habits. The level of evidence is following the general principle of evidence pyramid characterized as: (i) "theoretical" if mechanisms can be proposed; (ii) "low" if in vitro data are available; (iii) "moderate" if in vivo data are available; (iv) "good" if the DFI is observed clinically and (v) "high" if the level of evidence is adequate with randomized clinical studies and other data sources available [46–49]. The review is structured to present initially general pharmacological mechanisms of potentially clinically significant DFIs with some characteristic examples. Subsequently available scientific data and the level of evidence regarding DFIs between Med-D foods and drugs are presented and discussed. The potential impact of heavy metal or other pollutants is not included.

2. Pharmacological Mechanisms of Drug-Food Interactions

Pharmacological mechanisms of DFIs include the influence of PK processes and/or the PD effect of drugs in organs and secondary tissues (Figure 1). For a DFI to have a clinically significant impact it must modulate the pharmacological profile of a drug out of its therapeutic window. For example, in cases of PK-DFIs, one or more constituents of a food product must increase or decrease drug's concentration out of the minimum or maximum effective concentration, respectively, and/or change the time of drugs action. (Figure 2). Therefore, clinically significant DFIs occur when one or more food constituents are (i) contained in fair and consistent amounts in the product; (ii) can reach systemic circulation and biological ways in adequate concentrations and (iii) modulate meaningfully the drug's pharmacological action [50,51].

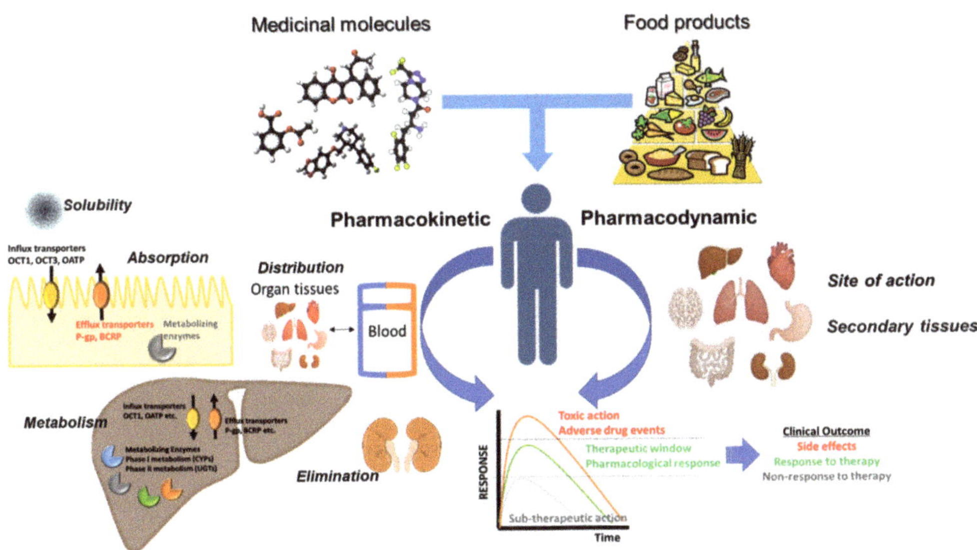

Figure 1. Representation of pharmacological mechanisms of DFIs and their impact on clinical outcome.

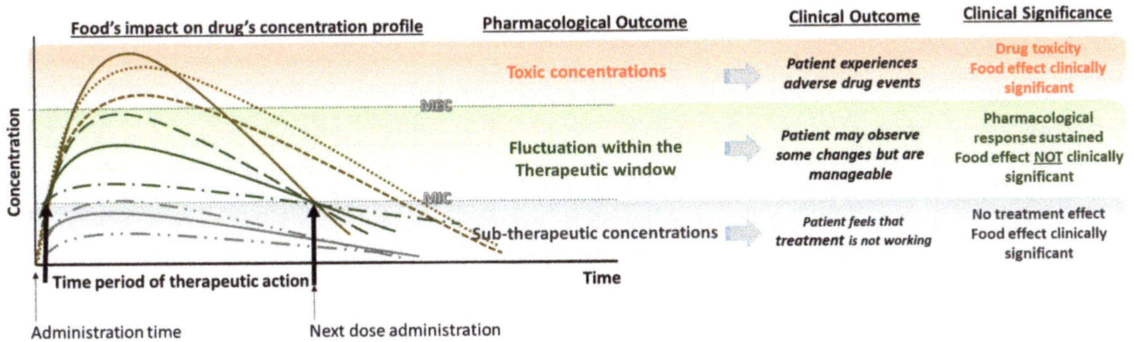

Figure 2. Graphical representation of food effect on the PK profile of a drug that leads in clinically significant DFIs (MEC maximum effective concentration, MIC: minimum effective concentration).

2.1. Pharmacokinetic Drug-Food Interactions (PK-DFIs)

2.1.1. Absorption

Drug's absorption—for orally administered drugs—is determined by its physicochemical properties, formulation, the capability of transportation (active or passive) across epithelial cells in the gastro-intestinal (GI) tract and the administration during fasted/fed state (Figure 3) [52]. Any modulation of intestinal transit time can change the drug's bioavailability—the fraction that reaches systemic circulation—and alter its systemic concentration. This is very important for drugs with reduced aqueous solubility but sufficient permeability [53,54]. Dietary compounds can modulate absorption either through biological activities or due to physicochemical mechanisms that change GI-environment, thus altering the biopharmaceutical properties of a drug formulation. For example, radish extract has been shown to stimulate GI motility in vitro through activation of muscarinic pathways [55]. Oat brans and fibers, although assisting in lowering cholesterol, also inhibit the intestinal absorption of statins, altering their PD properties, thus concurrent administration and consumption is proposed to be avoided or limited [56,57]. High-fat diets may lead in raised bile salt secretion which may increase the solubilization and absorption of lipophilic drugs such as anticancer drugs pazopanib, vemurafenib and lapatinib [58,59]. Generally the impact of fasted/fed state seems to be an important factor for orally administered drugs and especially for drugs used in oncology [60,61]. Thus, it is important relative information to be placed on drug labels since certain foods can change the pH in the stomach, alter the solubility, delay gastric emptying, change the stimulation of bile and blood flow, create unabsorbed chelate ligands, or change the gut microflora [50,62].

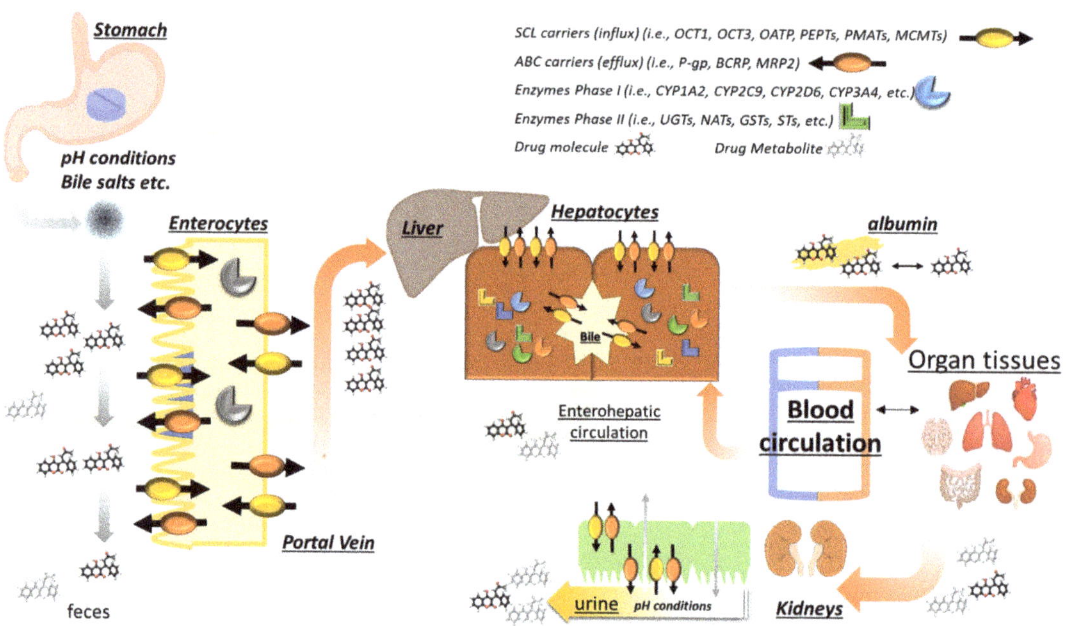

Figure 3. Pharmacokinetic processes for orally administered drugs and sites of potential PK-DFIs.

Following dispersion and dissolution in the GI track, a drug molecule is transported across epithelial GI-cells to the portal vein through active transport. Drug carriers can be classified into two major superfamilies, (i) ATP-binding cassette (ABC) transporters with efflux activity and (ii) solute carrier (SLC) transporters with influx activity [63–67]. P-glycoprotein (P-gp) is the most well-known efflux carrier regulating the absorption of drugs (i.e., anticancer, immunosuppressants, cardiovascular and lipid-modifying agents, antivirals, antibiotics, antiepileptics). Additional efflux transporters are: (i) the breast cancer resistance protein (BCRP) alleged to prevent absorption of toxic compounds and regulate the absorption of chemotherapeutics, and drugs such as prazosin, glyburide, cimetidine, sulfasalazine, rosuvastatin etc.; and (ii) the multidrug resistance-associated protein 2 (MRP2) that plays a role in the absorption of chemotherapeutic products [64,66,67]. Regarding influx transport, it is regulated by SLCs superfamily and includes a wide number of carriers such as organic anion transporters proteins (OATPs); organic cation transporters (OCTs); proton-dependent oligopeptide transporters (PEPTs); plasma membrane monoamine transporters (PMATs); and the monocarboxylate transporters (MCTs), all of which have been shown to play roles in drug transport across cell membranes [65,67].

Constituents of food products and DS can influence transporter's activity [68–71]. The efflux activity of P-gp has been shown in vitro to be inhibited from rosemary extracts, orange juices, strawberry and apricot extracts, dietary fatty acids, mint extracts, and spices constituents found in black pepper (piperine), chili peppers (capsaicin), and sesame oil (sesamin) [72]. Flavonoids such as quercetin, rutin, genistein, and silymarin, and terpenoids such as glycyrrhetinic acid are typical examples inhibiting activity in P-gp and in some cases for BCRP [73–76]. Green tea beverages have been shown in vivo to inhibit hepatic OATP2 reducing hepatic drug uptake and increase plasma exposure to atorvastatin [77].

2.1.2. Distribution

Drug distribution describes the reversible transfer of a drug from the blood to organs through vascular permeability into interstitial-intracellular space, and from there inside the cells [78]. Drug's physicochemical properties, plasma and/or tissue protein binding, tissue composition and the presence of several barriers (i.e., blood-brain barrier) are factors that play roles in distribution [79]. Drug's distribution is often related with the volume of distribution (Vd). It is defined as the apparent volume in which the existing amount of drug in the body must be dissolved to give its measured plasma concentration. Low Vd values are for drugs that remain mainly in plasma while high values are for lipophilic molecules that are distributed in tissues with high body-fat (i.e., CNS drugs). Regarding drugs in plasma, they are either bound (fb) or unbound (fu) to plasma proteins (i.e., human serum albumin (HAS), a1-acid-glycoprotein) depending on their affinity with them. This is important since only unbound drugs are distributed in the tissues and give the pharmacological action, as well as are eliminated. Protein binding is a site of potential drug interactions and alteration of protein binding can be significant in cases of fu < 0.05, low Vd (high plasma concentration) and narrow therapeutic index (NTI) [80]. For example, if a drug with fu = 0.05 changes it to fu = 0.1 and no added elimination occurs, the drug will double its action which can potentially lead in ADRs. Phenolic derivatives (i.e., flavonoids, phenolic acids, anthocyanidins, gallic acid derivatives, stilbene derivatives etc.) can form reversible complexes with plasma proteins such as albumin [81,82]. Flavonoids such as baicalin, rutin and quercetin that are found in several DS have been suggested to alter the albumin binding of drugs such as theophylline, nifedipine, promethazine, and ticagrelor [82]. As far as displacement of warfarin, experimental results vary from no risk up to clinically significant DFIs. For DFIs in protein binding with warfarin, an allosteric alteration of different binding regions is proposed, but the interaction is a minor one due also to the fact that S-warfarin (the active enantiomer) has lower protein binding [83].

2.1.3. Metabolism

Metabolic transformation of drugs occurs through a series of pathways in the liver and is divided in two phases (phase I and phase II) (Figure 2). Phase I reactions encompass the formation, or the modification of a functional group through oxidation, reduction, or hydrolysis while phase II are conjugation reactions.

Cytochromes P450 (CYPs) are responsible for 80% of the total phase I oxidation reactions [84]. Of the total 18 CYP families described in humans, the first three participate the most in drug metabolism. Variations in CYP genes among individuals results in phenotypes of different metabolic activity and usually classifying them as ultra-rapid, extensive, intermediate, and poor metabolizers [85–87]. CYP2D6 is the most recognized example regarding inter-individual and inter-population variability regarding phenotypes [88]. Pharmacogenetic testing for CYP2D6 has become an important clinical aspect of personalized therapy, especially since it is responsible for the metabolism of 20–25% of drugs such as CNS and cardiovascular medications [87,88]. This unique characteristic of CYP2D6 often makes it a focus point for potential drug interactions, including DFIs. CYP3A subfamily is the most abundant isoenzyme that plays role in drugs pharmacokinetics. CYP3A isoenzymes (CYP3A4 and CYP3A5) are expressed in the small intestine and the liver, and their metabolic activity accounts for 70–75% of drug metabolism [86]. Over the previous years it has become evident that CYP3A4 and P-gp play complementary roles for absorption, disposition, and metabolism processes. Hence, in many cases, interaction mechanisms for drugs often include both biological pathways of transport and metabolism [89].

The roles of food products in drug metabolism have been extensively evaluated [51]. St John's Wort (SJW) is a well-known example of DS involved in drug interactions through CYP3A4 and P-g. SJW has been promoted as a well-documented natural antidepressant supplement after long term use [90]. SJW's constituents (mainly hyperforin) can initially inhibit and subsequently induce the expression of CYP3A in the liver as well as induce the expression of P-pg [91,92]. The overall effect is an initial increase in drug exposure but the later induction of P-gp and CYP3A leads in reduced and sub-therapeutic concentration of CYP3A and P-gp substrates. The impact of SJW on P-gp and CYP3A is correlated with the hyperforin content in the product [93]. As of today there are several drug categories reported to interact with SJW such as anticancer drugs (i.e., imatinib, irinotecan, docetaxel), HIV drugs (i.e., indinavir), immunosuppressants (cyclosporine, tacrolimus), anticoagulants (i.e., coumarin analogues), cardiovascular medications and lipid modifying agents (i.e., digoxin, statins etc.), antiepileptics (i.e., carbamazepine), and oral contraceptives (i.e., levonorgestrel etc.) [92]. Hence, SJW, although promoted as a good natural antidepressant, needs special precautions when it is used form patients under therapies with NTI drugs [94].

Other important CYP enzymes are the CYP1A1 and CYP1A2, CYP2C9 and CYP2C19 that also metabolize a wide number of drugs. Polycyclic aromatic hydrocarbons (PAHs) present in tobacco are bio-transformed by CYP1A1 and CYP1A2 to oxy-derivatives, which can bind to DNA and initiate carcinogenesis. Tobacco can also induce CYP1A1 and CYP1A2 expression and modulate the systemic concentration of drugs that are substrates for these enzymes [95]. Isothiocyanates and other natural occurring polyphenols that are present in several vegetables have been shown to reduce the CYP1A enzyme activity, thus limiting the effect of PAHs [96]. CYP2C9 participates in the metabolism of coumarin anticoagulants (i.e., warfarin and acenocoumarol), thus is a site of potential interactions for anticoagulation treatment. A known case is that of gingko biloba, where ginkgolic acid I and II can inhibit CYP2C9, however the interaction is negligible [97,98].

In Phase II, metabolism transferases such as UDP-glucuronosyltransferases, sulfo-transferases, N-acetyltransferases, glutathione S-transferases and methyltransferases insert hydrophilic groups to the initial drug or the formed metabolite prior to renal elimination. Dietary polyphenols found in fruits, vegetables, wine, olive oil, tea and cocoa products are potential substrates and/or inhibitors of Phase II enzymes due to the presence of hydroxylic groups (-OH) that gives them structural similarity with drugs' metabolites. Until now, studies (mostly in vitro) have shown natural polyphenols' potential inhibitory activities on UDP-glucuronosyltransferases. For example, molecules such as glycyrrhetinic acid (licorice), chrysin (passiflora, mushrooms), silymarin (milk thistle), quercetin (onions, apples, grapes, berries, broccoli etc.), myricetin (tomatoes, oranges, berries, red wine etc.), naringenin (citrus fruits), luteolin (celery, parsley, broccoli, onion leaves), phloretin (apples), piceatannol (passion fruit and blueberries), resveratrol (grapes, wine, grape juice) etc. have been found to exhibit inhibitory activity against UGTs thus they can play role during Phase II metabolism of drugs and possibly contribute in DFIs [99–102].

2.1.4. Elimination

Drug's removal from the body is occurring through excretion processes for an unmetabolized drug or its metabolite. Kidneys are the main organ of drug elimination and drugs (or their metabolites) are filtered through them into the urine and excreted. A secondary pathway is through the bile where the bio-transformed drugs in the liver are excreted in the bile and eliminated through enterohepatic circulation in the feces (Figure 3). Renal elimination of drugs involves the processes of glomerular filtration (drug passive diffusion from blood stream into the urine), proximal tubular secretion, and distal tubular reabsorption from passive diffusion or active transport. For renal elimination, two factors are playing roles for potential DFIs. The first is the urinary pH since drug's ionization—depending on the alkaline or acidic environment—plays a role in its capability to passively diffuse through membranes, trapped in ionized form and excreted, whereas the non-ionized form is re-absorbed. The second factor is the transporter-mediated interactions which can impact drugs' elimination profile or accumulation in renal tubular cells, leading to drug-induced nephrotoxicity. Diet habits play a key role in the acidity or alkalinity of urine pH. Vegetables and fresh fruits are mostly related with alkaline diets, whereas high protein content diets usually acidify the urine [103]. An alkaline urine pH can reduce the excretion rate of weakly basic drugs and an acidic urine pH from diet may decrease the excretion amount for weakly acidic drugs. Memantine and flecainide are two characteristic examples where their elimination profile is changing based on the urine pH conditions [104,105].

2.2. Pharmacodynamic Drug-Food Interactions (PD-DFIs)

Pharmacodynamic interactions refer to modulation of the pharmacological action of a drug in the site of action or in secondary tissues that can induce ADRs and side effects. PD-DFIs are usually related with additive, synergistic, or antagonistic effects from food compounds on the pharmacological pathways of a drug. One of the most well-known DFIs is for foods containing tyramine (i.e., wine and cheese). These foods are to be avoided from patients treated with monoamine oxidase inhibitors (MAOIs) [42]. Another typical example is the potential modulation of anticoagulating action of coumarin analogues (warfarin, acenocoumarol, phenprocoumon) from vitamin K (Vit-K) [106]. Vit-K1 (dihydroquinone, KH2) is the necessary cofactor for activation of the clotting factors, thus coumarin analogues and Vit-K are antagonizing each other. Foods rich in Vit-K such as kale, collard greens, broccoli, spinach, cabbage, and lettuce should be consumed from patients under anticoagulation treatment keeping in mind potential alterations in international normalized ration (INR). The general guidance is that patients need to sustain a stable diet, so the daily intake of Vit-K must remain constant throughout their treatment [107,108]. Similarly to Vit-K, special precautions should be taken for diets rich in minerals such as Potassium (K^+) in cases of cardiovascular diseases treated with K^+ sparing diuretics (spironolactone, amiloride, triamterene etc.) or agents acting on the renin–angiotensin system (i.e., angiotensin-converting enzyme inhibitors, ACEs, angiotensin receptor blockers, ARBs, etc.). In this respect, fruits such as bananas, oranges, and apricots, and vegetables such as spinach, potatoes, mushrooms, and peas should be consumed keeping in mind that high intake may lead to hyperkaliemia resulting in cardiac side effects and arrhythmias [109,110].

3. Mediterranean Food Products and Potential DFIs

3.1. Med-D Food Products

Med-D allows the intake of all types of foods following a general food guidance pyramid (Figure 4) [2,3]. In its base (level 1) are vegetables, fruits, nuts, and cereals that should be consumed in greater amounts and at daily frequency. Above them (level-2), sea food proteins and omega-3-fatty acids (n-3 FAs) are suggested to be consumed biweekly. Animal food products (level3) such as dairy, cheese, and eggs can be consumed in moderate portions during the week. Red meat (level 4) should be consumed in less proportions during the week along with saturated fat products and sweets. Regarding alcohol, a moderate consumption of wine and other fermented beverages is recommended (one to two glasses with meals). The pyramid is complete considering daily activities and physical exercise. Tables 1–3 summarize available data regarding potential DFIs for food products of level1 in the Med-D pyramid.

Figure 4. The Mediterranean diet pyramid along with examples of characteristic constituents that can be found in Med-D's food products.

Table 1. DFIs, pharmacological mechanism, significance, and level of evidence for vegetables and herbs that are contained in the base of Med-D pyramid (PK: pharmacokinetic, PD: Pharmacodynamic, DFI: drug-food interaction).

Food	Type of DFIs	Suggested Mechanism	Clinical Significance	Level of Evidence
Artichoke	PK	CYP mediated drug metabolism	Moderate	Moderate
Arugula	PD	Anticoagulants—Vitamin K	Moderate	Theoretical
Asparagus	PD	Anticoagulants—Vitamin K	Moderate	Theoretical
Beetroots	-	-	-	-
Bell pepper	PD	Anticoagulants—Vitamin K	Moderate	Theoretical
Broccoli	PD	Anticoagulants—Vitamin K	Moderate	Theoretical
Brussel Sprouts	PK	CYP mediated drug metabolism	Moderate	Moderate
Cabbage	PK	CYP mediated drug metabolism	Moderate	Good
Carrots	PK	CYP mediated drug metabolism	Minor	Low
Cauliflower	PK	CYP mediated drug metabolism	Minor	Moderate
Celery	PD	Synergism sedatives	Minor	Good
Collard greens	PD	Synergism antidiabetic medications	Minor	Low
Cucumbers	-	-	-	-
Dandelion greens	PD	Lithium	Moderate	Low
Eggplant	-	-	-	-
Fennel	-	-	-	-
Fava beans	PD	synergism with anti-parkinson drugs (L-dopa content)	Moderate	Low

Table 1. *Cont.*

Food	Type of DFIs	Suggested Mechanism	Clinical Significance	Level of Evidence
Garlic	PK		Moderate	Good
Green beans	-	-	-	-
Kale	PD	Anticoagulants—Vitamin K	Moderate	Theoretical
Leeks	PD	Anticoagulants—Vitamin K	Minor	Theoretical
Mushrooms	-	-	-	-
Onions	PD	Anticoagulants—Vitamin K	Minor	Theoretical
Potatoes	-	-	-	-
Radish	PK	stimulate GI motility & transit time	Minor	Moderate
Scallions	PD	Anticoagulants—Vitamin K	Minor	Theoretical
Shallots	PD	Anticoagulants—Vitamin K	Minor	Theoretical
Spinach	PD	Anticoagulants—Vitamin K	Moderate	Theoretical
Squash	PD	stimulant laxative may alter K+	Minor	Theoretical
Sweet potatoes	-	-	-	-
Swiss chard	PD	Anticoagulants—Vitamin K	Minor	Theoretical
Tomatoes	PK	CYP mediated drug metabolism	Minor	Moderate
Turnips	PD	Synergism antidiabetic medications	Minor	Moderate
Yams	PD	Synergism with estrogens	Moderate	Good
Zucchini	-	-	-	-
Anise	PK	CYP mediated metabolism	Moderate	Good
Basil	PD	Anticoagulants, n-3 FAs	Moderate	Low
Bayleaf	PD	Synergism with antidiabetics & sedatives	Minor	Low
Cinnamon	PD	Synergism antidiabetic & hepatoxicity	Minor	Low
Cloves	PD	potentiate the effects of drugs that affect hemostasis	Moderate	Moderate
Cumin	PD	Synergism antidiabetic medications	Moderate	Moderate
Fennel	-	-	-	-
Lavender	PD	Synergism with sedatives	Moderate	Low
Marjoram	PD	Potentiate effects of drugs for hemostasis	Moderate	Low
Mint	-	-	-	-
Oregano	PD	Potentiate effects of drugs for hemostasis	Moderate	Low
Parsley	-	-	-	-
Pepper	PK	CYP mediated drug metabolism	Minor	Moderate
Rosemary	-	-	-	-
Sage	PD	Synergism with sedatives	Moderate	Low
Savory	PD	Anticoagulants, n-3 FAs	Moderate	Low
Sumac	PD	Induce nephrotoxicity of drugs	Minor	Low
Tarragon	PD	Anticoagulants, n-3 FAs	Moderate	Low
Thyme	PD	Anticoagulants, n-3 FAs	Moderate	Low

Level of evidence: (i) Theoretical (ii) low: in vitro data (iii) moderate (in vitro/in vivo) Good (in vitro/in vivo/clinical); (iv) High (in vitro/in vivo/clinical observations/clinical trials).

Table 2. DFIs, pharmacological mechanism, significance, and level of evidence for fruits that are contained in the base of Med-D pyramid (PK: pharmacokinetic, PD: Pharmacodynamic, DFI: drug-food interaction).

Fruit	Type of DFIs	Suggested Mechanism	Clinical Significance	Level of Evidence
Apples	PK	OAT transporter inhibition	Moderate	Good
Apricots	-	-	-	-
Avocados	PD	Anticoagulants—Vitamin K	Moderate	Theoretical
Bananas	PD	Hyperkaliemia, high K^+	Moderate	Theoretical
Cherries	-	-	-	-
Clementines	PK	CYP mediated drug metabolism	Moderate	Moderate
Grapefruit	PK	P-gp transport & CYP mediated drug metabolism	Serious-avoid	High
Grapes	PK	UGT mediated drug metabolism	Moderate	Good
Lime	PK			
Melons	PD	Synergism antidiabetic medications	Low	Theoretical
Oranges	PK	OAT transporter inhibition	Moderate	Good
Peaches	PD	Hyperkaliemia, high K^+	Moderate	Theoretical
Pears	-	-	-	-
Pomegranates	PK	CYP mediated drug metabolism	None	High
Pomelo	PK	P-gp transport & CYP mediated drug metabolism	Use with caution	Good
Strawberries	PK/PD	Metabolism & anticoagulation	Low	Theoretical
Tangerines	PK	CYP mediated drug metabolism	Moderate	Moderate

Level of evidence: (i) Theoretical (ii) low: in vitro data (iii) moderate (in vitro/in vivo) Good (in vitro/in vivo/clinical); (iv) High (in vitro/in vivo/clinical observations/clinical trials).

Table 3. DFIs, pharmacological mechanism, significance, and level of evidence for nuts, and cereals that are contained in the base of Med-D pyramid.

Food	Type of DFIs	Suggested Mechanism	Clinical Significance	Level of Evidence
Barley	PK	Modulation GI absorption	Moderate	Low
Buckwheat	-	-	-	-
Bulgur	-	-	-	-
Farro-flax seed	PD	Anticoagulants, n-3 FAs	Moderate	Low
Millet	-	-	-	-
Oats & fibers	PK	Modulation GI absorption	Moderate	Good
Polenta	-	-	-	-
Rice	-	-	-	-
Wheat Berries	PK	Modulation GI absorption	Moderate	Low
Breads	-	-	-	-
Couscous	-	-	-	-
Almonds	-	-	-	-
Cashews	PD	Synergism antidiabetic medications	Minor	Low
Chickpeas				
Fava beans	PD	Synergism with anti-Parkinson drugs (high L-dopa content)	Moderate	Low
Green peas	-	-	-	-
Hazelnuts				
Kidney beans	PK	Modulation GI absorption	Minor	Low
Lentils	PD	Hyperkaliemia, high K^+	Minor	Low
Peanuts	PD	Tyramine pressure effect (MAOIs)	Use with caution	Low
Pistachios	-	-	-	-
Sesame seeds (tachini)	PK	CYP mediated metabolism	None	Moderate
Split peas	PK	Modulation GI absorption	Minor	Theoretical
Walnuts	PK	Modulation GI absorption	Moderate	Theoretical

Level of evidence: (i) Theoretical (ii) low: in vitro data (iii) moderate (in vitro/in vivo) Good (in vitro/in vivo/clinical); (iv) High (in vitro/in vivo/clinical observations/clinical trials).

3.2. Drug-Med Diet Interactions

3.2.1. Vegetables, Herbals, Olive Oil, Cereals, and Nuts

The Med-D has its basis in foods of plant-origin with a wide variety of vegetables along with other herbals. It contains an extensive list of domestic and imported vegetables that reached the area through historical trading routes. The most common vegetables include artichokes, arugula, asparagus, beetroots, broccoli, brussel sprouts, cabbage, carrots, celery, collard greens, cucumbers, dandelion greens, eggplant, fennel, garlic leeks, lettuce, mushrooms, mustard greens, onions (all types), peas, peppers, potatoes, pumpkin, radishes, spinach, turnips, zucchini. Case reports and clinical data suggest that potential PK-DFIs can result from consumption of artichoke, broccoli, brussel sprouts, cabbage, cauliflower, and tomatoes [111–113]. A pharmacological mechanism can be attributed to the isothio-cyanate content (i.e., broccoli, cauliflower etc.) and their capability to modulate drugs' CYP-mediated metabolism or transport (ABC-transporters) [112]. Especially for CYP1A2, it has been shown through clinical trial that brassica vegetables can induce CYP1A2 metabolic activity modulating caffeine's pharmacokinetics [114]. In addition, celery and other api-aceous vegetables (i.e., carrot, celery, dill, cilantro, parsnip, parsley etc.) can decrease cytochrome CYP1A2 activity as has been shown through several studies [112,114]. Garlic components have shown inhibitory action for CYPs 2C, 2D, and 3A-mediated metabolism in vitro but in a later clinical pharmacokinetic study, long-term use of garlic caplets led to a significant decline in the plasma concentrations of saquinavir which is metabolized from CYP3A4 [115,116]. Tomato juice was shown in vitro to contain mechanism-based and competitive inhibitor(s) of CYP3A4 [117,118]. Cabbage and onion juices have also shown potential inhibiting activities on CYP3A4 in vitro [119]. Basil demonstrated in vitro potential reversible and time-dependent inhibition of CYP2B6 and CYP3A4 as well as esterase-mediated metabolism of rifampicin, but the concentrations were higher than the ones used in daily food consumption [120]. The significance of these potential PK-DFIs is currently unresolved and the level of evidence for most of the cases is low. The frequent consumption of these foods may contribute in an observed inter-individual variability within the treatment goals. A recent systematic review and meta-analysis of twenty-three dietary intervention trials in humans analyzed the effect of cruciferous vegetable-enriched diets on drug metabolism. The meta-analyses showed a significant effect on CYP1A2 and glutathione S-transferase-alpha (GSTa) [113]. Thus, healthcare advice is needed in case patients habitually consume excessive amounts of vegetables such as broccoli, brussel sprouts, cabbage, cauliflower, radish, and watercress and are under treatment with CYP1A2 substrates (i.e., clozapine, olanzapine, fluvoxamine, haloperidol, melatonin, ramelteon, tizanidine, and theophylline).

Regarding PD-DFIs, arugula, asparagus, bell peppers, broccoli, celery, collard greens, kale, onions and leeks, spinach, and chard due to their content of vitamin-K could modulate the INR for people treated with coumadin analogues such as warfarin or acenocoumarol. However, despite this, DFI represents a clinically significant case; it is estimated to be of moderate importance and has a low level of evidence with studies suggesting that a balanced consumption of vegetables does not interfere with INR in a clinically significant way [107,108]. Concerning other potential PD-DFIs, anise and aniseed's essential oil (used to enhance the flavor of Greek Ouzo and mastic) in vivo enhanced the effects of CNS drugs (codeine, diazepam, midazolam, pentobarbital, imipramine and fluoxetine) in mice suggesting potential synergism and a clinically significant PD-DFI if it is used in extensive doses [121]. Garlic has shown promising data as an antidiabetic agent; thus it may enhance the pharmacologic effect of antidiabetic medicines [122]. Turnips have demonstrated synergism with antidiabetic drugs towards hypoglycemia while yams have a good quality of evidence for synergy with estrogens and thus special precautions should be made for patients under estrogen therapy [123,124]. Naturally occurring levodopa and carbidopa have been quantified in fava beans in fair amounts, thus patients with Parkinson's under treatment should be aware of possible synergism with co-administered medications [125].

Concerning olive oil, its protective role against inflammation-related chronic non-communicable diseases (cardiovascular, diabetes, cancer etc.) have been described thoroughly [126]. Proposed mechanisms involve the action of bioactive constituents of olive oil on interleukin-6 (IL-6) and platelet activating factor (PAF) inflammation pathways. In particular, PAF, which is a class of lipid chemical mediators with messenger functions, has gained research attention as a potential drug target due to its involvement in inflammatory diseases such as allergies, asthma, atherosclerosis, diabetes etc. [127]. Moreover, it is a point of focus as a contributing biological mechanism for anti-inflammatory action of several food products with protective roles against inflammation [31]. Several bioactive phytochemical compounds such as terpenes and constituents in olive oil with PAF action have been related to protective mechanisms against atherosclerosis [128,129]. In addition, as of today there is no contribution of olive oil or its constituents in potential DFIs. On the other hand, for some terpenes with PAF action, i.e., cedrol, inhibiting properties against human P450 have been described in vitro and further studies are needed to clarify potential DFIs [130]. Regarding potential PD-DFIs of PAF modulators with anti-platelet medications, there are not any reports suggesting potential contribution in DFIs.

3.2.2. Fruits and Fruit Juices

The Mediterranean fruits are among the most famous and widely consumed food products globally. Apples, apricots, avocados, cherries, clementines, figs, grapefruits, grapes, melons, nectarines, oranges, peaches, pears, pomegranates, strawberries, tangerines are the most common ones that are consumed by people following the Med-D style. Regarding fruit juices, the fermentable but unfermented product obtained from the edible part of the fruit and preserved fresh, they sometimes contain (due to the extraction process) different constituents or quantities from the original fruit.

The most notorious DFI regarding fruits and/or fruit juice with medications is that of grapefruit and its juice (GFJ) [41,131–133]. GFJ constituents (i.e., furanocoumarins etc.) can inhibit CYP activity (mainly CYP3A) through mechanism-based inhibition as well as transporter proteins in the intestine and liver (i.e., P-gp) [131,133–136]. This can elevate drugs' bioavailability which, along with reduction in drugs' intrinsic clearance, can result in increased concentrations and potential side effects. Typical examples are the concomitant use of GFJ with: (i) Ca^{2+} channel antagonists that result in low blood pressure; (ii) HMG-CoA reductase inhibitors that may lead to rhabdomyolysis and renal impairment; and (iii) adverse pulmonary effects caused by amiodarone co-administration [41,70,131,137].

Grapefruit as a plant belongs to the plant family of Rutaceae within the genus of Citrus fruits such as lemons, oranges, limes, tangerines fruits that are widely consumed. As a result, the frequent consumption of these products alerted the previous years the scientific community to the need to examine if DFIs could be further observed [138–140]. Until today, the most often described mechanism for potential DFIs are related with modulation of the activity of OATP transporters and CYPs metabolic activities [141–143]. Citrus fruits such as orange, lemon, pomelo, and lime have been assessed for potential DFIs and compared with GFJ. Pomelo juice increased the bioavailability of cyclosporine in an open-label crossover PK study probably due to inhibition of CYP3A4 and P-gp [144]. Orange juice reduced the bioavailability of alendronate and aliskiren [145,146], whereas Seville orange juice interacts with felodipine in a similar way to GFJ [147]. Lime juice has been shown in vivo to increase (similarly to GFJ) the bioavailability and systemic concentrations of carbamazepine with a clinically significant risk for liver and kidney toxicity [148]. Although tangerine (fruit and juice) showed some effect in vitro on CYP3A4-mediated metabolism of midazolam, this was of no clinical significance [149]. Narirutin found in Citrus fruits has also been observed in vivo to inhibit OATP1A2 and OATP2B1 [150].

Regarding apple juice, there is evidence from in vitro, in vivo and clinical studies of inhibiting activity on OATPs and modulation of the PK profile of drugs such as fexofenadine, montelukast and aliskiren [142,143,146,151]. It also contributed to a DFI with atenolol in a dose-response relationship but with limited effect on the PD-profile of the drug [152].

Cranberry juice and its constituent avicularin inhibited uptake transporters OATP1A2 and AOATP2B1 in vitro [153]. In addition, in vitro data indicated inhibition of CYP-mediated metabolism (CYP2C9 and CYP3A4) similar to ketoconazole and fluconazole. The effect is mainly attributed to anthocyanins content but these compounds show poor bioavailability, thus in vitro data were not repeated in vivo or through clinical studies [154–156]. Pomegranate juice has demonstrated in vitro/in vivo inhibiting action against CYP2C9 and CYP3A4 but with no clinical impact based on the available clinical data [157].

Drug interactions with fruits, fruit juices or pulps can also be related with PD-DFIs, and especially for fruits that contain considerable amounts of potassium (K^+). Bananas, apricots, and oranges are some typical examples of high-K^+ fruits and in theory their over consumption can be implicated in potential PD-DFIs with ARBs and diuretics [109,110]. Although this effect is based in theoretical statements, an in vivo study with palm fruits and lisinopril demonstrated elevated serum K^+ levels [158]. The risk of potential hyperkaliemia is clinically significant, especially in cases of kidney diseases. Although observational studies suggest that adherence to Med-D improves survival for CKD patients, there is a lack of conclusive clinical data regarding DFIs and hyperkaliemia, hence vigilance should be advised from healthcare providers [159,160].

3.2.3. Fish and Sea Food

Fish and sea food (clams, cockles, crabs, groupers, lobsters, mackerel, mussels, octopuses, oysters, salmons, sardines, sea basses, shrimps, squids, sea breams, tunas, etc.) are the main sources of protein and fat within the Med-D diet. They contain high amounts of essential amino acids along with n-3 fatty acids (FA) (i.e., eicosapentaenoic acid and docosahexaenoic acid) especially the pelagic fishes (sardines, anchovies, mackerels etc.) [161]. Apart of the nutritional value in Med-D, marine n-3 FAs are known to have positive effects on human health such as a protecting role in cardiovascular diseases, inflammation, diabetes, neurocognitive disorders etc. [162]. Regarding DFIs, omega-3 FAs seem to reduce coagulation factors (i.e., fibrinogen and prothrombin), thus in theory can potentiate the effects of anticoagulants [163]. As of today there have been some case reports of interactions between warfarin co-administration with fish-oil supplements, but the results were not repeated in a retrospective study of a larger cohort of patients with atrial fibrillation and deep vein thrombosis [164–166]. Considering also that DS usually have a higher content of n-3 FAs than the consumed food, the potential DFI is of minor importance and negligible for patients that remain adherent in their treatment plan.

3.2.4. Milk Dairy Products, White and Red Meat

Milk and dairy products (yoghurt, cheese) are part of the traditional domestic livestock practices around the Mediterranean basin and are part of the historical heritage of the dietary habits for the region. In medicine, milk and dairy products are an old case of potential DFIs due to their content in Ca^{2+} and tyramine. The presence of Ca^{2+} in milk and dairy products can create unabsorbed chelate ligands with antibiotic classes of tetracyclines and quinolones resulting in reduced bioavailability (PK-DFIs) [62]. Tyramine is a precursor of catecholamines and the inhibition of their metabolism from MAOIs can lead to increased catecholamine levels which can cause hypertension. Tyramine is known to interact with mono-amino oxidase inhibitors (MAOIs) resulting in an effect known as tyramine pressor response with high blood pressure and risk of cerebral hemorrhage which can be fatal [42,167]. Finally, another important PK-DFI is the reduced bioavailability of ferrous from cow milk. Caseins in cow milk bind Fe^{2+} by clusters of phosphoserines, keeping it soluble in GI's alkaline pH, preventing its free form from being available for absorption, thus decreasing its bioavailability in cases of ferrous supplement co-administration [168]. In addition, co-administration of mercaptopurine and cow milk in patients with chronic myelogenous leukemias reduces the bioavailability of the drug due to milk's high content of xanthine oxidase, thus this co-administration should be avoided [169,170].

Meat products within Med-D, although consumed to a limited extent, are a valuable source of nutrients for a healthy and balanced diet. Their dietary value lies in their high protein content with essential amino acids, ferrous from red meat, vitamin B12 and other vitamins of B-complex, zinc, selenium, and phosphorus. Fat content is dependent on meat species, feeding system, as well as the meat part that is used in food [171]. As stated earlier, high-fat content may lead to raised salt and increase the solubilization of lipophilic drugs [38].

3.2.5. Wine and Other Beverages

Wines, except for their alcohol content (~11% for whites and 15% for reds), have a rich composition of bioactive compounds such as polyphenols. Resveratrol, anthocyanins, catechins, and tannins (proanthocyanins and ellagitannins) are some of the most often found polyphenols with a higher content in red wines which also explicate the beneficial effect of from wine consumption [172,173]. One of the most known examples is the French paradox, the epidemiological observation of low coronary heart disease death rates despite high intake of dietary cholesterol and saturated fat in southern France which is attributed in red wine consumption in those populations [174]. Regarding DFIs, polyphenols can modulate the phase I and II metabolism as stated earlier but for wine this effect can be considered minimal compared to the effects of alcohol in the case of regular or heavy drinking.

Alcohol can enhance the effects of medications, especially in cases of chronic conditions. PK-DFIs of alcohol are related mostly with induction of CYP2E1 and to a lesser extent CYP3A3 and CYP2A1. Heavy alcohol drinking can lead to PD-DFIs of alcohol such as sedation when combined with CNS acting drugs (sedatives, antihistamines, antidepressants, antipsychotics etc.), induce gastric bleeding when combined with aspirin and relative painkillers or anticoagulants, and hypoglycemia with antidiabetic drugs [43,175]. Regarding the vexed issue, the widespread opinion that concomitant drink of alcohol with antibiotics or other antimicrobials will cause toxicity or treatment failure, a recent review of the available evidence suggested that the data are poor and sometimes controversial [176]. The reduced efficacy refers mostly to erythromycin and doxycycline. Disulfiram-like reaction (distress, pain, flushes, irregular heartbeat) can occurs in co-administration of metronidazole, ketoconazole, griseofulvin and cephalosporines (i.e., cefuroxime, cefotetan, ceftriaxone, cefoperazone, ceftriazone). Ambiguous data for ADRs exist for trimethoprim sulfamethoxazole. On the other hand, penicillins, fluoroquinolones, azithromycin, tetracyclines, nitrofurantoin, secnidazole, tinidazole, and fluconazole have not been causally related to ADRs [176]. Another mechanism involved is the reduction in the immune response and epidemiological studies have shown alcohol abuse to be associated with an increased incidence of infectious diseases. But this is related mostly to cases of alcohol abuse, consumption of high content alcoholic drinks and overall, a poor quality of life regarding well-being and disease prevention. Thus, it is a good idea from the healthcare perspective, especially for heavy-drinking patients under treatment, to advise towards drinking cessation [177,178]. On the other hand, the moderate consumption of red wine seems to be beneficial in cases of immune protection due to its polyphenol content [177]. Thus, although the quality and quantity of data are vague, the avoidance of or reduction in consuming alcohol in low or moderate amounts (e.g., a social occasion with one glass of wine or beer) as is suggested through Med-D can be enjoyed.

4. Discussion

It is of no debate that a healthy diet and nutrition is essential for good health and shields from several chronic non-communicable diseases, such as cardiovascular heart disease, diabetes, CNS disorders, autoimmune disease and cancer [11,17,22,26,31,163,179,180] Thus, in the context of optimum healthcare provision is important to support patients as well as the general population to embrace a healthier balanced diet and remain adherent and compliant to it along with their medication [181]. Healthcare providers should be able to educate patients and in the context of precision medicine era, to individualize their consultation so as to empower patients to improve their health and well-being [182–184].

Med-D remains one of the most valuable and widely studied dietary habits for promoting well-being. The clinical evidence supports that long-term adherence to Med-D is valuable against metabolic disorders and cardiovascular diseases. Numerous studies demonstrate the anti-inflammatory and anti-oxidative properties of constituents found within the Med-D foods against risk factors and pathophysiological mechanisms lowering the risk for developing most of the major non-communicable disorders [2,6,8,23,28,185,186]. All those food products and nutraceuticals that promote well-being have been the focus of research the recent years for the identification of bioactive compounds that can be related with molecular mechanisms by which dietary components promote health and prevent diseases [187–189]. On the other hand, these bioactive dietary compounds can share same biological mechanisms with drug molecules and interfere with pharmacological mechanisms modulating the clinical outcome due to DFIs [45].

The DFIs, although seeming to be summarized in a simple clinical question "shall I take this drug with this food?" are more than that. DFIs depend on the type and amount of food consumed on daily basis, patient's health status, comorbidities, drug categories that are prescribed, time of drug administration and meal intake, the clinical significance of the DFI and whether it is beneficial or not [50,52]. Briguglio et al. in their work [45] described that DFIs can be categorized in three groups: (i) pharmaceutical, (ii) pharmacokinetic and (iii) pharmacodynamic based on the mechanism involved. Alternatively, four types can be considered based on the process prior to systemic concentration. Hence, pre-systemic interactions are referred to as type I ex-vivo interactions or type II modulation of enzyme (subtype a) or carriers (subtype b) and/or other deactivations in GI-track (subtype c). Post-systemic interactions are described as type III for modulation of distribution and metabolism in the body and type IV as modulation of drug's renal or enterohepatic clearance. The latter description omits the PD-DFIs which could be considered as a type V post-systemic phase interaction.

This work aimed to provide a literature review regarding DFIs focusing on examples of foods that are included in the Med-D and discuss cases in which they can be related with significant DFIs. DFIs for Med-D food products are associated with modulation of PK processes and primarily with drugs' first pass effect from the GI-track to reach the systemic circulation. The majority the cases discussed refer to type II and III (subtypes a and b) with the addition of PD-effects that can occur for some products (Figure 5). Absorption and metabolism are the two main processes that should be considered when the question comes to food effects for a drug and the main pathways involved are the carrier-mediated transport across membranes and/or the mediated metabolism from Phase I and Phase II enzymes [39,112]. As shown in Tables 1–3, there is a good quality of evidence that cruciferous vegetables may impact drug metabolism, and fruits (or juices) such as grapefruit, apples, pomelo, pomegranates, and Seville orange juice can modulate the PK profile of some drugs which can be clinically significant in cases of drugs with narrow therapeutic index [57,113,131,139,142]. In addition, oat brans and fibers can modulate the absorption of statins whereas the high-fat content foods can modulate the bioavailability of some orally administered cancer medications [56–59]. Regarding PD-effects, the high content in Vit-K and K^+ in some vegetables and fruits can modulate the pharmacological action for coumadin analogue anticoagulants (Vit-K) and contribute to hyperkaliemia with ACEs, ARBs, and diuretics [106,109]. Although there are case reports and general literature discussion, the currently accepted scientific point of view is that patients who remain adherent and compliant to a balanced diet prior to and during any therapy initiation can keep on consuming these food products in a similar way to before [45]. They key issue here is the communication of patient with healthcare provider in case any adverse event occurs and the healthcare provider's awareness to report it and examine whether a DFI is related with it.

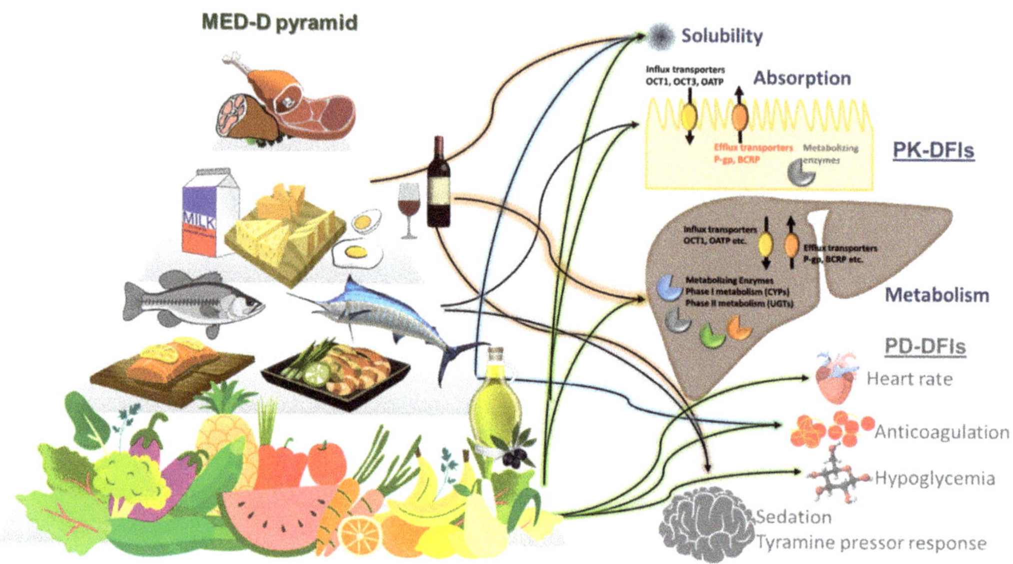

Figure 5. DFIs of Med-D food products regarding relation with type II and type III (subtypes a and b) interactions.

In addition, for cases where the food effect is negligible, or the pharmacological effect is sustained (Figure 2), it should be noted that Med-D diet can be considered as a collaborating partner with drugs [7,10,19,180,190]. For example, it may empower patients to better manage their health status and improve biochemical disease markers (i.e., blood glucose, cholesterol etc.) thus avoiding complicated therapeutic regimens with multiple medications that also raise the risk for adverse drug events from drug-drug interactions [32,191].

Apart from the well-documented cases (Tables 1–3) there are a lot of additional examples where data are scarce, theoretical, or poor. Hence, future studies are needed to further advance our knowledge in the field of DFIs. As the research is progressing on how dietary constituents engage in body homeostasis and health status, new knowledge is emerging towards prediction of food effects on drug's action. Increasing our understanding on PK and PD interaction mechanisms and utilization of novel experimental procedures will allow the gap on the available evidence regarding food effects on drugs to decrease. The incorporation of biomedical methods will allow the better prediction of potential interactions even through in silico methods during drug development or through big-data analysis [192–194]. From the citizen's perspective, the harnessing of digital evolution in relation to health promotion also cannot go unnoticed. Patient's oriented eHealth tools and apps are constantly emerging as means for patient empowerment to better manage their health status and well-being [195–200]. Even for Med-D, smartphone apps are available to assist people to remain adherent with the Med-D principles [201]. Moreover, a recent analysis of Android Google Play and Apple App Store Apps suggested that currently available apps can deliver information on the Med-D, but the integration of more behavioral change techniques within them is needed to expand the potential for improved adherence to Med-D [202]. Finally, eHealth evolution for patient empowerment can also assist in the dissemination of knowledge regarding DFIs, assisting patients to avoid potential interactions with DS and improve their communication with their healthcare providers [35,37,203,204].

5. Conclusions

The prediction, prevention, and management of DFIs is an essential element of optimum healthcare provision. Even for Med-D, one of the most effective diets, there can be cases where food habits may have negative impact on therapeutic outcome. Proper patient consultation is important, so they are aware to avoid any potential conflicts between administered treatment and dietary habits. Adherence to a diet program similar to the one that is proposed through Med-D and individualized patient consultation and education from healthcare providers for potential DFIs will help patients to manage their disease and their overall well-being.

Author Contributions: Conceptualization M.S.; Methodology, M.S. and A.P.; investigation M.S. and E.P.; Resources E.P.; writing—original draft preparation M.S., writing review and editing M.S. and A.P.; Visualization M.S.; Supervision, A.P.; project administration E.P. All authors have read and agreed to the published version of the manuscript.

Funding: This research received no external funding.

Institutional Review Board Statement: This study did not require ethical approval.

Informed Consent Statement: Not applicable.

Data Availability Statement: Not applicable.

Conflicts of Interest: The authors declare no conflict of interest.

References

1. Martínez-González, M.Á.; Hershey, M.S.; Zazpe, I.; Trichopoulou, A. Transferability of the Mediterranean Diet to Non-Mediterranean Countries. What Is and What Is Not the Mediterranean Diet. *Nutrients* **2017**, *9*, 1226. [CrossRef] [PubMed]
2. Willett, W.C.; Sacks, F.; Trichopoulou, A.; Drescher, G.; Ferro-Luzzi, A.; Helsing, E.; Trichopoulos, D. Mediterranean Diet Pyramid: A Cultural Model for Healthy Eating. *Am. J. Clin. Nutr.* **1995**, *61*, 1402S–1406S. [CrossRef] [PubMed]
3. Bach-Faig, A.; Berry, E.M.; Lairon, D.; Reguant, J.; Trichopoulou, A.; Dernini, S.; Medina, F.X.; Battino, M.; Belahsen, R.; Miranda, G.; et al. Mediterranean Diet Pyramid Today. Science and Cultural Updates. *Public Health Nutr.* **2011**, *14*, 2274–2284. [CrossRef] [PubMed]
4. United Nations Educational, S. and C.O. The Mediterranean Diet. Intangible Heritage. Available online: https://www.unesco.org/archives/multimedia/document-1680-eng-2 (accessed on 8 August 2022).
5. Wright, C.M. Biographical Notes on Ancel Keys and Salim Yusuf: Origins and Significance of the Seven Countries Study and the INTERHEART Study. *J. Clin. Lipidol.* **2011**, *5*, 434–440. [CrossRef]
6. Trichopoulou, A.; Martínez-González, M.A.; Tong, T.Y.N.; Forouhi, N.G.; Khandelwal, S.; Prabhakaran, D.; Mozaffarian, D.; de Lorgeril, M. Definitions and Potential Health Benefits of the Mediterranean Diet: Views from Experts around the World. *BMC Med.* **2014**, *12*, 112. [CrossRef]
7. El Amrousy, D.; Elashry, H.; Salamah, A.; Maher, S.; Abd-Elsalam, S.M.; Hasan, S. Adherence to the Mediterranean Diet Improved Clinical Scores and Inflammatory Markers in Children with Active Inflammatory Bowel Disease: A Randomized Trial. *J. Inflamm. Res.* **2022**, *15*, 2075–2086. [CrossRef]
8. Trichopoulou, A.; Costacou, T.; Bamia, C.; Trichopoulos, D. Adherence to a Mediterranean Diet and Survival in a Greek Population. *N. Engl. J. Med.* **2003**, *348*, 2599–2608. [CrossRef]
9. Iaccarino Idelson, P.; Scalfi, L.; Valerio, G. Adherence to the Mediterranean Diet in Children and Adolescents: A Systematic Review. *Nutr. Metab. Cardiovasc. Dis.* **2017**, *27*, 283–299. [CrossRef]
10. García-Fernández, E.; Rico-Cabanas, L.; Estruch, R.; Estruch, R.; Estruch, R.; Bach-Faig, A. Mediterranean Diet and Cardiodiabesity: A Review. *Nutrients* **2014**, *6*, 3474–3500. [CrossRef]
11. Dinu, M.; Pagliai, G.; Casini, A.; Sofi, F. Mediterranean Diet and Multiple Health Outcomes: An Umbrella Review of Meta-Analyses of Observational Studies and Randomised Trials. *Eur. J. Clin. Nutr.* **2017**, *72*, 30–43. [CrossRef]
12. Du, H.; Cao, T.; Lu, X.; Zhang, T.; Luo, B.; Li, Z. Mediterranean Diet Patterns in Relation to Lung Cancer Risk: A Meta-Analysis. *Front. Nutr.* **2022**, *9*, 844382. [CrossRef]
13. Bayán-Bravo, A.; Banegas, J.R.; Donat-Vargas, C.; Sandoval-Insausti, H.; Gorostidi, M.; Rodríguez-Artalejo, F.; Guallar-Castillón, P. The Mediterranean Diet Protects Renal Function in Older Adults: A Prospective Cohort Study. *Nutrients* **2022**, *14*, 432. [CrossRef]
14. Forsyth, C.; Kouvari, M.; D'Cunha, N.M.; Georgousopoulou, E.N.; Panagiotakos, D.B.; Mellor, D.D.; Kellett, J.; Naumovski, N. The Effects of the Mediterranean Diet on Rheumatoid Arthritis Prevention and Treatment: A Systematic Review of Human Prospective Studies. *Rheumatol. Int.* **2018**, *38*, 737–747. [CrossRef]
15. Morales-Ivorra, I.; Romera-Baures, M.; Roman-Viñas, B.; Serra-Majem, L. Osteoarthritis and the Mediterranean Diet: A Systematic Review. *Nutrients* **2018**, *10*, 30. [CrossRef]
16. Rees, K.; Takeda, A.; Martin, N.; Ellis, L.; Wijesekara, D.; Vepa, A.; Das, A.; Hartley, L.; Stranges, S. *Mediterranean-Style Diet for the Primary and Secondary Prevention of Cardiovascular Disease*; John Wiley and Sons Ltd.: Hoboken, NJ, USA, 2019; Volume 2019.

17. García-Casares, N.; Fuentes, P.G.; Barbancho, M.Á.; López-Gigosos, R.; García-Rodríguez, A.; Gutiérrez-Bedmar, M. Alzheimer's Disease, Mild Cognitive Impairment and Mediterranean Diet. A Systematic Review and Dose-Response Meta-Analysis. *J. Clin. Med.* **2021**, *10*, 4642. [CrossRef]
18. Bakaloudi, D.R.; Chrysoula, L.; Leonida, I.; Kotzakioulafi, E.; Theodoridis, X.; Chourdakis, M. Impact of the Level of Adherence to the Mediterranean Diet on Blood Pressure: A Systematic Review and Meta-Analysis of Observational Studies. *Clin. Nutr.* **2021**, *40*, 5771–5780. [CrossRef]
19. Bakaloudi, D.R.; Chrysoula, L.; Kotzakioulafi, E.; Theodoridis, X.; Chourdakis, M. Impact of the Level of Adherence to Mediterranean Diet on the Parameters of Metabolic Syndrome: A Systematic Review and Meta-Analysis of Observational Studies. *Nutrients* **2021**, *13*, 1514. [CrossRef]
20. Finicelli, M.; Di Salle, A.; Galderisi, U.; Peluso, G. The Mediterranean Diet: An Update of the Clinical Trials. *Nutrients* **2022**, *14*, 2956. [CrossRef]
21. Castro-Espin, C.; Agudo, A. The Role of Diet in Prognosis among Cancer Survivors: A Systematic Review and Meta-Analysis of Dietary Patterns and Diet Interventions. *Nutrients* **2022**, *14*, 348. [CrossRef]
22. Rufino-Palomares, E.E.; Pérez-Jiménez, A.; García-Salguero, L.; Mokhtari, K.; Reyes-Zurita, F.J.; Peragón-Sánchez, J.; Lupiáñez, J.A. Nutraceutical Role of Polyphenols and Triterpenes Present in the Extracts of Fruits and Leaves of Olea Europaea as Antioxidants, Anti-Infectives and Anticancer Agents on Healthy Growth. *Molecules* **2022**, *27*, 2341. [CrossRef]
23. Caponio, G.R.; Lippolis, T.; Tutino, V.; Gigante, I.; De Nunzio, V.; Milella, R.A.; Gasparro, M.; Notarnicola, M. Nutraceuticals: Focus on Anti-Inflammatory, Anti-Cancer, Antioxidant Properties in Gastrointestinal Tract. *Antioxidants* **2022**, *11*, 1274. [CrossRef]
24. Abenavoli, L.; Procopio, A.C.; Paravati, M.R.; Costa, G.; Milić, N.; Alcaro, S.; Luzza, F. Mediterranean Diet: The Beneficial Effects of Lycopene in Non-Alcoholic Fatty Liver Disease. *J. Clin. Med.* **2022**, *11*, 3477. [CrossRef]
25. Scoditti, E.; Capurso, C.; Capurso, A.; Massaro, M. Vascular Effects of the Mediterranean Diet-Part II: Role of Omega-3 Fatty Acids and Olive Oil Polyphenols. *Vascul. Pharmacol.* **2014**, *63*, 127–134. [CrossRef]
26. Massaro, M.; Scoditti, E.; Carluccio, M.A.; De Caterina, R. Nutraceuticals and Prevention of Atherosclerosis: Focus on Omega-3 Polyunsaturated Fatty Acids and Mediterranean Diet Polyphenols. *Cardiovasc. Ther.* **2010**, *28*, e13–e19. [CrossRef]
27. Augimeri, G.; Bonofiglio, D. The Mediterranean Diet as a Source of Natural Compounds: Does It Represent a Protective Choice against Cancer? *Pharmaceuticals* **2021**, *14*, 920. [CrossRef]
28. Vivancos, M.; Moreno, J.J. Effect of Resveratrol, Tyrosol and Beta-Sitosterol on Oxidised Low-Density Lipoprotein-Stimulated Oxidative Stress, Arachidonic Acid Release and Prostaglandin E2 Synthesis by RAW 264.7 Macrophages. *Br. J. Nutr.* **2008**, *99*, 1199–1207. [CrossRef]
29. Roman, G.C.; Jackson, R.E.; Gadhia, R.; Roman, A.N.; Reis, J. Mediterranean Diet: The Role of Long-Chain Omega-3 Fatty Acids in Fish; Polyphenols in Fruits, Vegetables, Cereals, Coffee, Tea, Cacao and Wine; Probiotics and Vitamins in Prevention of Stroke, Age-Related Cognitive Decline, and Alzheimer Disease. *Rev. Neurol.* **2019**, *175*, 724–741. [CrossRef]
30. Nadtochiy, S.M.; Redman, E.K. Mediterranean Diet and Cardioprotection: The Role of Nitrite, Polyunsaturated Fatty Acids, and Polyphenols. *Nutrition* **2011**, *27*, 733–744. [CrossRef]
31. Nomikos, T.; Fragopoulou, E.; Antonopoulou, S.; Panagiotakos, D.B. Mediterranean Diet and Platelet-Activating Factor: A Systematic Review. *Clin. Biochem.* **2018**, *60*, 1–10. [CrossRef]
32. Chatsisvili, A.; Sapounidis, I.; Pavlidou, G.; Zoumpouridou, E.; Karakousis, V.A.; Spanakis, M.; Teperikidis, L.; Niopas, I. Potential Drug-Drug Interactions in Prescriptions Dispensed in Community Pharmacies in Greece. *Pharm. World Sci.* **2010**, *32*, 187–193. [CrossRef]
33. Kohler, G.I.; Bode-Boger, S.M.; Busse, R.; Hoopmann, M.; Welte, T.; Boger, R.H. Drug-Drug Interactions in Medical Patients: Effects of in-Hospital Treatment and Relation to Multiple Drug Use. *Int. J. Clin. Pharmacol. Ther.* **2000**, *38*, 504–513. [CrossRef] [PubMed]
34. Dechanont, S.; Maphanta, S.; Butthum, B.; Kongkaew, C. Hospital Admissions/Visits Associated with Drug-Drug Interactions: A Systematic Review and Meta-Analysis. *Pharmacoepidemiol. Drug Saf.* **2014**, *23*, 489–497. [CrossRef] [PubMed]
35. Spanakis, M.; Spanakis, E.G.; Kondylakis, H.; Sfakianakis, S.; Genitsaridi, I.; Sakkalis, V.; Tsiknakis, M.; Marias, K. Addressing drug-drug and drug-food interactions through personalized empowerment services for healthcare. In Proceedings of the 38th Annual International Conference of the IEEE Engineering in Medicine and Biology Society (EMBS 2016), Orlando, FL, USA, 16–20 August 2016.
36. Vizirianakis, I.S.; Spanakis, M.; Termentzi, A.; Niopas, I.; Kokkalou, E. Clinical and Pharmacogenomic Assessment of Herb-Drug Interactions to Improve Drug Delivery and Pharmacovigilance. In *Plants in Traditional and Modern Medicine: Chemistry and Activity*; Kokkalou, E., Ed.; Transworld Research Network: Kerala, India, 2010; ISBN 978-81-7895-432-5.
37. Spanakis, M.; Sfakianakis, S.; Sakkalis, V.; Spanakis, E.G. PharmActa: Empowering Patients to Avoid Clinical Significant Drug(-)Herb Interactions. *Medicines* **2019**, *6*, 26. [CrossRef] [PubMed]
38. Lopes, M.; Coimbra, M.A.; Costa, M.D.C.; Ramos, F. Food supplement vitamins, minerals, amino-acids, fatty acids, phenolic and alkaloid-based substances: An overview of their interaction with drugs. *Crit. Rev. Food Sci. Nutr.* **2021**, 1–35. [CrossRef]
39. Won, C.S.; Oberlies, N.H.; Paine, M.F. Mechanisms Underlying Food-Drug Interactions: Inhibition of Intestinal Metabolism and Transport. *Pharmacol. Ther.* **2012**, *136*, 186. [CrossRef]
40. Frankel, E.H.; McCabe, B.J.; Wolfe, J.J. *Handbook of Food-Drug Interactions*; CRC Press: Boca Raton, FL, USA, 2003; ISBN 1135504571.
41. Kirby, B.J.; Unadkat, J.D. Grapefruit Juice, a Glass Full of Drug Interactions? *Clin. Pharmacol. Ther.* **2007**, *81*, 631–633. [CrossRef]

42. Brown, C.; Taniguchi, G.; Yip, K. The Monoamine Oxidase Inhibitor-Tyramine Interaction. *J. Clin. Pharmacol.* **1989**, *29*, 529–532. [CrossRef]
43. Chan, L.N.; Anderson, G.D. Pharmacokinetic and Pharmacodynamic Drug Interactions with Ethanol (Alcohol). *Clin. Pharmacokinet.* **2014**, *53*, 1115–1136. [CrossRef]
44. Amadi, C.N.; Mgbahurike, A.A. Selected Food/Herb-Drug Interactions: Mechanisms and Clinical Relevance. *Am. J. Ther.* **2018**, *25*, e423–e433. [CrossRef]
45. Briguglio, M.; Hrelia, S.; Malaguti, M.; Serpe, L.; Canaparo, R.; Dell'Osso, B.; Galentino, R.; De Michele, S.; Dina, C.Z.; Porta, M.; et al. Food Bioactive Compounds and Their Interference in Drug Pharmacokinetic/Pharmacodynamic Profiles. *Pharmaceutics* **2018**, *10*, 277. [CrossRef]
46. Spanakis, M.; Patelarou, A.; Patelarou, E.; Tzanakis, N. Drug Interactions for Patients with Respiratory Diseases Receiving COVID-19 Emerged Treatments. *Int. J. Environ. Res. Public Health* **2021**, *18*, 1711. [CrossRef]
47. Spanakis, M.; Roubedaki, M.; Tzanakis, I.; Zografakis-Sfakianakis, M.; Patelarou, E.; Patelarou, A. Impact of Adverse Drug Reactions in Patients with End Stage Renal Disease in Greece. *Int. J. Environ. Res. Public Health* **2020**, *17*, 9101. [CrossRef]
48. Spanakis, M.; Melissourgaki, M.; Lazopoulos, G.; Patelarou, A.E.; Patelarou, E. Prevalence and Clinical Significance of Drug–Drug and Drug–Dietary Supplement Interactions among Patients Admitted for Cardiothoracic Surgery in Greece. *Pharmaceutics* **2021**, *13*, 239. [CrossRef]
49. Murad, M.H.; Asi, N.; Alsawas, M.; Alahdab, F. New Evidence Pyramid. *Evid. Based Med.* **2016**, *21*, 125–127. [CrossRef]
50. Deng, J.; Zhu, X.; Chen, Z.; Fan, C.H.; Kwan, H.S.; Wong, C.H.; Shek, K.Y.; Zuo, Z.; Lam, T.N. A Review of Food-Drug Interactions on Oral Drug Absorption. *Drugs* **2017**, *77*, 1833–1855. [CrossRef]
51. Koziolek, M.; Alcaro, S.; Augustijns, P.; Basit, A.W.; Grimm, M.; Hens, B.; Hoad, C.L.; Jedamzik, P.; Madla, C.M.; Maliepaard, M.; et al. The Mechanisms of Pharmacokinetic Food-Drug Interactions—A Perspective from the UNGAP Group. *Eur. J. Pharm. Sci.* **2019**, *134*, 31–59. [CrossRef]
52. Schmidt, L.E.; Dalhoff, K. Food-Drug Interactions. *Drugs* **2012**, *62*, 1481–1502. [CrossRef]
53. Wu, C.Y.; Benet, L.Z. Predicting Drug Disposition via Application of BCS: Transport/Absorption/Elimination Interplay and Development of a Biopharmaceutics Drug Disposition Classification System. *Pharm. Res.* **2005**, *22*, 11–23. [CrossRef]
54. Sharma, S.; Prasad, B. Meta-Analysis of Food Effect on Oral Absorption of Efflux Transporter Substrate Drugs: Does Delayed Gastric Emptying Influence Drug Transport Kinetics? *Pharmaceutics* **2021**, *13*, 1035. [CrossRef]
55. Jung, K.Y.; Choo, Y.K.; Kim, H.M.; Choi, B.K. Radish Extract Stimulates Motility of the Intestine via the Muscarinic Receptors. *J. Pharm. Pharmacol.* **2000**, *52*, 1031–1036. [CrossRef]
56. Eussen, S.R.B.M.; Rompelberg, C.J.M.; Andersson, K.E.; Klungel, O.H.; Hellstrand, P.; Öste, R.; Van Kranen, H.; Garssen, J. Simultaneous Intake of Oat Bran and Atorvastatin Reduces Their Efficacy to Lower Lipid Levels and Atherosclerosis in LDLr-/- Mice. *Pharmacol. Res.* **2011**, *64*, 36–43. [CrossRef]
57. Vaquero, M.P.; Muniz, F.J.S.; Redondo, S.J.; Oliván, P.P.; Higueras, F.J.; Bastida, S. Major Diet-Drug Interactions Affecting the Kinetic Characteristics and Hypolipidaemic Properties of Statins. *Nutr. Hosp.* **2010**, *25*, 193–206.
58. Willemsen, A.E.C.A.B.; Lubberman, F.J.E.; Tol, J.; Gerritsen, W.R.; Van Herpen, C.M.L.; Van Erp, N.P. Effect of Food and Acid-Reducing Agents on the Absorption of Oral Targeted Therapies in Solid Tumors. *Drug Discov. Today* **2016**, *21*, 962–976. [CrossRef]
59. Lewis, L.D.; Koch, K.M.; Reddy, N.J.; Cohen, R.B.; Lewis, N.L.; Whitehead, B.; Mackay, K.; Stead, A.; Beelen, A.P. Effects of Food on the Relative Bioavailability of Lapatinib in Cancer Patients. *J. Clin. Oncol.* **2009**, *27*, 1191–1196. [CrossRef]
60. Kang, S.P.; Ratain, M.J. Inconsistent Labeling of Food Effect for Oral Agents across Therapeutic Areas: Differences between Oncology and Non-Oncology Products. *Clin. Cancer Res.* **2010**, *16*, 4446–4451. [CrossRef]
61. Omachi, F.; Kaneko, M.; Iijima, R.; Watanabe, M.; Itagaki, F. Relationship between the Effects of Food on the Pharmacokinetics of Oral Antineoplastic Drugs and Their Physicochemical Properties. *J. Pharm. Health Care Sci.* **2019**, *5*, 26. [CrossRef]
62. Jung, H.; Perergina, A.A.; Rodriguez, J.M.; Moreno- Esparza, R. The Influence of Coffee with Milk and Tea with Milk on the Bioavailability of Tetracycline. *Biopharm. Drug Dispos.* **1997**, *18*, 459–463. [CrossRef]
63. Estudante, M.; Morais, J.G.; Soveral, G.; Benet, L.Z. Intestinal Drug Transporters: An Overview. *Adv. Drug Deliv. Rev.* **2013**, *65*, 1340–1356. [CrossRef]
64. Terada, T.; Hira, D. Intestinal and Hepatic Drug Transporters: Pharmacokinetic, Pathophysiological, and Pharmacogenetic Roles. *J. Gastroenterol.* **2015**, *50*, 508–519. [CrossRef]
65. Lin, L.; Yee, S.W.; Kim, R.B.; Giacomini, K.M. SLC transporters as therapeutic targets: Emerging opportunities. *Nat. Rev. Drug Discov.* **2015**, *14*, 543–560. [CrossRef]
66. Giacomini, K.M.; Huang, S.M.; Tweedie, D.J.; Benet, L.Z.; Brouwer, K.L.R.; Chu, X.; Dahlin, A.; Evers, R.; Fischer, V.; Hillgren, K.M.; et al. Membrane Transporters in Drug Development. *Nat. Rev. Drug Discov.* **2010**, *9*, 215–236. [CrossRef] [PubMed]
67. Liu, Z.; Liu, K. The Transporters of Intestinal Tract and Techniques Applied to Evaluate Interactions between Drugs and Transporters. *Asian J. Pharm. Sci.* **2013**, *8*, 151–158. [CrossRef]
68. Glaeser, H.; Bailey, D.G.; Dresser, G.K.; Gregor, J.C.; Schwarz, U.I.; McGrath, J.S.; Jolicoeur, E.; Lee, W.; Leake, B.F.; Tirona, R.G.; et al. Intestinal Drug Transporter Expression and the Impact of Grapefruit Juice in Humans. *Clin. Pharmacol. Ther.* **2007**, *81*, 362–370. [CrossRef]

69. Bailey, D.G. Fruit Juice Inhibition of Uptake Transport: A New Type of Food-Drug Interaction. *Br. J. Clin. Pharmacol.* **2010**, *70*, 645–655. [CrossRef]

70. Zhou, S.; Lim, L.Y.; Chowbay, B. Herbal Modulation of P-Glycoprotein. *Drug Metab. Rev.* **2004**, *36*, 57–104. [CrossRef]

71. Nakanishi, T.; Tamai, I. Interaction of Drug or Food with Drug Transporters in Intestine and Liver. *Curr. Drug Metab.* **2015**, *16*, 753–764. [CrossRef]

72. Deferme, S.; Augustijns, P. The Effect of Food Components on the Absorption of P-Gp Substrates: A Review. *J. Pharm. Pharmacol.* **2003**, *55*, 153–162. [CrossRef]

73. Kim, T.H.; Shin, S.; Yoo, S.D.; Shin, B.S. Effects of Phytochemical P-Glycoprotein Modulators on the Pharmacokinetics and Tissue Distribution of Doxorubicin in Mice. *Molecules* **2018**, *23*, 349. [CrossRef]

74. Yu, J.; Zhou, P.; Asenso, J.; Yang, X.D.; Wang, C.; Wei, W. Advances in Plant-Based Inhibitors of P-Glycoprotein. *J. Enzyme Inhib. Med. Chem.* **2016**, *31*, 867–881. [CrossRef]

75. Alvarez, A.I.; Real, R.; Pérez, M.; Mendoza, G.; Prieto, J.G.; Merino, G. Modulation of the Activity of ABC Transporters (P-Glycoprotein, MRP2, BCRP) by Flavonoids and Drug Response. *J. Pharm. Sci.* **2010**, *99*, 598–617. [CrossRef]

76. Katayama, K.; Masuyama, K.; Yoshioka, S.; Hasegawa, H.; Mitsuhashi, J.; Sugimoto, Y. Flavonoids Inhibit Breast Cancer Resistance Protein-Mediated Drug Resistance: Transporter Specificity and Structure-Activity Relationship. *Cancer Chemother. Pharmacol.* **2007**, *60*, 789–797. [CrossRef]

77. Yao, H.T.; Hsu, Y.R.; Li, M.L. Beverage–Drug Interaction: Effects of Green Tea Beverage Consumption on Atorvastatin Metabolism and Membrane Transporters in the Small Intestine and Liver of Rats. *Membranes* **2020**, *10*, 233. [CrossRef]

78. Onetto, A.J.; Shariff, S. *Drug Distribution*; StatPearls Publishing: Treasure Island, FL, USA, 2022.

79. Caldwell, J.; Gardner, I.; Swales, N. An introduction to drug disposition: The basic principles of absorption, distribution, metabolism, and excretion. *Toxicol. Pathol.* **1995**, *23*, 102–114. [CrossRef]

80. McElnay, J.C.; D'Arcy, P.F. Protein Binding Displacement Interactions and Their Clinical Importance. *Drugs* **1983**, *25*, 495–513. [CrossRef]

81. Xiao, J.; Kai, G. A Review of Dietary Polyphenol-Plasma Protein Interactions: Characterization, Influence on the Bioactivity, and Structure-Affinity Relationship. *Crit. Rev. Food Sci. Nutr.* **2012**, *52*, 85–101. [CrossRef]

82. López-Yerena, A.; Perez, M.; Vallverdú-Queralt, A.; Escribano-Ferrer, E. Insights into the Binding of Dietary Phenolic Compounds to Human Serum Albumin and Food-Drug Interactions. *Pharmaceutics* **2020**, *12*, 1123. [CrossRef]

83. Rimac, H.; Dufour, C.; Debeljak, Ž.; Zorc, B.; Bojić, M. Warfarin and Flavonoids Do Not Share the Same Binding Region in Binding to the IIA Subdomain of Human Serum Albumin. *Molecules* **2017**, *22*, 1153. [CrossRef]

84. Sim, S.C.; Ingelman-Sundberg, M. The Human Cytochrome P450 (CYP) Allele Nomenclature Website: A Peer-Reviewed Database of CYP Variants and Their Associated Effects. *Hum. Genomics* **2010**, *4*, 278–281. [CrossRef]

85. Zhou, S.F.; Liu, J.P.; Chowbay, B. Polymorphism of Human Cytochrome P450 Enzymes and Its Clinical Impact. *Drug Metab. Rev.* **2009**, *41*, 89–295. [CrossRef]

86. Zanger, U.M.; Schwab, M. Cytochrome P450 Enzymes in Drug Metabolism: Regulation of Gene Expression, Enzyme Activities, and Impact of Genetic Variation. *Pharmacol. Ther.* **2013**, *138*, 103–141. [CrossRef]

87. Ingelman-Sundberg, M.; Sim, S.C.; Gomez, A.; Rodriguez-Antona, C. Influence of Cytochrome P450 Polymorphisms on Drug Therapies: Pharmacogenetic, Pharmacoepigenetic and Clinical Aspects. *Pharmacol. Ther.* **2007**, *116*, 496–526. [CrossRef] [PubMed]

88. Daly, A.K.; Brockmoller, J.; Broly, F.; Eichelbaum, M.; Evans, W.E.; Gonzalez, F.J.; Huang, J.D.; Idle, J.R.; Ingelman-Sundberg, M.; Ishizaki, T.; et al. Nomenclature for Human CYP2D6 Alleles. *Pharmacogenetics* **1996**, *6*, 193–201. [CrossRef] [PubMed]

89. Zhang, L.; Zhang, Y.; Huang, S.M. Scientific and Regulatory Perspectives on Metabolizing Enzyme-Transporter Interplay and Its Role in Drug Interactions: Challenges in Predicting Drug Interactions. *Mol. Pharm.* **2009**, *6*, 1766–1774. [CrossRef] [PubMed]

90. Di Carlo, G.; Borrelli, F.; Ernst, E.; Izzo, A.A. St John's Wort: Prozac from the Plant Kingdom. *Trends Pharmacol. Sci.* **2001**, *22*, 292–297. [CrossRef]

91. Chrubasik-Hausmann, S.; Vlachojannis, J.; McLachlan, A.J. Understanding Drug Interactions with St John's Wort (Hypericum Perforatum L.): Impact of Hyperforin Content. *J. Pharm. Pharmacol.* **2019**, *71*, 129–138. [CrossRef]

92. Madabushi, R.; Frank, B.; Drewelow, B.; Derendorf, H.; Butterweck, V. Hyperforin in St. John's Wort Drug Interactions. *Eur. J. Clin. Pharmacol.* **2006**, *62*, 225–233. [CrossRef]

93. Nicolussi, S.; Drewe, J.; Butterweck, V.; Meyer zu Schwabedissen, H.E. Clinical Relevance of St. John's Wort Drug Interactions Revisited. *Br. J. Pharmacol.* **2020**, *177*, 1212–1226. [CrossRef]

94. Soleymani, S.; Bahramsoltani, R.; Rahimi, R.; Abdollahi, M. Clinical Risks of St John's Wort (Hypericum Perforatum) Co-Administration. *Expert Opin. Drug Metab. Toxicol.* **2017**, *13*, 1047–1062. [CrossRef]

95. Van Der Weide, J.; Steijns, L.S.W.; Van Weelden, M.J.M. The Effect of Smoking and Cytochrome P450 CYP1A2 Genetic Polymorphism on Clozapine Clearance and Dose Requirement. *Pharmacogenetics* **2003**, *13*, 169–172. [CrossRef]

96. Skupinska, K.; Misiewicz-Krzeminska, I.; Lubelska, K.; Kasprzycka-Guttman, T. The effect of isothiocyanates on CYP1A1 and CYP1A2 activities induced by polycyclic aromatic hydrocarbons in Mcf7 cells. *Toxicol. In Vitro* **2009**, *23*, 763–771. [CrossRef]

97. Jiang, X.; Williams, K.M.; Liauw, W.S.; Ammit, A.J.; Roufogalis, B.D.; Duke, C.C.; Day, R.O.; McLachlan, A.J. Effect of Ginkgo and Ginger on the Pharmacokinetics and Pharmacodynamics of Warfarin in Healthy Subjects. *Br. J. Clin. Pharmacol.* **2005**, *59*, 425. [CrossRef]

98. Von Moltke, L.L.; Weemhoff, J.L.; Bedir, E.; Khan, I.A.; Harmatz, J.S.; Goldman, P.; Greenblatt, D.J. Inhibition of Human Cytochromes P450 by Components of Ginkgo Biloba. *J. Pharm. Pharmacol.* **2004**, *56*, 1039–1044. [CrossRef]

99. Liu, D.; Zhang, L.; Duan, L.; Wu, J.J.; Hu, M.; Liu, Z.Q.; Wang, C. yan Potential of Herb-Drug/Herb Interactions between Substrates and Inhibitors of UGTs Derived from Herbal Medicines. *Pharmacol. Res.* **2019**, *150*, 104510. [CrossRef]

100. Li, X.; Wang, C.; Chen, J.; Hu, X.; Zhang, H.; Li, Z.; Lan, B.; Zhang, W.; Su, Y.; Zhang, C. Potential Interactions among Myricetin and Dietary Flavonols through the Inhibition of Human UDP-Glucuronosyltransferase in Vitro. *Toxicol. Lett.* **2022**, *358*, 40–47. [CrossRef]

101. Jiang, L.; Wang, Z.; Wang, X.; Wang, S.; Wang, Z.; Liu, Y. Piceatannol exhibits potential food-drug interactions through the inhibition of human UDP-glucuronosyltransferase (UGT) in Vitro. *Toxicol. In Vitro* **2020**, *67*, 104890. [CrossRef]

102. Chen, J.; Zhang, H.; Hu, X.; Xu, M.; Su, Y.; Zhang, C.; Yue, Y.; Zhang, X.; Wang, X.; Cui, W.; et al. Phloretin exhibits potential food-drug interactions by inhibiting human UDP-glucuronosyltransferases in vitro. *Toxicol. In Vitro* **2022**, *84*, 105447. [CrossRef]

103. Remer, T.; Manz, F. Potential Renal Acid Load of Foods and Its Influence on Urine PH. *J. Am. Diet. Assoc.* **1995**, *95*, 791–797. [CrossRef]

104. Freudenthaler, S.; Meineke, I.; Schreeb, K.H.; Boakye, E.; Gundert-Remy, U.; Gleiter, C.H. Influence of Urine PH and Urinary Flow on the Renal Excretion of Memantine. *Br. J. Clin. Pharmacol.* **1998**, *46*, 541. [CrossRef]

105. Hertrampf, R.; Gundert-Remy, U.; Beckmann, J.; Hoppe, U.; Elsäßer, W.; Stein, H. Elimination of Flecainide as a Function of Urinary Flow Rate and PH. *Eur. J. Clin. Pharmacol.* **1991**, *41*, 61–63. [CrossRef]

106. Couris, R.; Tataronis, G.; McCloskey, W.; Oertel, L.; Dallal, G.; Dwyer, J.; Blumberg, J.B. Dietary Vitamin K Variability Affects International Normalized Ratio (INR) Coagulation Indices. *Int. J. Vitam. Nutr. Res.* **2006**, *76*, 65–74. [CrossRef]

107. Kim, K.H.; Choi, W.S.; Lee, J.H.; Lee, H.; Yang, D.H.; Chae, S.C. Relationship between Dietary Vitamin K Intake and the Stability of Anticoagulation Effect in Patients Taking Long-Term Warfarin. *Thromb. Haemost.* **2010**, *104*, 755–759. [CrossRef]

108. Mahtani, K.R.; Heneghan, C.J.; Nunan, D.; Roberts, N.W. Vitamin K for improved anticoagulation control in patients receiving warfarin. *Cochrane Database Syst. Rev.* **2014**, *5*, CD009917. [CrossRef]

109. Mohamed Pakkir Maideen, N.; Balasubramanian, R.; Muthusamy, S.; Nallasamy, V. An Overview of Clinically Imperative and Pharmacodynamically Significant Drug Interactions of Renin-Angiotensin-Aldosterone System (RAAS) Blockers. *Curr. Cardiol. Rev.* **2022**, *18*, e110522204611. [CrossRef]

110. Batra, V.; Villgran, V. Hyperkalemia from Dietary Supplements. *Cureus* **2016**, *8*, e859. [CrossRef] [PubMed]

111. Campos, M.G.; Machado, J.; Costa, M.L.; Lino, S.; Correia, F.; Maltez, F. Case Report: Severe Hematological, Muscle and Liver Toxicity Caused by Drugs and Artichoke Infusion Interaction in an Elderly Polymedicated Patient. *Curr. Drug Saf.* **2018**, *13*, 44–50. [CrossRef] [PubMed]

112. Rodríguez-Fragoso, L.; Martínez-Arismendi, J.L.; Orozco-Bustos, D.; Reyes-Esparza, J.; Torres, E.; Burchiel, S.W. Potential Risks Resulting from Fruit/Vegetable–Drug Interactions: Effects on Drug-Metabolizing Enzymes and Drug Transporters. *J. Food Sci.* **2011**, *76*, R112–R124. [CrossRef] [PubMed]

113. Eagles, S.K.; Gross, A.S.; McLachlan, A.J. The Effects of Cruciferous Vegetable-Enriched Diets on Drug Metabolism: A Systematic Review and Meta-Analysis of Dietary Intervention Trials in Humans. *Clin. Pharmacol. Ther.* **2020**, *108*, 212–227. [CrossRef] [PubMed]

114. Lampe, J.W.; King, I.B.; Li, S.; Grate, M.T.; Barale, K.V.; Chen, C.; Feng, Z.; Potter, J.D. Brassica Vegetables Increase and Apiaceous Vegetables Decrease Cytochrome P450 1A2 Activity in Humans: Changes in Caffeine Metabolite Ratios in Response to Controlled Vegetable Diets. *Carcinogenesis* **2000**, *21*, 1157–1162. [CrossRef]

115. Foster, B.C.; Foster, M.S.; Vandenhoek, S.; Krantis, A.; Budzinski, J.W.; Arnason, J.T.; Gallicano, K.D.; Choudri, S. An in Vitro Evaluation of Human Cytochrome P450 3A4 and P-Glycoprotein Inhibition by Garlic. *J. Pharm. Pharm. Sci.* **2001**, *4*, 176–184.

116. Piscitelli, S.C.; Burstein, A.H.; Welden, N.; Gallicano, K.D.; Falloon, J. The Effect of Garlic Supplements on the Pharmacokinetics of Saquinavir. *Clin. Infect. Dis.* **2002**, *34*, 234–238. [CrossRef]

117. Sunaga, K.; Ohkawa, K.; Nakamura, K.; Ohkubo, A.; Harada, S.; Tsuda, T. Mechanism-Based Inhibition of Recombinant Human Cytochrome P450 3A4 by Tomato Juice Extract. *Biol. Pharm. Bull.* **2012**, *35*, 329–334. [CrossRef]

118. Ohkubo, A.; Chida, T.; Kikuchi, H.; Tsuda, T.; Sunaga, K. Effects of Tomato Juice on the Pharmacokinetics of CYP3A4 Substrate Drugs. *Asian J. Pharm. Sci.* **2017**, *12*, 464. [CrossRef]

119. Tsujimoto, M.; Agawa, C.; Ueda, S.; Yamane, T.; Kitayama, H.; Terao, A.; Fukuda, T.; Minegaki, T.; Nishiguchi, K. Inhibitory Effects of Juices Prepared from Individual Vegetables on CYP3A4 Activity in Recombinant CYP3A4 and LS180 Cells. *Biol. Pharm. Bull.* **2017**, *40*, 1561–1565. [CrossRef]

120. Kumar, S.; Bouic, P.J.; Rosenkranz, B. In Vitro Assessment of the Interaction Potential of Ocimum Basilicum (L.) Extracts on CYP2B6, 3A4, and Rifampicin Metabolism. *Front. Pharmacol.* **2020**, *11*, 517. [CrossRef]

121. Samojlik, I.; Mijatović, V.; Petković, S.; Škrbić, B.; Božin, B. The Influence of Essential Oil of Aniseed (*Pimpinella Anisum*, L.) on Drug Effects on the Central Nervous System. *Fitoterapia* **2012**, *83*, 1466–1473. [CrossRef]

122. Gupta, R.C.; Chang, D.; Nammi, S.; Bensoussan, A.; Bilinski, K.; Roufogalis, B.D. Interactions between Antidiabetic Drugs and Herbs: An Overview of Mechanisms of Action and Clinical Implications. *Diabetol Metab Syndr* **2017**, *9*, 59. [CrossRef]

123. Hassanzadeh-Taheri, M.; Hassanpour-Fard, M.; Doostabadi, M.; Moodi, H.; Vazifeshenas-Darmiyan, K.; Hosseini, M. Co-Administration Effects of Aqueous Extract of Turnip Leaf and Metformin in Diabetic Rats. *J. Tradit. Complement. Med.* **2018**, *8*, 178. [CrossRef]

124. Zeng, M.; Zhang, L.; Li, M.; Zhang, B.; Zhou, N.; Ke, Y.; Feng, W.; Zheng, X. Estrogenic Effects of the Extracts from the Chinese Yam (Dioscorea Opposite Thunb.) and Its Effective Compounds in Vitro and in Vivo. *Molecules* **2018**, *23*, 11. [CrossRef]

125. Mohseni, M.S.M.; Golshani, B. Simultaneous Determination of Levodopa and Carbidopa from Fava Bean, Green Peas and Green Beans by High Performance Liquid Gas Chromatography. *J. Clin. Diagn. Res.* **2013**, *7*, 1004. [CrossRef]

126. Fernandes, J.; Fialho, M.; Santos, R.; Peixoto-Plácido, C.; Madeira, T.; Sousa-Santos, N.; Virgolino, A.; Santos, O.; Vaz Carneiro, A. Is olive oil good for you? A systematic review and meta-analysis on anti-inflammatory benefits from regular dietary intake. *Nutrition* **2020**, *69*, 110559. [CrossRef]

127. Papakonstantinou, V.D.; Lagopati, N.; Tsilibary, E.C.; Demopoulos, C.A.; Philippopoulos, A.I. A Review on Platelet Activating Factor Inhibitors: Could a New Class of Potent Metal-Based Anti-Inflammatory Drugs Induce Anticancer Properties? *Bioinorg. Chem. Appl.* **2017**, *2017*, 6947034. [CrossRef]

128. Schwingshackl, L.; Krause, M.; Schmucker, C.; Hoffmann, G.; Rücker, G.; Meerpohl, J.J. Impact of Different Types of Olive Oil on Cardiovascular Risk Factors: A Systematic Review and Network Meta-Analysis. *Nutr. Metab. Cardiovasc. Dis.* **2019**, *29*, 1030–1039. [CrossRef]

129. Albadawi, D.A.I.; Ravishankar, D.; Vallance, T.M.; Patel, K.; Osborn, H.M.I.; Vaiyapuri, S. Impacts of Commonly Used Edible Plants on the Modulation of Platelet Function. *Int. J. Mol. Sci.* **2022**, *23*, 605. [CrossRef]

130. Jeong, H.U.; Kwon, S.S.; Kong, T.Y.; Kim, J.H.; Lee, H.S. Inhibitory Effects of Cedrol, β-Cedrene, and Thujopsene on Cytochrome P450 Enzyme Activities in Human Liver Microsomes. *J. Toxicol. Environ. Health A* **2014**, *77*, 1522–1532. [CrossRef]

131. Bailey, D.G. Grapefruit-Medication Interactions. *CMAJ* **2013**, *185*, 507–508. [CrossRef]

132. Hanley, M.J.; Cancalon, P.; Widmer, W.W.; Greenblatt, D.J. The Effect of Grapefruit Juice on Drug Disposition. *Expert Opin. Drug Metab. Toxicol.* **2011**, *7*, 267–286. [CrossRef]

133. Seden, K.; Dickinson, L.; Khoo, S.; Back, D. Grapefruit-Drug Interactions. *Drugs* **2010**, *70*, 2373–2407. [CrossRef]

134. Guo, L.Q.; Yamazoe, Y. Inhibition of Cytochrome P450 by Furanocoumarins in Grapefruit Juice and Herbal Medicines. *Acta Pharmacol. Sin.* **2004**, *25*, 129–136. [PubMed]

135. Lown, K.S.; Bailey, D.G.; Fontana, R.J.; Janardan, S.K.; Adair, C.H.; Fortlage, L.A.; Brown, M.B.; Guo, W.; Watkins, P.B. Grapefruit Juice Increases Felodipine Oral Availability in Humans by Decreasing Intestinal CYP3A Protein Expression. *J. Clin. Investig.* **1997**, *99*, 2545–2553. [CrossRef]

136. Bailey, D.G.; Dresser, G.; Arnold, J.M. Grapefruit-Medication Interactions: Forbidden Fruit or Avoidable Consequences? *CMAJ* **2013**, *185*, 309–316. [CrossRef] [PubMed]

137. Bailey, D.G. Better to Avoid Grapefruit with Certain Statins. *Am. J. Med.* **2016**, *129*, e301. [CrossRef] [PubMed]

138. Bailey, D.G.; Dresser, G.K.; Bend, J.R. Bergamottin, Lime Juice, and Red Wine as Inhibitors of Cytochrome P450 3A4 Activity: Comparison with Grapefruit Juice. *Clin. Pharmacol. Ther.* **2003**, *73*, 529–537. [CrossRef]

139. Chen, M.; Zhou, S.Y.; Fabriaga, E.; Zhang, P.H.; Zhou, Q. Food-Drug Interactions Precipitated by Fruit Juices Other than Grapefruit Juice: An Update Review. *J. Food Drug Anal.* **2018**, *26*, S61–S71. [CrossRef] [PubMed]

140. Petric, Z.; Žuntar, I.; Putnik, P.; Kovačević, D.B. Food–Drug Interactions with Fruit Juices. *Foods* **2021**, *10*, 33. [CrossRef]

141. An, G.; Mukker, J.K.; Derendorf, H.; Frye, R.F. Enzyme- and transporter-mediated beverage-drug interactions: An update on fruit juices and green tea. *J. Clin. Pharmacol.* **2015**, *55*, 1313–1331. [CrossRef]

142. Dresser, G.K.; Bailey, D.G.; Leake, B.F.; Schwarz, U.I.; Dawson, P.A.; Freeman, D.J.; Kim, R.B. Fruit Juices Inhibit Organic Anion Transporting Polypeptide-Mediated Drug Uptake to Decrease the Oral Availability of Fexofenadine. *Clin. Pharmacol. Ther.* **2002**, *71*, 11–20. [CrossRef]

143. Kamath, A.V.; Yao, M.; Zhang, Y.; Chong, S. Effect of Fruit Juices on the Oral Bioavailability of Fexofenadine in Rats. *J. Pharm. Sci.* **2005**, *94*, 233–239. [CrossRef]

144. Grenier, J.; Fradette, C.; Morelli, G.; Merritt, G.J.; Vranderick, M.; Ducharme, M.P. Pomelo Juice, but Not Cranberry Juice, Affects the Pharmacokinetics of Cyclosporine in Humans. *Clin. Pharmacol. Ther.* **2006**, *79*, 255–262. [CrossRef]

145. Gertz, B.J.; Holland, S.D.; Kline, W.F.; Matuszewski, B.K.; Freeman, A.; Quan, H.; Lasseter, K.C.; Mucklow, J.C.; Porras, A.G. Studies of the Oral Bioavailability of Alendronate. *Clin. Pharmacol. Ther.* **1995**, *58*, 288–298. [CrossRef]

146. Tapaninen, T.; Neuvonen, P.J.; Niemi, M. Orange and Apple Juice Greatly Reduce the Plasma Concentrations of the OATP2B1 Substrate Aliskiren. *Br. J. Clin. Pharmacol.* **2011**, *71*, 718–726. [CrossRef]

147. Malhotra, S.; Bailey, D.G.; Paine, M.F.; Watkins, P.B. Seville Orange Juice-Felodipine Interaction: Comparison with Dilute Grapefruit Juice and Involvement of Furocoumarins. *Clin. Pharmacol. Ther.* **2001**, *69*, 14–23. [CrossRef]

148. Karmakar, S.; Biswas, S.; Bera, R.; Mondal, S.; Kundu, A.; Ali, M.A.; Sen, T. Beverage-Induced Enhanced Bioavailability of Carbamazepine and Its Consequent Effect on Antiepileptic Activity and Toxicity. *J. Food Drug Anal.* **2015**, *23*, 327–334. [CrossRef]

149. Backman, J.T.; Mäenpää, J.; Belle, D.J.; Wrighton, S.A.; Kivistö, K.T.; Neuvonen, P.J. Lack of Correlation between in Vitro and m Vivo Studies on the Effects of Tangeretin and Tangerine Juice on Midazolam Hydroxylation. *Clin. Pharmacol. Ther.* **2000**, *67*, 382–390. [CrossRef]

150. Morita, T.; Akiyoshi, T.; Sato, R.; Uekusa, Y.; Katayama, K.; Yajima, K.; Imaoka, A.; Sugimoto, Y.; Kiuchi, F.; Ohtani, H. Citrus Fruit-Derived Flavanone Glycoside Narirutin Is a Novel Potent Inhibitor of Organic Anion-Transporting Polypeptides. *J. Agric. Food Chem.* **2020**, *68*, 14182–14191. [CrossRef]

151. Akamine, Y.; Miura, M.; Komori, H.; Saito, S.; Kusuhara, H.; Tamai, I.; Ieiri, I.; Uno, T.; Yasui-Furukori, N. Effects of One-Time Apple Juice Ingestion on the Pharmacokinetics of Fexofenadine Enantiomers. *Eur. J. Clin. Pharmacol.* **2014**, *70*, 1087–1095. [CrossRef]

152. Jeon, H.; Jang, I.J.; Lee, S.; Ohashi, K.; Kotegawa, T.; Ieiri, I.; Cho, J.Y.; Yoon, S.H.; Shin, S.G.; Yu, K.S.; et al. Apple Juice Greatly Reduces Systemic Exposure to Atenolol. *Br. J. Clin. Pharmacol.* **2013**, *75*, 172–179. [CrossRef]

153. Morita, T.; Akiyoshi, T.; Tsuchitani, T.; Kataoka, H.; Araki, N.; Yajima, K.; Katayama, K.; Imaoka, A.; Ohtani, H. Inhibitory Effects of Cranberry Juice and Its Components on Intestinal OATP1A2 and OATP2B1: Identification of Avicularin as a Novel Inhibitor. *J. Agric. Food Chem.* **2022**, *70*, 3310–3320. [CrossRef]

154. Srinivas, N.R. Cranberry Juice Ingestion and Clinical Drug-Drug Interaction Potentials; Review of Case Studies and Perspectives. *J. Pharm. Pharm. Sci.* **2013**, *16*, 289–303. [CrossRef]

155. Greenblatt, D.J.; Von Moltke, L.L.; Perloff, E.S.; Luo, Y.; Harmatz, J.S.; Zinny, M.A. Interaction of Flurbiprofen with Cranberry Juice, Grape Juice, Tea, and Fluconazole: In Vitro and Clinical Studies. *Clin. Pharmacol. Ther.* **2006**, *79*, 125–133. [CrossRef]

156. Lilja, J.J.; Backman, J.T.; Neuvonen, P.J. Effects of Daily Ingestion of Cranberry Juice on the Pharmacokinetics of Warfarin, Tizanidine, and Midazolam—Probes of CYP2C9, CYP1A2, and CYP3A4. *Clin. Pharmacol. Ther.* **2007**, *81*, 833–839. [CrossRef]

157. Srinivas, N.R. Is Pomegranate Juice a Potential Perpetrator of Clinical Drug-Drug Interactions? Review of the in Vitro, Preclinical and Clinical Evidence. *Eur. J. Drug Metab. Pharmacokinet.* **2013**, *38*, 223–229. [CrossRef]

158. Yusuff, K.B.; Emeka, P.M.; Attimarad, M. Concurrent Administration of Date Palm Fruits with Lisinopril Increases Serum Potassium Level in Male Rabbits. *Int. J. Pharmacol.* **2018**, *14*, 93–98. [CrossRef]

159. St-Jules, D.E.; Goldfarb, D.S.; Sevick, M.A. Nutrient Non-Equivalence: Does Restricting High-Potassium Plant Foods Help to Prevent Hyperkalemia in Hemodialysis Patients? *J. Ren. Nutr.* **2016**, *26*, 282–287. [CrossRef]

160. Huang, X.; Jiménez-Molén, J.J.; Lindholm, B.; Cederholm, T.; Ärnlöv, J.; Risérus, U.; Sjögren, P.; Carrero, J.J. Mediterranean Diet, Kidney Function, and Mortality in Men with CKD. *Clin. J. Am. Soc. Nephrol.* **2013**, *8*, 1548–1555. [CrossRef]

161. Molina-Vega, M.; Gómez-Pérez, A.M.; Tinahones, F.J. Fish in the Mediterranean diet. In *The Mediterranean Diet: An Evidence-Based Approach*, 2nd ed.; Preedy, V.R., Watson, R.R., Eds.; Elsevier Science: Amsterdam, The Netherlands, 2020; pp. 275–284. [CrossRef]

162. Méndez, L.; Dasilva, G.; Taltavull, N.; Romeu, M.; Medina, I. Marine Lipids on Cardiovascular Diseases and Other Chronic Diseases Induced by Diet: An Insight Provided by Proteomics and Lipidomics. *Mar. Drugs* **2017**, *15*, 258. [CrossRef]

163. Vanschoonbeek, K.; Feijge, M.A.; Paquay, M.; Rosing, J.; Saris, W.; Kluft, C.; Giesen, P.L.; de Maat, M.P.; Heemskerk, J.W. Variable Hypocoagulant Effect of Fish Oil Intake in Humans: Modulation of Fibrinogen Level and Thrombin Generation. *Arter. Thromb Vasc Biol* **2004**, *24*, 1734–1740. [CrossRef]

164. Buckley, M.S.; Goff, A.D.; Knapp, W.E. Fish Oil Interaction with Warfarin. *Ann. Pharmacother.* **2004**, *38*, 50–53. [CrossRef]

165. Jalili, M.; Dehpour, A.R. Extremely Prolonged INR Associated with Warfarin in Combination with Both Trazodone and Omega-3 Fatty Acids. *Arch. Med. Res.* **2007**, *38*, 901–904. [CrossRef]

166. Pryce, R.; Bernaitis, N.; Davey, A.K.; Badrick, T.; Anoopkumar-Dukie, S. The Use of Fish Oil with Warfarin Does Not Significantly Affect Either the International Normalised Ratio or Incidence of Adverse Events in Patients with Atrial Fibrillation and Deep Vein Thrombosis: A Retrospective Study. *Nutrients* **2016**, *8*, 578. [CrossRef]

167. KANEKO, J.; TANIMUKAI, H. MONOAMINE OXIDASE INHIBITORS (MAO). *Sogo. Rinsho.* **1964**, *13*, 833–846. [CrossRef]

168. Kibangou, I.B.; Bouhallab, S.; Henry, G.; Bureau, F.; Allouche, S.; Blais, A.; Guérin, P.; Arhan, P.; Bouglé, D.L. Milk Proteins and Iron Absorption: Contrasting Effects of Different Caseinophosphopeptides. *Pediatr. Res.* **2005**, *58*, 731–734. [CrossRef] [PubMed]

169. De Lemos, M.L.; Hamata, L.; Jennings, S.; Leduc, T. Interaction between Mercaptopurine and Milk. *J. Oncol. Pharm. Pract.* **2007**, *13*, 237–240. [CrossRef] [PubMed]

170. Sofianou-Katsoulis, A.; Khakoo, G.; Kaczmarski, R. Reduction in Bioavailability of 6-Mercaptopurine on Simultaneous Administration with Cow's Milk. *Pediatr. Hematol. Oncol.* **2006**, *23*, 485–487. [CrossRef] [PubMed]

171. Pereira, P.M.C.C.; Vicente, A.F.R.B. Meat Nutritional Composition and Nutritive Role in the Human Diet. *Meat Sci.* **2013**, *93*, 586–592. [CrossRef] [PubMed]

172. Fernandes, I.; Pérez-Gregorio, R.; Soares, S.; Mateus, N.; De Freitas, V.; Santos-Buelga, C.; Feliciano, A.S. Wine Flavonoids in Health and Disease Prevention. *Molecules* **2017**, *22*, 292. [CrossRef] [PubMed]

173. Snopek, L.; Mlcek, J.; Sochorova, L.; Baron, M.; Hlavacova, I.; Jurikova, T.; Kizek, R.; Sedlackova, E.; Sochor, J. Contribution of Red Wine Consumption to Human Health Protection. *Molecules* **2018**, *23*, 1684. [CrossRef] [PubMed]

174. Renaud, S.; de Lorgeril, M. Wine, Alcohol, Platelets, and the French Paradox for Coronary Heart Disease. *Lancet* **1992**, *339*, 1523–1526. [CrossRef]

175. Weathermon, R.; Crabb, D.W. Alcohol and Medication Interactions. *Alcohol Res. Health* **1999**, *23*, 40.

176. Mergenhagen, K.A.; Wattengel, B.A.; Skelly, M.K.; Clark, C.M.; Russo, T.A. Fact versus Fiction: A Review of the Evidence behind Alcohol and Antibiotic Interactions. *Antimicrob. Agents Chemother.* **2020**, *64*, e02167-19. [CrossRef]

177. Romeo, J.; Würnberg, J.; Nova, E.; Díaz, L.E.; Gómez-Martinez, S.; Marcos, A. Moderate Alcohol Consumption and the Immune System: A Review. *Br. J. Nutr.* **2007**, *98*, S111–S115. [CrossRef]

178. Foster, J.H.; Powell, J.E.; Marshall, E.J.; Peters, T.J. Quality of Life in Alcohol-Dependent Subjects–a Review. *Qual. Life Res.* **1999**, *8*, 255–261. [CrossRef]

179. Pauwels, E.K.J. The Protective Effect of the Mediterranean Diet: Focus on Cancer and Cardiovascular Risk. *Med. Princ. Pract.* **2011**, *20*, 103–111. [CrossRef]

180. Tsigalou, C.; Konstantinidis, T.; Paraschaki, A.; Stavropoulou, E.; Voidarou, C.; Bezirtzoglou, E. Mediterranean diet as a tool to combat inflammation and chronic diseases. An overview. *Biomedicines* **2020**, *8*, 201. [CrossRef]
181. Baute, V.; Sampath-Kumar, R.; Nelson, S.; Basil, B. Nutrition Education for the Health-care Provider Improves Patient Outcomes. *Glob. Adv. Health Med.* **2018**, *7*, 2164956118795995. [CrossRef]
182. Spanakis, M.; Patelarou, A.E.; Patelarou, E. Nursing Personnel in the Era of Personalized Healthcare in Clinical Practice. *J. Pers. Med.* **2020**, *10*, 56. [CrossRef]
183. Thorpe, M.G.; Milte, C.M.; Crawford, D.; McNaughton, S.A. Education and lifestyle predict change in dietary patterns and diet quality of adults 55 years and over. *Nutr. J.* **2019**, *18*, 67. [CrossRef]
184. Tuttolomondo, A.; Simonetta, I.; Daidone, M.; Mogavero, A.; Ortello, A.; Pinto, A. Metabolic and Vascular Effect of the Mediterranean Diet. *Int. J. Mol. Sci.* **2019**, *20*, 4716. [CrossRef]
185. Aljefree, N.M.; Almoraie, N.M.; Shatwan, I.M. Association of two types of dietary pattern scores with cardiovascular disease risk factors and serum 25 hydroxy vitamin D levels in Saudi Arabia. *Food Nutr. Res.* **2021**, *65*. [CrossRef]
186. Herrera-Marcos, L.V.; Lou-Bonafonte, J.M.; Arnal, C.; Navarro, M.A.; Osada, J. Transcriptomics and the Mediterranean Diet: A Systematic Review. *Nutrients* **2017**, *9*, 472. [CrossRef]
187. Badimon, L.; Vilahur, G.; Padro, T. Systems biology approaches to understand the effects of nutrition and promote health. *Br. J. Clin. Pharmacol.* **2017**, *83*, 38. [CrossRef]
188. Yeh, C.T.; Yen, G.C. Effect of Vegetables on Human Phenolsulfotransferases in Relation to Their Antioxidant Activity and Total Phenolics. *Free Radic. Res.* **2005**, *39*, 893–904. [CrossRef] [PubMed]
189. Harasym, J.; Oledzki, R. Effect of Fruit and Vegetable Antioxidants on Total Antioxidant Capacity of Blood Plasma. *Nutrition* **2014**, *30*, 511–517. [CrossRef] [PubMed]
190. Ravera, A.; Carubelli, V.; Sciatti, E.; Bonadei, I.; Gorga, E.; Cani, D.; Vizzardi, E.; Metra, M.; Lombardi, C. Nutrition and Cardiovascular Disease: Finding the Perfect Recipe for Cardiovascular Health. *Nutrients* **2016**, *8*, 363. [CrossRef] [PubMed]
191. Bjerrum, L.; Lopez-Valcarcel, B.G.; Petersen, G. Risk Factors for Potential Drug Interactions in General Practice. *Eur. J. Gen. Pract.* **2008**, *14*, 23–29. [CrossRef] [PubMed]
192. Guttman, Y.; Kerem, Z. Computer-Aided (In Silico) Modeling of Cytochrome P450-Mediated Food-Drug Interactions (FDI). *Int. J. Mol. Sci.* **2022**, *23*, 8498. [CrossRef] [PubMed]
193. Rahman, M.M.; Vadrev, S.M.; Magana-Mora, A.; Levman, J.; Soufan, O. A novel graph mining approach to predict and evaluate food-drug interactions. *Sci. Rep.* **2022**, *12*, 1061. [CrossRef] [PubMed]
194. Ryu, J.Y.; Kim, H.U.; Lee, S.Y. Deep Learning Improves Prediction of Drug–Drug and Drug–Food Interactions. *Proc. Natl. Acad. Sci. USA* **2018**, *115*, E4304–E4311. [CrossRef]
195. Car, J.; Tan, W.S.; Huang, Z.; Sloot, P.; Franklin, B.D. eHealth in the future of medications management: Personalisation, monitoring and adherence. *BMC Med.* **2017**, *15*, 73. [CrossRef]
196. Spanakis, E.G.; Santana, S.; Tsiknakis, M.; Marias, K.; Sakkalis, V.; Teixeira, A.; Janssen, J.H.; de Jong, H.; Tziraki, C. Technology-Based Innovations to Foster Personalized Healthy Lifestyles and Well-Being: A Targeted Review. *J. Med. Internet Res.* **2016**, *18*, e128. [CrossRef]
197. Ware, P.; Bartlett, S.J.; Paré, G.; Symeonidis, I.; Tannenbaum, C.; Bartlett, G.; Poissant, L.; Ahmed, S. Using EHealth Technologies: Interests, Preferences, and Concerns of Older Adults. *Interact. J. Med. Res.* **2017**, *6*, e3. [CrossRef]
198. Ossebaard, H.C.; Van Gemert-Pijnen, L. EHealth and Quality in Health Care: Implementation Time. *Int. J. Qual. Health Care* **2016**, *28*, 415–419. [CrossRef]
199. Black, A.D.; Car, J.; Pagliari, C.; Anandan, C.; Cresswell, K.; Bokun, T.; McKinstry, B.; Procter, R.; Majeed, A.; Sheikh, A. The Impact of EHealth on the Quality and Safety of Health Care: A Systematic Overview. *PLoS Med.* **2011**, *8*, e1000387. [CrossRef]
200. Maniadi, E.; Kondylakis, H.; Spanakis, E.G.; Spanakis, M.; Tsiknakis, M.; Marias, K.; Dong, F. Designing a digital patient avatar in the context of the MyHealthAvatar project initiative. In Proceedings of the 13th IEEE International Conference on BioInformatics and BioEngineering (BIBE 2013), Chania, Greece, 10–13 November 2013.
201. Vasiloglou, M.F.; Lu, Y.; Stathopoulou, T.; Papathanail, I.; Faeh, D.; Ghosh, A.; Baumann, M.; Mougiakakou, S. Assessing Mediterranean Diet Adherence with the Smartphone: The Medipiatto Project. *Nutrients* **2020**, *12*, 3763. [CrossRef]
202. McAleese, D.; Linardakis, M.; Papadaki, A. Quality and Presence of Behaviour Change Techniques in Mobile Apps for the Mediterranean Diet: A Content Analysis of Android Google Play and Apple App Store Apps. *Nutrients* **2022**, *14*, 1290. [CrossRef]
203. Spanakis, M.; Sfakianakis, S.; Spanakis, E.G.; Kallergis, G.; Sakkalis, V. PDCA: An EHealth Service for the Management of Drug Interactions with Complementary and Alternative Medicines. In Proceedings of the 2018 IEEE EMBS International Conference on Biomedical and Health Informatics (BHI), Las Vegas, NV, USA, 4–7 March 2018.
204. Spanakis, M.; Sfakianakis, S.; Kallergis, G.; Spanakis, E.G.; Sakkalis, V. PharmActa: Personalized Pharmaceutical Care EHealth Platform for Patients and Pharmacists. *J. Biomed. Inform.* **2019**, *100*, 103336. [CrossRef]

MDPI AG
Grosspeteranlage 5
4052 Basel
Switzerland
Tel.: +41 61 683 77 34

Applied Sciences Editorial Office
E-mail: applsci@mdpi.com
www.mdpi.com/journal/applsci

www.ingramcontent.com/pod-product-compliance
Lightning Source LLC
LaVergne TN
LVHW070904170726
843515LV00005B/702